Environmentalism and Sustainability

Edited by
Johnny McDougall

Larsen & Keller
www.larsen-keller.com

Environmentalism and Sustainability
Edited by Johnny McDougall
ISBN: 978-1-63549-114-2 (Hardback)

☰ Larsen & Keller

Published by Larsen and Keller Education,
5 Penn Plaza,
19th Floor,
New York, NY 10001, USA

Cataloging-in-Publication Data

Environmentalism and sustainability / edited by Johnny McDougall.
    p. cm.
Includes bibliographical references and index.
ISBN 978-1-63549-114-2
1. Environmentalism. 2. Sustainability. 3. Environmental
management. 4. Natural resources. 5. Conservation of natural resources.
I. McDougall, Johnny.
GE195 .E58 2017
363.7--dc23

The publisher's policy is to use permanent paper from mills that operate a sustainable forestry policy. Furthermore, the publisher ensures that the text paper and cover boards used have met acceptable environmental accreditation standards.

Printed and bound in the United States of America.

For more information regarding Larsen and Keller Education and its products, please visit the publisher's website www.larsen-keller.com

Table of Contents

VI   Contents

Permissions

Index

Preface

This book outlines the processes and applications of environmentalism and sustainability in detail. It will explain broadly the various techniques and methods of sustainable use of environmental resources. Environmentalism refers to a philosophy and world-view which stresses on the optimum and sustainable use of natural resources. It includes concepts like biodiversity preservation, ecology conservation, land ethics and biophilia hypothesis, etc. This book elucidates the concepts and innovative models around prospective development with respect to environmentalism and sustainability. Also included in this book is a detailed explanation of the various applications related to this field. Selected concepts that redefine the field have been presented in this text. The book is appropriate for those seeking detailed information in this area.

To facilitate a deeper understanding of the contents of this book a short introduction of every chapter is written below:

Chapter 1- Environmentalism involves the preservation of the environment and a recovery in the health of the environment. Environmentalism tries to strike a balance between humans and natural systems. This chapter offers an insightful focus, keeping in mind the subject matter.

Chapter 2- The essential elements of environmentalism are free-market environmentalism, evangelical environmentalism, conservation ethic, new urbanism, ecological footprint and green theory. Free-market environmentalism considers free markets and property rights to be the appropriate means of preserving the environment. The section strategically encompasses and incorporates the important elements of environmentalism, providing a complete understanding.

Chapter 3- Natural resources are resources that exist without the help of humankind, such as sunlight, atmosphere, land, water and animal life. Natural resource management, natural capital, natural capital accounting, non-renewable resource, exploitation of natural resources etc. are some of the topics included in this section. This text serves as a source to understand the concept of natural resources.

Chapter 4- Biological systems are diverse in nature and can be indefinitely productive. This ability of biological systems is known as sustainability. Aspects of sustainability include circles of sustainability, sustainable living, sustainable design, sustainable development, sustainability and environmental management and sustainability and systemic change resistance. The topics discussed in the section are of great importance to broaden the existing knowledge on sustainability.

Chapter 5- Permaculture is a system of agriculture that appropriately implements the patterns and features observed in the natural ecosystem. Holzer permaculture is an important branch of permaculture. The text also focuses on aspects such as farmer-managed natural regeneration, keyline design, managed intensive rotational grazing, natural farming and regenerative agriculture. The aim of this chapter is to explore all the techniques and concepts related to permaculture.

Chapter 6- Environmental and conservation movements have increased in number in the past couple of years. The increase in global warming and environmental degradation has caused several NGOs and governments to take steps regarding environmental degradation. Some of the movements explained in this section are conservation movement, Chipko movement and Anarcho-naturism. This section helps the reader in understanding the importance and the history of conserving our environment.

Finally, I would like to thank the entire team involved in the inception of this book for their valuable time and contribution. This book would not have been possible without their efforts. I would also like to thank my friends and family for their constant support.

Editor

Introduction to Environmentalism

Environmentalism involves the preservation of the environment and a recovery in the health of the environment. Environmentalism tries to strike a balance between humans and natural systems. This chapter offers an insightful focus, keeping in mind the subject matter.

Environmentalism

Environmentalism or environmental rights is a broad philosophy, ideology, and social movement regarding concerns for environmental protection and improvement of the health of the environment, particularly as the measure for this health seeks to incorporate the concerns of non-human elements. Environmentalism advocates the lawful preservation, restoration and/or improvement of the natural environment, and may be referred to as a movement to control pollution or protect plant and animal diversity. For this reason, concepts such as a land ethic, environmental ethics, biodiversity, ecology, and the biophilia hypothesis figure predominantly.

At its crux, environmentalism is an attempt to balance relations between humans and the various natural systems on which they depend in such a way that all the components are accorded a proper degree of sustainability. The exact measures and outcomes of this balance is controversial and there are many different ways for environmental concerns to be expressed in practice. Environmentalism and environmental concerns are often represented by the color green, but this association has been appropriated by the marketing industries for the tactic known as greenwashing. Environmentalism is opposed by anti-environmentalism, which says that the Earth is less fragile than some environmentalists maintain, and portrays environmentalism as overreacting to the human contribution to climate change or opposing human advancement.

Definitions

Environmentalism denotes a social movement that seeks to influence the political process by lobbying, activism, and education in order to protect natural resources and ecosystems. The word was first coined in 1922.

An *environmentalist* is a person who may speak out about our natural environment and the sustainable management of its resources through changes in public policy or individual behavior. This may include supporting practices such as informed consumption, conservation initiatives, investment in renewable resources, improved efficiencies in the materials economy, transitioning to new accounting paradigms such as Ecological economics and renewing and revitalizing our connections with non-human life.

In various ways (for example, grassroots activism and protests), environmentalists and environmental organizations seek to give the natural world a stronger voice in human affairs.

In general terms, environmentalists advocate the sustainable management of resources, and the protection (and restoration, when necessary) of the natural environment through changes in public policy and individual behavior. In its recognition of humanity as a participant in ecosystems, the movement is centered around ecology, health, and human rights.

While the term *environmentalism* focuses more on the environmental and nature-related aspects of green ideology and politics, *ecologism* as a term combines the ideology of social ecology and environmentalism. *Ecologism* as a term is more commonly used in continental European languages while *environmentalism* is more commonly used in English but the words have slightly different connotations.

History

A concern for environmental protection has recurred in diverse forms, in different parts of the world, throughout history. For example, in Europe, King Edward I of England banned the burning of sea-coal by proclamation in London in 1272, after its smoke had become a problem. The fuel was so common in England that this earliest of names for it was acquired because it could be carted away from some shores by the wheelbarrow.

Earlier in the Middle East, the Caliph Abu Bakr in the 630s commanded his army to "Bring no harm to the trees, nor burn them with fire," and "Slay not any of the enemy's flock, save for your food." Arabic medical treatises during the 9th to 13th centuries dealing with environmentalism and environmental science, including pollution, were written by Al-Kindi, Qusta ibn Luqa, Al-Razi, Ibn Al-Jazzar, al-Tamimi, al-Masihi, Avicenna, Ali ibn Ridwan, Ibn Jumay, Isaac Israeli ben Solomon, Abd-el-latif, Ibn al-Quff, and Ibn al-Nafis. Their works covered a number of subjects related to pollution, such as air pollution, water pollution, soil contamination, municipal solid waste mishandling, and environmental impact assessments of certain localities.

Early Environmental Legislation

At the advent of steam and electricity the muse of history holds her nose and shuts her eyes (Herbert Wells 1918).

Levels of air pollution rose during the Industrial Revolution, sparking the first modern environmental laws to be passed in the mid-19th century.

The origins of the environmental movement lay in the response to increasing levels of smoke pollution in the atmosphere during the Industrial Revolution. The emergence of great factories and the concomitant immense growth in coal consumption gave rise to an unprecedented level of air pollution in industrial centers; after 1900 the large volume of industrial chemical discharges added to the growing load of untreated human waste. The first large-scale, modern environmental laws came in the form of Britain's Alkali Acts, passed in 1863, to regulate the deleterious air pollution (gaseous hydrochloric acid) given off by the Leblanc process, used to produce soda ash. An Alkali inspector and four sub-inspectors were appointed to curb this pollution. The responsibilities of the inspectorate were gradually expanded, culminating in the Alkali Order 1958 which placed all major heavy industries that emitted smoke, grit, dust and fumes under supervision.

In industrial cities local experts and reformers, especially after 1890, took the lead in identifying environmental degradation and pollution, and initiating grass-roots movements to demand and achieve reforms. Typically the highest priority went to water and air pollution. The Coal Smoke Abatement Society was formed in 1898 making it one of the oldest environmental NGOs. It was founded by artist Sir William Blake Richmond, frustrated with the pall cast by coal smoke. Although there were earlier pieces of legislation, the Public Health Act 1875 required all furnaces and fireplaces to consume their own smoke. It also provided for sanctions against factories that emitted large amounts of black smoke. The provisions of this law were extended in 1926 with the Smoke Abatement Act to include other emissions, such as soot, ash and gritty particles and to empower local authorities to impose their own regulations.

It was, however, only under the impetus of the Great Smog of 1952 in London, which almost brought the city to a standstill and may have caused upward of 6,000 deaths that the Clean Air Act 1956 was passed and pollution in the city was finally brought to an end. Financial incentives were offered to householders to replace open coal fires with alternatives (such as installing gas fires), or for those who preferred, to burn coke instead (a byproduct of town gas production) which produces minimal smoke. 'Smoke control areas' were introduced in some towns and cities in which only smokeless fuels could be burnt and power stations were relocated away from cities. The act formed an important impetus to modern environmentalism, and caused a rethinking of the dangers of environmental degradation to people's quality of life.

The late 19th century also saw the passage of the first wildlife conservation laws. The zoologist Alfred Newton published a series of investigations into the *Desirability of establishing a 'Close-time' for the preservation of indigenous animals* between 1872 and 1903. His advocacy for legislation to protect animals from hunting during the mating season led to the formation of the Royal Society for the Protection of Birds and influenced the passage of the Sea Birds Preservation Act in 1869 as the first nature protection law in the world.

First Environmental Movements

Early interest in the environment was a feature of the Romantic movement in the early 19th century. The poet William Wordsworth travelled extensively in the Lake District and wrote that it is a "sort of national property in which every man has a right and interest who has an eye to perceive and a heart to enjoy".

John Ruskin an influential thinker who articulated the Romantic ideal of environmental protection and conservation.

Systematic efforts on behalf of the environment only began in the late 19th century; it grew out of the amenity movement in Britain in the 1870s, which was a reaction to industrialization, the growth of cities, and worsening air and water pollution. Starting with the formation of the Commons Preservation Society in 1865, the movement championed rural preservation against the encroachments of industrialisation. Robert Hunter, solicitor for the society, worked with Hardwicke Rawnsley, Octavia Hill, and John Ruskin to lead a successful campaign to prevent the construction of railways to carry slate from the quarries, which would have ruined the unspoilt valleys of Newlands and Ennerdale. This success led to the formation of the Lake District Defence Society (later to become The Friends of the Lake District).

In 1893 Hill, Hunter and Rawnsley agreed to set up a national body to coordinate environmental conservation efforts across the country; the "National Trust for Places of Historic Interest or Natural Beauty" was formally inaugurated in 1894. The organisation obtained secure footing through the 1907 National Trust Bill, which gave the trust the status of a statutory corporation. and the bill was passed in August 1907.

An early "Back-to-Nature" movement, which anticipated the romantic ideal of modern environmentalism, was advocated by intellectuals such as John Ruskin, William Morris, George Bernard Shaw and Edward Carpenter, who were all against consumerism, pollution and other activities that were harmful to the natural world. The movement was a reaction to the urban conditions of the industrial towns, where sanitation was awful, pollution levels intolerable and housing terribly cramped. Idealists championed the rural life as a mythical Utopia and advocated a return to it. John Ruskin argued that people should return to a *small piece of English ground, beautiful, peaceful, and fruitful. We will have no steam engines upon it . . . we will have plenty of flowers and vegetables . . . we will have some music and poetry; the children will learn to dance to it and sing it.*

Practical ventures in the establishment of small cooperative farms were even attempted and old rural traditions, without the "taint of manufacture or the canker of artificiality", were enthusiastically revived, including the Morris dance and the maypole.

These ideas also inspired various environmental groups in the UK, such as the Royal Society for the Protection of Birds, established in 1889 by Emily Williamson as a protest group to campaign for greater protection for the indigenous birds of the island. The Society attracted growing support from the suburban middle-classes as well as support from many other influential figures, such as the ornithologist Professor Alfred Newton. By 1900, public support for the organisation had grown, and it had over 25,000 members. The Garden city movement incorporated many environmental concerns into its urban planning manifesto; the Socialist League and The Clarion movement also began to advocate measures of nature conservation.

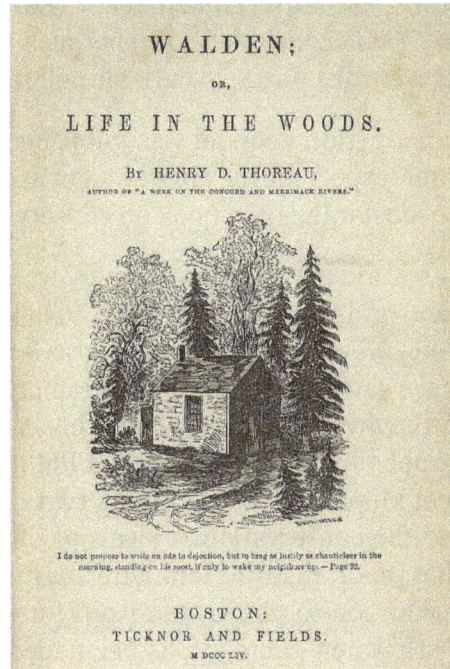

Original title page of *Walden* by Henry David Thoreau.

The movement in the United States began in the late 19th century, out of concerns for protecting the natural resources of the West, with individuals such as John Muir and Henry David Thoreau making key philosophical contributions. Thoreau was interested in peoples' relationship with nature and studied this by living close to nature in a simple life. He published his experiences in the book *Walden*, which argues that people should become intimately close with nature. Muir came to believe in nature's inherent right, especially after spending time hiking in Yosemite Valley and studying both the ecology and geology. He successfully lobbied congress to form Yosemite National Park and went on to set up the Sierra Club in 1892. The conservationist principles as well as the belief in an inherent right of nature were to become the bedrock of modern environmentalism.

In the 20th century, environmental ideas continued to grow in popularity and recognition. Efforts were starting to be made to save some wildlife, particularly the American bison. The death of the last passenger pigeon as well as the endangerment of the American bison helped to focus the minds of conservationists and popularize their concerns. In 1916 the National Park Service was founded by US President Woodrow Wilson.

The Forestry Commission was set up in 1919 in Britain to increase the amount of woodland in Britain by buying land for afforestation and reforestation. The commission was also tasked with promoting forestry and the production of timber for trade. During the 1920s the Commission focused on acquiring land to begin planting out new forests; much of the land was previously used for agricultural purposes. By 1939 the Forestry Commission was the largest landowner in Britain.

During the 1930s the Nazis had elements that were supportive of animal rights, zoos and wildlife, and took several measures to ensure their protection. In 1933 the government created a stringent animal-protection law and in 1934, *Das Reichsjagdgesetz* (The Reich Hunting Law) was enacted which limited hunting. Several Nazis were environmentalists (notably Rudolf Hess), and species protection and animal welfare were significant issues in the regime. In 1935, the regime enacted the "Reich Nature Protection Act" (*Reichsnaturschutzgesetz*). The concept of the *Dauerwald* (best translated as the "perpetual forest") which included concepts such as forest management and protection was promoted and efforts were also made to curb air pollution.

In 1949, *A Sand County Almanac* by Aldo Leopold was published. It explained Leopold's belief that humankind should have moral respect for the environment and that it is unethical to harm it. The book is sometimes called the most influential book on conservation.

Throughout the 1950s, 1960s, 1970s and beyond, photography was used to enhance public awareness of the need for protecting land and recruiting members to environmental organizations. David Brower, Ansel Adams and Nancy Newhall created the Sierra Club Exhibit Format Series, which helped raise public environmental awareness and brought a rapidly increasing flood of new members to the Sierra Club and to the environmental movement in general. "This Is Dinosaur" edited by Wallace Stegner with photographs by Martin Litton and Philip Hyde prevented the building of dams within Dinosaur National Monument by becoming part of a new kind of activism called environmentalism that combined the conservationist ideals of Thoreau, Leopold and Muir with hard-hitting advertising, lobbying, book distribution, letter writing campaigns, and more. The powerful use of photography in addition to the written word for conservation dated back to the creation of Yosemite National Park, when photographs persuaded Abraham Lincoln to preserve the beautiful glacier carved landscape for all time. The Sierra Club Exhibit Format Series galvanized public opposition to building dams in the Grand Canyon and protected many other national treasures. The Sierra Club often led a coalition of many environmental groups including the Wilderness Society and many others. After a focus on preserving wilderness in the 1950s and 1960s, the Sierra Club and other groups broadened their focus to include such issues as air and water pollution, population concern, and curbing the exploitation of natural resources.

Post-war Expansion

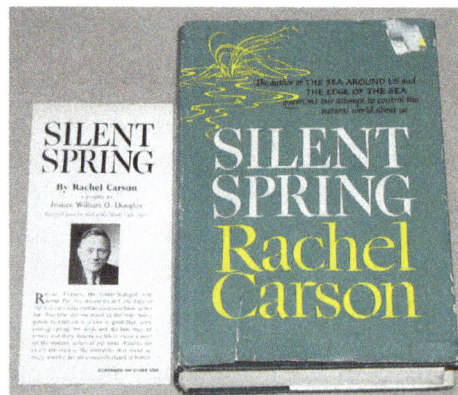

Silent Spring by Rachael Carson, published in 1962, included an endorsement by William O. Douglas.

In 1962, *Silent Spring* by American biologist Rachel Carson was published. The book cataloged the environmental impacts of the indiscriminate spraying of DDT in the US and questioned the logic of releasing large amounts of chemicals into the environment without fully understanding their effects on ecology or human health. The book suggested that DDT and other pesticides may cause cancer and that their agricultural use was a threat to wildlife, particularly birds. The resulting public concern led to the creation of the United States Environmental Protection Agency in 1970 which subsequently banned the agricultural use of DDT in the US in 1972. The limited use of DDT in disease vector control continues to this day in certain parts of the world and remains controversial. The book's legacy was to produce a far greater awareness of environmental issues and interest into how people affect the environment. With this new interest in environment came interest in problems such as air pollution and petroleum spills, and environmental interest grew. New pressure groups formed, notably Greenpeace and Friends of the Earth (US), as well as notable local organizations such as the Wyoming Outdoor Council, which was founded in 1967.

In the 1970s, the environmental movement gained rapid speed around the world as a productive outgrowth of the counterculture movement.

The world's first political parties to campaign on a predominantly environmental platform were the United Tasmania Group Tasmania, Australia and the Values Party of New Zealand. The first green party in Europe was the Popular Movement for the Environment, founded in 1972 in the Swiss canton of Neuchâtel. The first national green party in Europe was PEOPLE, founded in Britain in February 1973, which eventually turned into the Ecology Party, and then the Green Party.

Protection of the environment also became important in the developing world; the Chipko movement was formed in India under the influence of Mohandas Gandhi and they set up peaceful resistance to deforestation by literally hugging trees (leading to the term "tree huggers"). Their peaceful methods of protest and slogan "ecology is permanent economy" were very influential.

Another milestone in the movement was the creation of an Earth Day. Earth Day was first observed in San Francisco and other cities on March 21, 1970, the first day of spring. It was created to give awareness to environmental issues. On March 21, 1971, United Nations Secretary-General U Thant spoke of a spaceship Earth on Earth Day, hereby referring to the ecosystem services the earth supplies to us, and hence our obligation to protect it (and with it, ourselves). Earth Day is now coordinated globally by the Earth Day Network, and is celebrated in more than 175 countries every year.

The UN's first major conference on international environmental issues, the United Nations Conference on the Human Environment (also known as the Stockholm Conference), was held on June 5–16, 1972. It marked a turning point in the development of international environmental politics.

By the mid-1970s, many felt that people were on the edge of environmental catastrophe. The Back-to-the-land movement started to form and ideas of environmental ethics joined with anti-Vietnam War sentiments and other political issues. These individuals lived outside normal society and started to take on some of the more radical environmental theories such as deep ecology. Around this time more mainstream environmentalism was starting to show force with the signing of the Endangered Species Act in 1973 and the formation of CITES in 1975. Significant amendments were also enacted to the United States Clean Air Act and Clean Water Act.

In 1979, James Lovelock, a British scientist, published *Gaia: A new look at life on Earth*, which put forth the Gaia hypothesis; it proposes that life on earth can be understood as a single organism.

This became an important part of the Deep Green ideology. Throughout the rest of the history of environmentalism there has been debate and argument between more radical followers of this Deep Green ideology and more mainstream environmentalists.

Today

Environmentalism continues to evolve to face up to new issues such as global warming, overpopulation and genetic engineering.

Recent research demonstrates a precipitous decline in the public's interest in 19 different areas of environmental concern. Americans are less likely be actively participating in an environmental movement or organization and more likely to identify as "unsympathetic" to an environmental movement then in 2000. This is likely a lingering factor of the Great Recession in 2008. Since 2005 the percentage of Americans agreeing that the environment should be given priority over economic growth has dropped 10 points, in contrast, those feeling that growth should be given priority "even if the environment suffers to some extent" has risen 12 percent. These numbers point to the growing complexity of environmentalism and its relationship to economics.

Environmental Movement

The *environmental movement* (a term that sometimes includes the conservation and green movements) is a diverse scientific, social, and political movement. Though the movement is represented by a range of organizations, because of the inclusion of environmentalism in the classroom curriculum, the environmental movement has a younger demographic than is common in other social movements.

Before flue-gas desulfurization was installed, the air-polluting emissions from this power plant in New Mexico contained excessive amounts of sulfur dioxide.

Environmentalism as a movement covers broad areas of institutional oppression, including for example: consumption of ecosystems and natural resources into waste, dumping waste into disadvantaged communities, air pollution, water pollution, weak infrastructure, exposure of organic life to toxins, mono-culture, anti-polythene drive (jhola movement) and various other focuses. Because of these divisions, the environmental movement can be categorized into these primary focuses: environmental science, environmental activism, environmental advocacy, and environmental justice.

Free Market Environmentalism

Free market environmentalism is a theory that argues that the free market, property rights, and tort law provide the best tools to preserve the health and sustainability of the environment. It considers environmental stewardship to be natural, as well as the expulsion of polluters and other aggressors through individual and class action. It has been supported by libertarians and many conservatives.

Evangelical Environmentalism

Evangelical environmentalism is an environmental movement in the United States in which some Evangelicals have emphasized biblical mandates concerning humanity's role as steward and subsequent responsibility for the caretaking of Creation. While the movement has focused on different environmental issues, it is best known for its focus of addressing climate action from a biblically grounded theological perspective. The Evangelical Climate Initiative argues that human-induced climate change will have severe consequences and impact the poor the hardest, and that God's mandate to Adam to care for the Garden of Eden also applies to evangelicals today, and that it is therefore a moral obligation to work to mitigate climate impacts and support communities in adapting to change.

Preservation and Conservation

Environmental preservation in the United States and other parts of the world, including Australia, is viewed as the setting aside of natural resources to prevent damage caused by contact with humans or by certain human activities, such as logging, mining, hunting, and fishing, often to replace them with new human activities such as tourism and recreation. Regulations and laws may be enacted for the preservation of natural resources.

Organizations and Conferences

Environmental organizations can be global, regional, national or local; they can be government-run or private (NGO). Environmentalist activity exists in almost every country. Moreover, groups dedicated to community development and social justice also focus on environmental concerns.

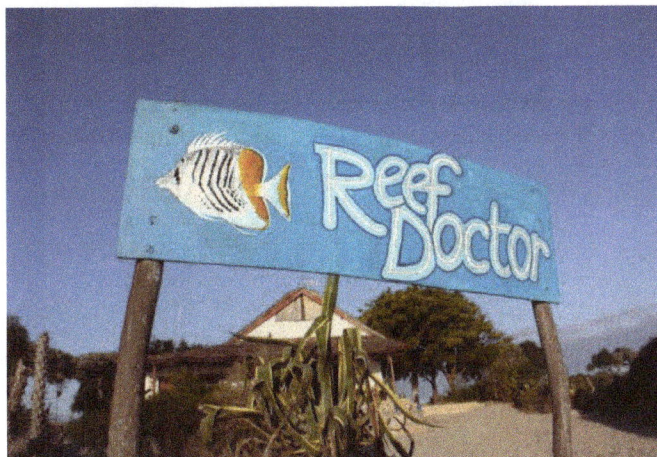

Reef doctor work station in Ifaty, Madagascar

There are some volunteer organizations. For example, Ecoworld and Paryawaran Sachetak Samiti are environmental organizations which are based on teamwork and volunteer work. Some US environmental organizations, among them the Natural Resources Defense Council and the Environmental Defense Fund, specialize in bringing lawsuits (a tactic seen as particularly useful in that country). Other groups, such as the US-based National Wildlife Federation, the Nature Conservancy, and The Wilderness Society, and global groups like the World Wide Fund for Nature and Friends of the Earth, disseminate information, participate in public hearings, lobby, stage demonstrations, and may purchase land for preservation. Statewide nonprofit organizations such as the Wyoming Outdoor Council often collaborate with these national organizations and employ similar strategies. Smaller groups, including Wildlife Conservation International, conduct research on endangered species and ecosystems. More radical organizations, such as Greenpeace, Earth First!, and the Earth Liberation Front, have more directly opposed actions they regard as environmentally harmful. While Greenpeace is devoted to nonviolent confrontation as a means of bearing witness to environmental wrongs and bringing issues into the public realm for debate, the underground *Earth Liberation Front* engages in the clandestine destruction of property, the release of caged or penned animals, and other criminal acts. Such tactics are regarded as unusual within the movement, however.

On an international level, concern for the environment was the subject of a United Nations Conference on the Human Environment in Stockholm in 1972, attended by 114 nations. Out of this meeting developed UNEP (United Nations Environment Programme) and the follow-up United Nations Conference on Environment and Development in 1992. Other international organizations in support of environmental policies development include the Commission for Environmental Cooperation (as part of NAFTA), the European Environment Agency (EEA), and the Intergovernmental Panel on Climate Change (IPCC).

Environmental Protests

Climate activists blockade British Airports Authority's headquarters for day of action

"March Against Monsanto", Vancouver, Canada, May 25, 2013

Assassination of Environmental Activists

Every year, more than 100 environmentalist activists are assassinated throughout the world.

Usage in Popular Culture

- The popular media have been used to convey conservation messages in the U.S. For instance, the U.S. Forest Service created Smokey the Bear in 1944; he appeared in countless posters, radio and television programs, movies, press releases, and other guises to warn about forest fires. The comic strip Mark Trail, by environmentalist Ed Dodd, began in 1946; it still appears weekly in 175 newspapers. Another example is the children's animated show *Captain Planet and the Planeteers*, created by Ted Turner and Barbara Pyle in 1989 to inform kids on environmental issues. The show aired for six seasons and 113 episodes, in 100 countries worldwide from 1990 to 1996.

- Miss Earth is one of the three largest international beauty pageants alongside Miss Universe and Miss World that promotes Environmental Awareness. The reigning titleholders dedicate their year to promote specific projects and often address issues concerning the environment and other global issues through school tours, tree planting activities, street campaigns, coastal clean ups, speaking engagements, shopping mall tours, media guesting, environmental fair, storytelling programs, eco-fashion shows, and other environmental activities. The Miss Earth winner is the spokesperson for the Miss Earth Foundation, the United Nations Environment Programme (UNEP) and other environmental organizations. The Miss Earth Foundation also works with the environmental departments and ministries of participating countries, various private sectors and corporations, as well as Greenpeace and the World Wildlife Foundation (WWF).

An Alternative View

Many environmentalists believe that human interference with 'nature' should be restricted or minimised as a matter of urgency (for the sake of life, or the planet, or just for the benefit of the human species), whereas environmental skeptics and anti-environmentalists do not believe that there is such a need. One can also regard oneself as an environmentalist and believe that human 'interference' with 'nature' should be *increased*. Nevertheless, there is a risk that the shift from emotional environmentalism into the technical management of natural resources and hazards could decrease the touch of humans with nature, leading to less concern with environment preservation.

Environmental History

Environmental history is the study of human interaction with the natural world over time. In contrast to other historical disciplines, it emphasizes the active role nature plays in influencing human affairs. Environmental historians study how humans both shape their environment and are shaped by it.

The city of Machu Picchu was constructed c. 1450 AD, at the height of the Inca Empire. It has commanding views down two valleys and a nearly impassable mountain at its back. There is an ample supply of spring water and enough land for a plentiful food supply. The hillsides leading to it have been terraced to provide farmland for crops, reduce soil erosion, protect against landslides, and create steep slopes to discourage potential invaders.

Environmental history emerged in the United States out of the environmental movement of the 1960s and 1970s, and much of its impetus still stems from present-day global environmental concerns. The field was founded on conservation issues but has broadened in scope to include more general social and scientific history and may deal with cities, population or sustainable development. As all history occurs in the natural world, environmental history tends to focus on particular time-scales, geographic regions, or key themes. It is also a strongly multidisciplinary subject that draws widely on both the humanities and natural science.

The subject matter of environmental history can be divided into three main components. The first, nature itself and its change over time, includes the physical impact of humans on the Earth's land, water, atmosphere and biosphere. The second category, how humans use nature, includes the environmental consequences of increasing population, more effective technology and changing patterns of production and consumption. Other key themes are the transition from nomadic hunter-gatherer communities to settled agriculture in the neolithic revolution, the effects of colonial expansion and settlements, and the environmental and human consequences of the industrial and technological revolutions. Finally, environmental historians study how people think about nature - the way attitudes, beliefs and values influence interaction with nature, especially in the form of myths, religion and science.

Origin of Name and Early Works

In 1967 Roderick Nash published "*Wilderness and the American Mind*", a work that has become a classic text of early environmental history. In an address to the *Organization of American Historians* in 1969 (published in 1970) Nash used the expression "environmental history", although 1972 is generally taken as the date when the term was first coined. The 1959 book by Samuel P. Hays, *Conservation and the Gospel of Efficiency: The Progressive Conservation Movement, 1890-1920*, while being a major contribution to American political history, is now also regarded as a founding document in the field of environmental history. Hays is Professor Emeritus of History at the University of Pittsburgh.

Historiography

Brief overviews of the field of environmental history have been given by John McNeill in 2003, Richard White in 1985, and J. Donald Hughes in 2006. In 2014 Oxford University Press published a volume of 25 essays entitled *The Oxford Handbook of Environmental History*. This collection was edited by Andrew C. Isenberg.

Definition

There is no universally accepted definition of environmental history. In general terms it is a history that tries to explain why our environment is like it is and how humanity has influenced its current condition, as well as commenting on the problems and opportunities of tomorrow. Donald Worster's widely quoted 1988 definition states: *"Environmental history is the interaction between human cultures and the environment in the past."*

In 2001 J. Donald Hughes defined the subject as *"The study of human relationships through time with the natural communities of which they are a part in order to explain the processes of change that affect that relationship."* and, in 2006, as *"... history that seeks understanding of human beings as they have lived, worked and thought in relationship to the rest of nature through the changes brought by time"* ... *"As a method, environmental history is the use of ecological analysis as a means of understanding human history ... an account of changes in human societies as they relate to changes in the natural environment."* Environmental historians are also *"interested in what people think about nature, and how they have expressed those ideas in folk religions, popular culture, literature and art."* In 2003 McNeill suggested that environmental history was *"... the history of the mutual relations between humankind and the rest of nature"*.

Subject Matter

Traditional historical analysis has over time extended its range of study from the activities and influence of a few significant people to a much broader social, political, economic and cultural analysis. Environmental history further broadens the subject matter of conventional history. In 1988, Donald Worster stated that environmental history *"attempts to make history more inclusive in its narratives"* by examining the *"role and place of nature in human life"*, and in 1993, that *"Environmental history explores the ways in which the biophysical world has influenced the course of human history and the ways in which people have thought about and tried to transform their surroundings"*. The interdependency of human and environmental factors in the creation of landscapes is expressed through the notion of the cultural landscape. Worster also questioned the scope of the discipline, asking: *"We study humans and nature; therefore can anything human or natural be outside our enquiry?"*

Environmental history is generally treated as a subfield of history, an established discipline. But some environmental historians challenge this assumption, arguing that while traditional history is human history – the story of people and their institutions, *"humans cannot place themselves outside the principles of nature."* In this sense environmental history is a version of human history within a larger context, one less dependent on anthropocentrism (even though anthropogenic change is at the center of its narrative).

Dimensions

J. Donald Hughes responded to the view that environmental history is *"light on theory"* or lacking theoretical structure by viewing the subject through the lens of three "dimensions": nature and culture, history and science, and scale. This advances beyond Worster's recognition of three broad clusters of issues to be addressed by environmental historians although both historians recognize that the emphasis of their categories might vary according to the particular study as, clearly, some studies will concentrate more on society and human affairs and others more on the environment.

General view of Funkville in 1864, Oil Creek, Pennsylvania, USA

Themes

Several themes are used to express these historical dimensions. A more traditional historical approach is to analyse the transformation of the globe's ecology through themes like the separation of man from nature during the neolithic revolution, imperialism and colonial expansion, exploration, agricultural change, the effects of the industrial and technological revolution, and urban expansion. More environmental topics include human impact through influences on forestry, fire, climate change, sustainability and so on. According to Paul Warde, "*the increasingly sophisticated history of colonization and migration can take on an environmental aspect, tracing the pathways of ideas and species around the globe and indeed is bringing about an increased use of such analogies and 'colonial' understandings of processes within European history.*" The importance of the colonial enterprise in Africa, the Caribbean and Indian Ocean has been detailed by Richard Grove. Much of the literature consists of case-studies targeted at the global, national and local levels.

Scale

Although environmental history can cover billions of years of history over the whole Earth, it can equally concern itself with local scales and brief time periods. Many environmental historians are occupied with local, regional and national histories. Some historians link their subject exclusively to the span of human history – "*every time period in human history*" while others include the period before human presence on Earth as a legitimate part of the discipline. Ian Simmons's *Environmental History of Great Britain* covers a period of about 10,000 years. There is a tendency to difference in time scales between natural and social phenomena: the causes of environmental change that stretch back in time may be dealt with socially over a comparatively brief period.

Although at all times environmental influences have extended beyond particular geographic regions and cultures, during the 20th and early 21st centuries anthropogenic environmental change has assumed global proportions, most prominently with climate change but also as a result of settlement, the spread of disease and the globalization of world trade.

History of the Subject

The questions posed and themes covered by environmental history date back to antiquity: historians have always included the effects of natural phenomena on human affairs. Hippocrates, ancient Greek father of medicine, in his *Airs, Waters, Places*, asserted that different cultures and human temperaments could be related to the surroundings in which peoples lived. During the Enlightenment there was a rising awareness of the environment as concept and early environmental scientists addressed themes of sustainability via the subjects of natural history and medicine. However, the origins of the subject in its present form are generally traced to the twentieth century.

Nature preservationist John Muir with US President Theodore Roosevelt (left) on Glacier Point in Yosemite National Park

In 1929 a group of French historians founded the journal *Annales*, in many ways a forerunner of modern environmental history since it took as its subject matter the reciprocal global influences of the environment and human society. The idea of the impact of the physical environment on civilizations was espoused by this Annales School to describe the long term developments that shape human history by focusing away from political and intellectual history, toward agriculture, demography, and geography. Emmanuel Le Roy Ladurie, a pupil of the Annales School, was the first to really embrace, in the 1950s, environmental history in a more contemporary form. One of the most influential members of the Annales School was Lucien Febvre (1878–1956), whose book *A Geographical Introduction to History* is now a classic in the field.

The most influential empirical and theoretical work in the subject has been done in the United States where teaching programs first emerged and a generation of trained environmental historians is now active. In the United States environmental history as an independent field of study emerged in the general cultural reassessment and reform of the 1960s and 1970s along with environmentalism, "conservation history", and a gathering awareness of the global scale of some environmental issues. This was in large part a reaction to the way nature was represented in history at the time, which "portrayed the advance of culture and technology as releasing humans from dependence on the natural world and providing them with the means to manage it [and] celebrated human mastery over other forms of life and the natural environment, and expected technological improvement and economic growth to accelerate". Environmental historians intended to develop a post-colonial historiography that was "*more inclusive in its narratives*".

Moral and political inspiration

Moral and political inspiration to environmental historians has come from American writers and activists Henry Thoreau (1817 – 1862), John Muir (1838 – 1914), Aldo Leopold (1887 – 1948), and Rachel Carson (1907 – 1964). Environmental history frequently promoted a moral and political agenda although it steadily became a more scholarly enterprise." Early attempts to define the field were made in the United States by Roderick Nash in "The State of Environmental History" and in other works by frontier historians Frederick Jackson Turner, James Malin, and Walter Prescott Webb who analysed the process of settlement. Their work was expanded by a second generation of more specialized environmental historians such as Alfred Crosby, Samuel P. Hays, Donald Worster, William Cronon, Richard White, Carolyn Merchant, John McNeill, Donald Hughes, Chad Montrie, and Europeans Paul Warde, Sverker Sorlin, Robert A. Lambert, T.C. Smout and Peter Coates.

British Empire

Although environmental history was growing rapidly after 1970, it only reached historians of the British Empire in the 1990s. Gregory Barton argues that the concept of environmentalism emerged from forestry studies, and emphasizes the British imperial role in that research. He argues that imperial forestry movement in India around 1900 included government reservations, new methods of fire protection, and attention to revenue-producing forest management. The result eased the fight between romantic preservationists and laissez-faire businessmen, thus giving the compromise from which modern environmentalism emerged.

In recent years numerous scholars cited by James Beattie have examined the environmental impact of the Empire. Beinart and Hughes argue that the discovery and commercial or scientific use of new plants was an important concern in the 18th and 19th centuries. The efficient use of rivers through dams and irrigation projects was an expensive but important method of raising agricultural productivity. Searching for more efficient ways of using natural resources, the British moved flora, fauna and commodities around the world, sometimes resulting in ecological disruption and radical environmental change. Imperialism also stimulated more modern attitudes toward nature and subsidized botany and agricultural research. Scholars have used the British Empire to examine the utility of the new concept of eco-cultural networks as a lens for examining interconnected, wide-ranging social and environmental processes.

Current practice

Frontier historian
Frederick Jackson Turner (1861–1932)

In the United States the American Society for Environmental History was founded in 1975 while the first institute devoted specifically to environmental history in Europe was established in 1991, based at the University of St. Andrews in Scotland. In 1986, the Dutch foundation for the history of environment and hygiene *Net Werk* was founded and publishes four newsletters per year. In the UK the White Horse Press in Cambridge has, since 1995, published the journal *Environment and History* which aims to bring scholars in the humanities and biological sciences closer together in constructing long and well-founded perspectives on present day environmental problems and a similar publication *Tijdschrift voor Ecologische Geschiedenis* is a combined Flemish-Dutch initiative mainly dealing with topics in the Netherlands and Belgium although it also has an interest in European environmental history. Each issue contains abstracts in English, French and German. In 1999 the Journal was converted into a yearbook for environmental history. In Canada the Network in Canadian History and Environment facilitates the growth of environmental history through numerous workshops and a significant digital infrastructure including their website and podcast.

Communication between European nations is restricted by language difficulties. In April 1999 a meeting was held in Germany to overcome these problems and to co-ordinate environmental history in Europe. This meeting resulted in the creation of the European Society for Environmental History in 1999. Only two years after its establishment, ESEH held its first international conference in St. Andrews, Scotland. Around 120 scholars attended the meeting and 105 papers were presented on topics covering the whole spectrum of environmental history. The conference showed that environmental history is a viable and lively field in Europe and since then ESEH has expanded to over 400 members and continues to grow and attracted international conferences in 2003 and 2005. In 1999 the *Centre for Environmental History* was established at the University of Stirling. Some history departments at European universities are now offering introductory courses in environmental history and postgraduate courses in Environmental history have been established at the Universities of Nottingham, Stirling and Dundee and more recently a Graduierten Kolleg was created at the University of Göttingen in Germany. In 2009, the Rachel Carson Center for Environment and Society (RCC), an international, interdisciplinary center for research and education in the environmental humanities and social sciences, was founded as a joint initiative of Munich's Ludwig-Maximilians-Universität and the Deutsches Museum, with the generous support of the German Federal Ministry of Education and Research. The Environment & Society Portal (environmentandsociety.org) is the Rachel Carson Center's open access digital archive and publication platform.

Related Disciplines

The 77 km long Panama Canal, opened in 1914, connects the Caribbean Sea to the Pacific Ocean, replacing a long and treacherous shipping route passing via the Drake Passage and Cape Horn at the tip of South America. Construction was plagued by problems, including disease (particularly malaria and yellow fever) and landslides. By the time the canal was completed, a total of 27,500 French and American workmen are estimated to have died.

Environmental history prides itself in bridging the gap between the arts and natural sciences although to date the scales weigh on the side of science. A definitive list of related subjects would be lengthy indeed and singling out those for special mention a difficult task. However, those frequently quoted include, historical geography, the history and philosophy of science, history of technology and climate science. On the biological side there is, above all, ecology and historical ecology, but also forestry and especially forest history, archaeology and anthropology. When the subject engages in environmental advocacy it has much in common with environmentalism.

With increasing globalization and the impact of global trade on resource distribution, concern over never-ending economic growth and the many human inequities environmental history is now gaining allies in the fields of ecological and environmental economics.

Engagement with sociological thinkers and the humanities is limited but cannot be ignored through the beliefs and ideas that guide human action. This has been seen as the reason for a perceived lack of support from traditional historians.

Issues

The subject has a number of areas of lively debate. These include discussion concerning: what subject matter is most appropriate; whether environmental advocacy can detract from scholarly objectivity; standards of professionalism in a subject where much outstanding work has been done by non-historians; the relative contribution of nature and humans in determining the passage of history; the degree of connection with, and acceptance by, other disciplines - but especially mainstream history. For Paul Warde the sheer scale, scope and diffuseness of the environmental history endeavour calls for an analytical toolkit "*a range of common issues and questions to push forward collectively*" and a "*core problem*". He sees a lack of "*human agency*" in its texts and suggest it be written more to act: as a source of information for environmental scientists; incorporation of the notion of risk; a closer analysis of what it is we mean by "environment"; confronting the way environmental history is at odds with the humanities because it emphasises the division between "materialist, and cultural or constructivist explanations for human behaviour".

Global sustainability

Achieving sustainability will enable the Earth to continue supporting human life as we know it. Blue Marble NASA composite images: 2001 (left), 2002 (right)

Many of the themes of environmental history inevitably examine the circumstances that produced the environmental problems of the present day, a litany of themes that challenge global sustainability including: population, consumerism and materialism, climate change, waste disposal, deforestation and loss of wilderness, industrial agriculture, species extinction, depletion of natural resources, invasive organisms and urban development. The simple message of sustainable use of renewable resources is frequently repeated and early as 1864 George Perkins Marsh was pointing out that the changes we make in the environment may later reduce the environments usefulness to humans so any changes should be made with great care - what we would nowadays call enlightened self-interest. Richard Grove has pointed out that *States will act to prevent environmental degradation only when their economic interests are threatened*.

Advocacy

It is not clear whether environmental history should promote a moral or political agenda. The strong emotions raised by environmentalism, conservation and sustainability can interfere with historical objectivity: polemical tracts and strong advocacy can compromise objectivity and professionalism. Engagement with the political process certainly has its academic perils although accuracy and commitment to the historical method is not necessarily threatened by environmental involvement: environmental historians have a reasonable expectation that their work will inform policy-makers.

Declensionist narratives

Narratives of environmental history tend to be declensionist, that is, accounts of progressive decline under human activity. Thus environmental history, like environmentalism, is perceived as entrenched pessimism, a litany of degeneration, failure, loss, decline and decay often portrayed as proceeding from some halcyon golden age of the past.

Presentism and culpability

Under the accusation of "presentism" it is sometimes claimed that, with its genesis in the late 20th century environmentalism and conservation issues, environmental history is simply a reaction to contemporary problems, an "attempt to read late twentieth century developments and concerns back into past historical periods in which they were not operative, and certainly not conscious to human participants during those times". This is strongly related to the idea of culpability. In environmental debate blame can always be apportioned, but it is more constructive for the future to understand the values and imperatives of the period under discussion so that causes are determined and the context explained. An awareness of presentism can help us to be wary of the easy wisdom of hindsight.

Environmental Determinism

Ploughing farmer in ancient Egypt. Mural in the burial chamber of artisan Sennedjem c. 1200 BCE

For some environmental historians "*the general conditions of the environment, the scale and arrangement of land and sea, the availability of resources, and the presence or absence of animals available for domestication, and associated organisms and disease vectors, that makes the development of human cultures possible and even predispose the direction of their development*" and that "*history is inevitably guided by forces that are not of human origin or subject to human choice*". This approach has been attributed to American environmental historians Webb and Turner and, more recently to Jared Diamond in his book "*Guns, Germs, and Steel*", where the presence or absence of disease vectors and resources such as plants and animals that are amenable to domestication that may not only stimulate the development of human culture but even determine, to some extent, the direction of that development. The claim that the path of history has been forged by environmental rather than cultural forces is referred to as environmental determinism while, at the other extreme, is what may be called cultural determinism. An example of cultural determinism would be the view that human influence is so pervasive that the idea of pristine nature has little validity - that there is no way of relating to nature without culture.

Methodology

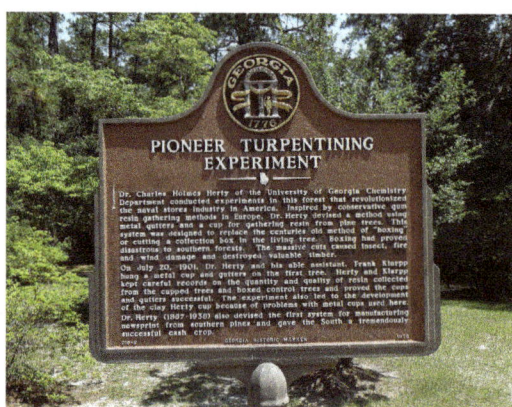

Recording historical events

Useful guidance on the process of doing environmental history has been given by Donald Worster, Carolyn Merchant, William Cronon and Ian Simmons. Worster's three core subject areas (the environment itself, human impacts on the environment, and human thought about the environment) are generally taken as a starting point for the student as they encompass many of the different skills required. The tools are those of both history and science with a requirement for fluency in the language of natural science and especially ecology. In fact methodologies and insights from a range of physical and social sciences is required, there seeming to be universal agreement that environmental history is indeed a multidisciplinary subject.

Future

Environmental history, like all historical studies, shares the hope that through an examination of past events it may be possible to forge a more considered future. In particular a greater depth of historical knowledge can inform environmental controversies and guide policy decisions.

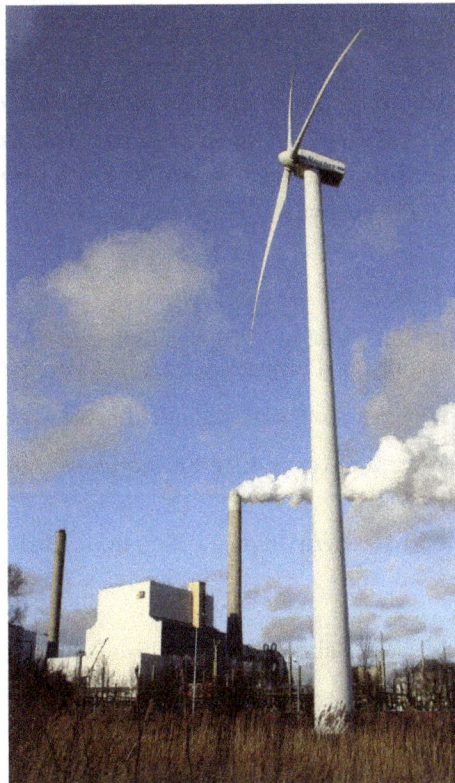

Old and new human uses of the atmosphere

The subject continues to provide new perspectives, offering cooperation between scholars with different disciplinary backgrounds and providing an improved historical context to resource and environmental problems. There seems little doubt that, with increasing concern for our environmental future, environmental history will continue along the path of environmental advocacy from which it originated as *"human impact on the living systems of the planet bring us no closer to utopia, but instead to a crisis of survival"* with key themes being population growth, climate change, conflict over environmental policy at different levels of human organization, extinction, biological invasions, the environmental consequences of technology especially biotechnology, the reduced supply of resources - most notably energy, materials and water. Hughes comments that environmental historians *"will find themselves increasingly challenged by the need to explain the background of the world market economy and its effects on the global environment. Supranational instrumentalities threaten to overpower conservation in a drive for what is called sustainable development, but which in fact envisions no limits to economic growth"*. Hughes also notes that *"environmental history is notably absent from nations that most adamantly reject US, or Western influences"*.

Michael Bess sees the world increasingly permeated by potent technologies in a process he calls "artificialization" which has been accelerating since the 1700s, but at a greatly accelerated rate after 1945. Over the next fifty years, this transformative process stands a good chance of turning our physical world, and our society, upside-down. Environmental historians can *"play a vital role in helping humankind to understand the gale-force of artifice that we have unleashed on our planet and on ourselves"*.

Against this background *"environmental history can give an essential perspective, offering knowledge of the historical process that led to the present situation, give examples of past problems and solutions, and an analysis of the historical forces that must be dealt with"* or, as expressed by William Cronon, *"The viability and success of new human modes of existing within the constraints of the environment and its resources requires both an understanding of the past and an articulation of a new ethic for the future."*

References

- Aboul-Enein, H. Yousuf; Zuhur, Sherifa (2004), Islamic Rulings on Warfare, Strategic Studies Institute, US Army War College, Diane Publishing Co., Darby PA, p. 22, ISBN 9781584871774

- Sylvie Nail (July 2008). Forest Policies and Social Change in England. Springer. p. 332. ISBN 9781402083648.

- Thomas R. DeGregori (2002). Bountiful Harvest: Technology, Food Safety, and the Environment. Cato Institute. p. 153. ISBN 1-930865-31-7.

- Martin Kitchen (2006). A History of Modern Germany, 1800-2000. Blackwell Publishing. p. 278. ISBN 1-4051-0040-0.

- Huesemann, Michael H., and Joyce A. Huesemann (2011). Technofix: Why Technology Won't Save Us or the Environment, New Society Publishers, Gabriola Island, British Columbia, Canada, ISBN 0865717044, 464 pp.

- https://www.theguardian.com/environment/2016/jun/20/environmental-activist-murders-global-witness-report

- James Beattie, Edward Melillo, and Emily O'Gorman. "Rethinking the British Empire through eco-cultural networks: materialist-cultural environmental history, relational connections and agency." Environment and History 20#4 (2014): 561-575.

- "Vasconcelos, Vitor Vieira "The Environment Professional and the Touch with Nature." Qualit@s, v 1, p. 1-10, 2011". Pt.scribd.com. 2013-04-28. Retrieved 2013-05-15.

- McCallum, M.L. and G.W. Bury (2013). "Google search patterns suggest declining interest in the environment". Biodiversity and Conservation. 22 (6): 1355–1367. doi:10.1007/s10531-013-0476-6.

- Neil Paul Cummins "An Evolutionary Perspective on the Relationship Between Humans and Their Surroundings: Geoengineering, the Purpose of Life & the Nature of the Universe", Cranmore Publications, 2012.

- James Beattie, "Recent Themes in the Environmental History of the British Empire," History Compass (Feb 2012) 10#2 pp 129-139

- "An Act to incorporate and confer powers upon the National Trust for Places of Historic Interest or Natural Beauty", The National Trust, accessed 4 June 2012

- "Environmentalism - Definition and More from the Free Merriam-Webster Dictionary". Merriam-webster. com. 2010-08-13. Retrieved 2012-06-20.

2

Essential Elements of Environmentalism

The essential elements of environmentalism are free-market environmentalism, evangelical environmentalism, conservation ethic, new urbanism, ecological footprint and green theory. Free-market environmentalism considers free markets and property rights to be the appropriate means of preserving the environment. The section strategically encompasses and incorporates the important elements of environmentalism, providing a complete understanding.

Free-market Environmentalism

Free-market environmentalism argues that the free market, property rights, and tort law provide the best means of preserving the environment, internalizing pollution costs, and conserving resources.

While environmental problems may be viewed as market failures, free market environmentalists argue that environmental problems arise because:

1. The state encodes, provides and enforces laws which override or obscure property rights and thus fail to protect them adequately.

2. Given the technological and legal context in which people operate, transaction costs are too high to allow parties to negotiate to a solution better for the environment.

3. Laws governing class or individual tort claims provide polluters with immunity from tort claims, or interfere with those claims in such a way as to make it difficult to legally sustain them.

Free-market environmentalists therefore argue that the best way to protect the environment is to clarify and protect property rights. This allows parties to negotiate improvements in environmental quality. It also allows them to use tortes to stop environmental harm. If affected parties can compel polluters to compensate them they will reduce or eliminate the externality. Market proponents advocate changes to the legal system that empower affected parties to obtain such compensation. They further claim that governments have limited affected parties' ability to do so by complicating the tort system to benefit producers over others.

Government Failure

Though many environmentalists blame markets for many of today's environmental problems, free-market environmentalists blame many of these problems on distortions of the market and the lack of markets. Government actions are blamed for a number of environmental detriments.

- Tragedy of the Commons is seen as a fundamental problem for the environment. When land is held in common, anybody may use it. Since resources are consumable, this creates the incentive for entrepreneurs to use common resources before somebody else does. Many environmental resources are held by the government or in common, such as air, water, forests. The problem with regulation is that it puts property into a political commons, where individuals try to appropriate public resources for their own gain, a phenomenon called rent-seeking.

- Tenure - Renters do not benefit from value accrued during their tenure and thus face an incentive to extract as much value as possible without conservation.

- Political Allocation - Political information does not have the incentives that markets do to seek superior information (profit and loss). Though many participants provide input to governments, they can only make one decision. This means that governments create rules that are not well crafted for local situations. The government's strategy is that of anticipation, to hide from danger through regulations. A healthier society would use resilience, facing and overcoming risks.

- Perverse Subsidies - Governments offer cross subsidies that distort price systems. This means that underconsumers and overconsumers are paying the same rates, so the underconsumer is overpaying and the overconsumer is underpaying. The incentive leads to more overconsumers and fewer underconsumers.

- Increased Transaction Costs - Governments may create rules that make it difficult to transfer rights in ways that benefit the environment. For example, in the US West, many states have laws over water rights that make it difficult for environmental groups to purchase instream flows from farmers.

Market Tools

Markets are not perfect, and free-market environmentalists assert that market-based solutions will have their mistakes. Through strong feedback mechanisms such as risk, profit and loss, market-driven have strong incentives to learn from mistakes.

- Individual Choice - Consumers have the incentive to maximize their satisfaction and try to find low cost, high value options. Markets allocate resources to the highest bidder. Producers make purchases on behalf of the consumer. Due to many actors in the market, there is no one-size-fits-all solution and entrepreneurs will seek to fulfill many values of society, including conservation.

- Entrepreneurship - Entrepreneurs seek value, problem-solve, and coordinate resources.

- Price System - When resources become scarce, prices rise. Rising prices incentivize entrepreneurs to find substitutions for these resources. These resources are often conserved. E.g. as prices for coal rise, consumers will use less and higher prices will drive substitution for different energy sources.

- Property Rights - Owners face a strong incentive to take care of and protect their property. They must decide how much to use today and how much to use tomorrow. Everybody is trying to grow value. Corporate value and share price is based on their anticipated future profits. Owners with the possibility of transferring their property, either to an heir or through sale want their property to grow in value. Property rights encourage conservation and defend resources against depletion, since there is a strong incentive to maximize the value of the resource for the future.

- Common Law - In order to have working property rights, you need a good system to defend them. When rights are weak, people will violate them. By creating a strong system, where common resources can be homesteaded, transferred, and defended against harm, resources can be protected, managed, allocated with the results that aggregate and balance humanity's needs and wants.

The market is a non-political allocation device. Many environmentalists proposals call to return resources from markets to become political problems.

Issues

Coasian Bargaining

Some economists argue that, if industries internalized the costs of negative externalities, they would face an incentive to reduce them, perhaps even becoming enthusiastic about taking advantage of opportunities to improve profitability through lower costs. Moreover, economists claim this would lead to the optimal balance between the marginal benefits of pursuing an activity and the marginal cost of its environmental consequences. One well-known means of internalizing a negative consequence is to establish a property right over some phenomenon formerly in the public domain.

The Coase Theorem is one extreme version of this logic. If property rights are well defined and if there are no transaction costs, then market participants can negotiate to a solution that internalizes the externality. Moreover, this solution will not depend on who is allocated the property right. For example, a paper mill and a resort might be on the same lake. Suppose the benefits to the resort of a clean lake outweigh the benefits to the mill of being able to pollute. It the mill has the right to pollute, the resort will pay it not to. If the resort has the right to a pollution-free lake, it will keep that right, as the mill will be unable to compensate it for its pollution. However, critics have charged that the "theorem" attributed to Coase is of extremely limited practicability because of its assumptions, including no transaction costs, and is ill-suited to real world externalities which have high bargaining costs due to many factors.

More generally, free-market environmentalists argue that transaction costs "count" as real costs. If the cost of re-allocating property rights exceeds the benefits of doing so, then it is actually optimal to stay in the status quo. This means the initial allocation of property rights is not neutral and also that it has important implications for efficiency. Nevertheless, given the existing property rights regime, costly changes to it are not necessarily efficient, even if in hindsight an alternative regime would have been better. But if there are opportunities for property rights to evolve, entrepreneurs can find them to create new wealth.

Geolibertarianism

Libertarian Georgists (or Geolibertarians) maintain a strong essential commitment to free markets but reject the Coasian solution in favor of land value taxation, wherein the economic rent of land is collected by the community and either equally distributed to adult residents in the form of universal basic income, called the *Citizen's Dividend*, or used to fund necessary functions of a minimal government. Under the LVT system, only landholders are taxed and on the basis of the market value of the earth in its unimproved state, that is to say, apart from the value of any structures or products of human labor. Geolibertarians regard the LVT as just compensation for a legal land title granting exclusive access to that which logically precedes and generates private capital, whose supply is inelastic, which properly belongs to all, and to which all have an equal right because it is vital to human existence and economic activity—the ground itself—and thus consider land value capture both morally imperative and a natural source of revenue.

Taxation of land values has been advocated by many classical economists and theorists of classical liberalism, but this approach was popularized as the *Single Tax* by political economist and public intellectual Henry George in the late 19th century. Geolibertarians generally also support Pigouvian taxes on pollution and fees as compensation for natural resource extraction, negative externalities which adversely affect land values in particular. Many argue the monopolization of land promotes idle land speculation, real estate bubbles, urban sprawl and artificially severe wealth inequality, while violating the Lockean proviso and denying others rightful access to the earth.

Anarcho-Capitalism

Rothbardians also reject the proposed Coasian solution as making invalid assumptions about the purely subjective notion of costs being measurable in monetary terms, and also of making unexamined and invalid value judgments (i.e., ethical judgments). (PDF) The Rothbardians' solution is to recognize individuals' Lockean property rights, of which the Rothbardians maintain that *Wertfreiheit* (i.e., value-free) economic analysis demonstrates that this arrangement necessarily maximizes social utility. (PDF) Rothbard himself however believed the term free-market environmentalism to be oxymoronic, because the environment, being undeveloped and unowned, is in itself not truly property until it is transformed via the Lockean proviso. On environmentalism Rothbard said: "The problem is that environmentalists are not interested in efficiency or preserving private property....The environmentalists are acolytes and prisoners of a monstrous literally anti-human philosophy. They despise and condemn the human race, which by its very nature and in contrast to other creatures, changes and transforms the environment instead of being passively subjected to it....I have come to the conclusion that a 'free-market environmentalist' is an oxymoron. Scratch one and you get...an environmentalist."

Markets and Ecosystems as Spontaneous Orders

Recent arguments in the academic literature have used Friedrich Hayek's concept of a spontaneous order to defend a broadly non-interventionist environmental policy. Hayek originally used the concept of a spontaneous order to argue against government intervention in the market. Like the market, ecosystems contain complex networks of information, involve an ongoing dynamic process, contain orders within orders, and the entire system operates without being directed by a conscious mind. On this analysis, species takes the place of price as a visible element of the system formed by a complex set of largely unknowable elements. Human ignorance about the countless interactions between the organisms of an ecosystem limits our ability to manipulate nature. Since humans rely on the ecosystem to sustain themselves, it is argued that we have an obligation to not disrupt such systems. This analysis of ecosystems as spontaneous orders does not rely on markets qualifying as spontaneous orders. As such, one need not endorse Hayek's analysis of markets to endorse ecosystems as spontaneous orders.

Others

Proponents of free-market environmentalism use the example of the recent destruction of the once prosperous Grand Banks fishery off Newfoundland. Once one of the world's most abundant fisheries, it has been almost completely depleted of fish. Those primarily responsible were large "factory-fishing" enterprises driven by the imperative to realize profits in a competitive global market. It is contended that if the fishery had been owned by a single entity, the owner would have had an interest in keeping a renewable supply of fish to maintain profits over the long term. The owner would thus have charged high fees to fish in the area, sharply reducing how many fish were caught. The owner also would have closely enforced rules on not catching young fish. Instead commercial ships from around the world raced to get the fish out of the water before competitors could, including catching fish that had not yet reproduced.

Another example is in the 19th century early gold miners in California developed a trade in rights to draw from water courses based on the doctrine of prior appropriation. This was curtailed in 1902 by the Newlands Reclamation Act which introduced subsidies for irrigation projects. This had the effect of sending a signal to farmers that water was inexpensive and abundant, leading to uneconomic use of a scarce resource. Increasing difficulties in meeting demand for water in the western United States have been blamed on the continuing establishment of governmental control and a return to tradable property rights has been proposed.

According to Richard L. Stroup, markets in the environmental field, in order to function well, require "3-D" property rights to each important resource — i.e., rights that are clearly defined, easily defended against invasion, and divestible (transferable) by owners on terms agreeable to buyer and seller. The first two rights prevent property owners from being forced to accept pollution, and the third right provides an incentive for owners to be good stewards.

Criticisms

Some critics argue that free-market environmentalists have no method of dealing with collective problems like environmental degradation and natural resource depletion because of their rejection of collective regulation and control. They see natural resources as too difficult to privatize (e.g. water), as well as legal responsibility for pollution and degrading biodiversity as too hard to trace.

Evangelical Environmentalism

Evangelical environmentalism is an environmental movement in the United States in which some Evangelicals have emphasized biblical mandates concerning humanity's role as steward and subsequent responsibility for the care taking of Creation. While the movement has focused on different environmental issues, it is best known for its focus of addressing climate action from a biblically grounded theological perspective. The Evangelical Climate Initiative argues that human-induced climate change will have severe consequences and impact the poor the hardest, and that God's mandate to Adam to care for the Garden of Eden also applies to evangelicals today, and that it is therefore a moral obligation to work to mitigate climate impacts and support communities in adapting to change.

Some Evangelical groups have allied with environmentalists in teaching knowledge and developing awareness of global warming. The National Association of Evangelicals, a non profit organization, is working to encourage lawmakers to pass a law that would put restrictions on carbon emissions in the U.S.

Overview

The Evangelical environmental movement is rooted in the idea that humanity is engaging in sinfulness and disobedience to God by ignoring the mandate to "tend and keep" the land in which they were originally placed (Garden of Eden).

Evangelical environmentalists are committed to the authority of the Bible. Drawing from Genesis (Gen.) 2:15, humans are seen as caretakers (stewards) of God's Creation. Genesis 2:15 states: "And the LORD God took the man, and put him into the garden of Eden to dress it and to keep it." Richard T. Ritenbaugh writes, "Tend means "to work or serve", and thus referring to the ground or a garden, it can be defined as "to till or cultivate". It possesses the nuance seen in the King James Version's choice in its translation: "dress", implying adornment, embellishment, and improvement. Keep (Hebrew shamar) means "to exercise great care over". In the context of Genesis 2:15, it expresses God's wish that mankind, in the person of Adam, "take care of", "guard", or "watch over" the garden. A caretaker maintains and protects his charge so that he can return it to its owner in as good or better condition than when he received it.

From an Evangelical Environmentalist perspective, the response to the ecological crisis involves the restoration of correct doctrine, the restoration of Christianity as guide, and a balancing of the bible and biology. It is important to Evangelical Environmentalists that they are not seen as worshiping nature; they feel obligated to the stewardship of creation because of their focus on the creator of nature. Many Evangelical environmentalists prefer terms such as "creation care" or "stewardship of Creation" instead of environmentalism. The main reason for this preference is to emphasize the biblical basis for their engagement.

- "The love of God and Jesus require Christians to love Creation. The world is the property of God and as his stewards we are called to a high standard to tend and care for it" (Gen. 1; Ps. 24; Col. 1:16).

- "God commanded humanity to be stewards of His Creation" (Gen. 1:26-28).

In *Green Like God: Unlocking the Divine Plan for Our Planet* Merritt states the *Noah Covenant* is God entering a Covenant with all the Earth by citing Gen 9:9-10. Merritt continues from Gen 2:15 ...

- "Cultivate it and keep it" New American Standard Bible

- "Tend it and watch over it" New Living Translation

- "Work it and keep it" English Standard Version

- "Take care of it and to look after it" Holman Christian Standard Bible

- "Till it and watch keep it" Revised Standard Version

- "Work the ground and keep it in order" The Message (Bible)

Specific Actions

Many American religious organizations have a long record of opposing nuclear weapons. Rejecting the development and use of nuclear weapons is "...one of the most widely shared convictions across faith traditions". In the 1980s religious groups organized large anti-nuclear protests involving hundreds of thousands of people, and specific groups involved included the Southern Baptist Convention, and the Episcopal Church. The Protestant, Catholic, and Jewish communities published explicitly anti-nuclear statements, and in 2000 Muslims also began to take a stance against nuclear weaponry.

In February 2006, a group of 86 notable U.S. evangelical Christian leaders launched the Evangelical Climate Initiative, a campaign for environmental reform, calling on all Christians to push for federal legislation that would reduce carbon dioxide emissions in an effort to stem global warming.

The initiative's organizers intended to lobby federal legislators, hold creation care talks at churches and colleges, and air television and radio advertisements that link drought, starvation, and hurricanes to global warming. The president of the NAE (National Association of Evangelicals), Rev. Ted Haggard, did not join the leaders in the statement on global warming. He declined because it would have been interpreted as an endorsement by the entire NAE. Speaking just for himself, he said, "There is no doubt about it in my mind that climate change is happening, and there is no doubt about it that it would be wise for us to stop doing the foolish things we're doing that could potentially be causing this. In my mind there is no downside to being cautious."

Televangelist Pat Robertson changed his stance on global warming. In October 2005, Robertson accused the Evangelical Climate Initiative of teaming up with "far-left environmentalists," but in the summer of 2006 on his 700 Club television show, Robertson stated that "they're making a convert out of me." He also said "We really need to address the burning of fossil fuels. If we are contributing to the destruction of this planet, we need to do something about it.

In March 2008, a group of Southern Baptist leaders issued a statement that their denomination had been timid when it came to environmental issues and that they have a duty to stop global warming. This declaration was signed by the President of the congregation. The Southern Baptist Convention is the largest Protestant denomination in the United States with 16.3 million members.

The Evangelical Environmental Network (EEN) has received much attention for actions such as lobbying the United States Congress and producing biblical educational materials which were sent out to Evangelical congregations across the country. Also, in 1996 the EEN produced television advertisements warning Christians, "don't let the special interests sink the Endangered Species Act." The EEN also mailed out 38,000 "Let the Earth Be Glad" packets and enlisted 1000 churches as "Noah Congregations".(Kearns, 1997)

Michael Banach, the Vatican representative to the International Atomic Energy Agency, told a Vienna conference in 2011 that the Japanese Fukushima nuclear disaster created new safety concerns about nuclear plants globally. Auxiliary bishop of Osaka Michael Goro Matsuura said this nuclear power disaster should encourage Japan and other countries to abandon nuclear projects. He called on worldwide Christian solidarity to provide wide support for an anti-nuclear campaign. Statements from bishops' conferences in Korea and the Philippines called on their governments to abandon atomic power. Columban priest Fr Seán McDonagh has written a book entitled *Is Fukushima the Death Knell for Nuclear Energy?*. Nobel laureate Kenzaburō Ōe has said Japan should quickly abandon its nuclear reactors.

Criticism

A criticism of Creation Care, stemming from Jan Markell (who believes that evangelicals are "called by God to win souls for Jesus, not to exercise stewardship over creation"), is that Creation Care is primarily a push for a massive "Cap and Trade" tax increase under the veil of helping the environment.

In January 2006, a group of evangelicals opposed the Evangelical Climate Initiative's stance and issued a letter to the NAE which stated that "global warming is not a consensus issue, and our love for the Creator and respect for His Creation does not require us to take a position [supporting a cap and trade tax increase]". In 2007 the New York Times reported, "leaders of the conservative Christian wing of the Republican Party, including James Dobson, Gary Bauer and Paul Weyrich, told the policy director of the NAE, the Rev. Richard Cizik, to shut up already about global warming". In 2008, the NAE clarified that it does not "have a specific position on global warming or emissions. [Cizik] has spoken as an individual on that."

Ann Coulter focuses on Genesis 1:27-28 which gives dominion to humanity over nature. Ann Coulter claims: "God gave us the earth. We have dominion over the plants, the animals, the trees. God said, 'Earth is yours. Take it. Rape it. It's yours.'" Lynn White (1967) implies that this is a common view among Christians, but the accuracy of this statement is debatable.

In 2008, a survey conducted by the Pew Forum on Religion and Public Life, found that 31% of white Evangelical Protestants did not believe that there is solid evidence showing that the Earth is warming. 34% of white Evangelical Protestants did believe that there was evidence, but 17% didn't believe that warming was due to human impacts. 47% of the total U.S. population does believe that the Earth is warming because of human influences and 58% of unaffiliated Americans believe that global warming due to human impacts is real.

Conservation (Ethic)

Conservation is an ethic of resource use, allocation, and protection. Its primary focus is upon maintaining the health of the natural world, its fisheries, habitats, and biological diversity. Secondary focus is on materials conservation, including non-renewable resources such as metals, minerals and fossil fuels, and energy conservation, which is important to protect the natural world. Those who follow the conservation ethic and, especially, those who advocate or work toward conservation goals are termed conservationists.

Satellite photograph of industrial deforestation in the Tierras Bajas project in eastern Bolivia, using skyline logging and replacement of forests by agriculture

Much attention has been given to preserving the natural characteristics of Hopetoun Falls, Australia, while allowing access for visitors

The terms *conservation* and *preservation* are frequently conflated outside of the academic, scientific, and professional literatures. The US National Park Service offers the following explanation of the important ways in which these two terms represent very different conceptions of environmental protection ethics:

"Conservation and preservation are closely linked and may indeed seem to mean the same thing. Both terms involve a degree of protection, but how that is protection is carried out is the key difference. Conservation is generally associated with the protection of natural resources, while preservation is associated with the protection of buildings, objects, and landscapes. Put simply, *conservation seeks the proper use of nature, while preservation seeks protection of nature from use.*

During the environmental movement of the early 20th century, two opposing factions emerged: conservationists and preservationists. Conservationists sought to regulate human use while preservationists sought to eliminate human impact altogether."

Introduction

To conserve habitat in terrestrial ecoregions and to stop deforestation is a goal widely shared by many groups with a wide variety of motivations.

To protect sea life from extinction due to overfishing or climate change is another commonly stated goal of conservation — ensuring that "some will be available for future generations" to continue a way of life.

The consumer conservation ethic is sometimes expressed by the *four R's*: " Rethink, Reduce, Recycle, Repair" This social ethic primarily relates to local purchasing, moral purchasing, the sustained, and efficient use of renewable resources, the moderation of destructive use of finite resources, and the prevention of harm to common resources such as air and water quality, the natural functions of a living earth, and cultural values in a built environment.

The principal value underlying most expressions of the conservation ethic is that the natural world has intrinsic and intangible worth along with utilitarian value — a view carried forward by the scientific conservation movement and some of the older Romantic schools of ecology movement.

More Utilitarian schools of conservation seek a proper valuation of local and global impacts of human activity upon nature in their effect upon human well being, now and to posterity. How such values are assessed and exchanged among people determines the social, political, and personal restraints and imperatives by which conservation is practiced. This is a view common in the modern environmental movement.

These movements have diverged but they have deep and common roots in the conservation movement.

In the United States of America, the year 1864 saw the publication of two books which laid the foundation for Romantic and Utilitarian conservation traditions in America. The posthumous publication of Henry David Thoreau's *Walden* established the grandeur of unspoiled nature as a citadel to nourish the spirit of man. From George Perkins Marsh a very different book, *Man and Nature*, later subtitled "The Earth as Modified by Human Action", catalogued his observations of man exhausting and altering the land from which his sustenance derives.

Terminology

The conservation of natural resources is the fundamental problem. Unless we solve that problem, it will avail us little to solve all others.

Theodore Roosevelt

In common usage, the term refers to the activity of systematically protecting natural resources such as forests, including biological diversity. Carl F. Jordan defines the term as:

biological conservation as being a philosophy of managing the environment in a manner that does not despoil, exhaust or extinguish.

While this usage is not new, the idea of biological conservation has been applied to the principles of ecology, biogeography, anthropology, economy and sociology to maintain. biodiversity.

The term "conservation" itself may cover the concepts such as cultural diversity, genetic diversity and the concept of movements environmental conservation, seedbank (preservation of seeds). These are often summarized as the priority to respect diversity, especially by Greens.

Much recent movement in conservation can be considered a resistance to commercialism and globalization. Slow food is a consequence of rejecting these as moral priorities, and embracing a slower and more locally focused lifestyle.

Practice

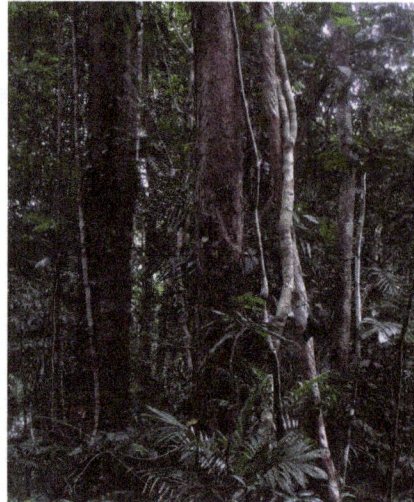

The Daintree Rainforest in Queensland, Australia

Distinct trends exist regarding conservation development. While many countries' efforts to preserve species and their habitats have been government-led, those in the North Western Europe tended to arise out of the middle-class and aristocratic interest in natural history, expressed at the level of the individual and the national, regional or local learned society. Thus countries like Britain, the Netherlands, Germany, etc. had what we would today term NGOs — in the shape of the RSPB, National Trust and County Naturalists' Trusts (dating back to 1889, 1895 and 1912 respectively) Natuurmonumenten, Provincial conservation Trusts for each Dutch province, Vogelbescherming, etc. — a long time before there were national parks and national nature reserves. This in part reflects the absence of wilderness areas in heavily cultivated Europe, as well as a longstanding interest in laissez-faire government in some countries, like the UK, leaving it as no coincidence that John Muir, the Scottish-born founder of the National Park movement (and hence of government-sponsored conservation) did his sterling work in the USA, where he was the motor force behind the establishment of such NPs as Yosemite and Yellowstone. Nowadays, officially more than 10 percent of the world is legally protected in some way or the other, and in practice private fundraising is insufficient to pay for the effective management of so much land with protective status.

Protected areas in developing countries, where probably as many as 70–80 percent of the species of the world live, still enjoy very little effective management and protection. Some countries, such as Mexico, have non-profit civil organizations and land owners dedicated to protect vast private property, such is the case of Hacienda Chichen's Maya Jungle Reserve and Bird Refuge in Chichen Itza, Yucatán. The Adopt A Ranger Foundation has calculated that worldwide about 140,000 rangers are needed for the protected areas in developing and transition countries. There are no data on how many rangers are employed at the moment, but probably less than half the protected areas in developing and transition countries have any rangers at all and those that have them are at least 50% short This means that there would be a worldwide ranger deficit of 105,000 rangers in the developing and transition countries.

One of the world's foremost conservationists, Dr. Kenton Miller, stated about the importance of rangers: "The future of our ecosystem services and our heritage depends upon park rangers. With the rapidity at which the challenges to protected areas are both changing and increasing, there has never been more of a need for well prepared human capacity to manage. Park rangers are the backbone of park management. They are on the ground. They work on the front line with scientists, visitors, and members of local communities."

Adopt A Ranger, fears that the ranger deficit is the greatest single limiting factor in effectively conserving nature in 75% of the world. Currently, no conservation organization or western country or international organization addresses this problem. Adopt A Ranger has been incorporated to draw worldwide public attention to the most urgent problem that conservation is facing in developing and transition countries: protected areas without field staff. Very specifically, it will contribute to solving the problem by fund raising to finance rangers in the field. It will also help governments in developing and transition countries to assess realistic staffing needs and staffing strategies.

Others, including Survival International, have advocated instead for cooperation with local tribal peoples, who are natural allies of the conservation movement and can provide cost-effective protection.

New Urbanism

New Urbanism is an urban design movement which promotes environmentally friendly habits by creating walkable neighborhoods containing a wide range of housing and job types. It arose in the United States in the early 1980s, and has gradually influenced many aspects of real estate development, urban planning, and municipal land-use strategies.

Seaside, Florida

Market Street, Celebration, Florida

New Urbanism is strongly influenced by urban design practices that were prominent until the rise of the automobile prior to World War II; it encompasses ten basic principles such as traditional neighborhood design (TND) and transit-oriented development (TOD). These ideas can all be circled back to two concepts: building a sense of community and the development of ecological practices.

The organizing body for New Urbanism is the Congress for the New Urbanism, founded in 1993. Its foundational text is the *Charter of the New Urbanism*, which begins:

We advocate the restructuring of public policy and development practices to support the following principles: neighborhoods should be diverse in use and population; communities should be designed for the pedestrian and transit as well as the car; cities and towns should be shaped by physically defined and universally accessible public spaces and community institutions; urban places should be framed by architecture and landscape design that celebrate local history, climate, ecology, and building practice.

New Urbanists support: regional planning for open space; context-appropriate architecture and planning; adequate provision of infrastructure such as sporting facilities, libraries and community centres; and the balanced development of jobs and housing. They believe their strategies can reduce traffic congestion by encouraging the population to ride bikes, walk, or take the train. They also hope that this set up will increase the supply of affordable housing and rein in suburban sprawl. The *Charter of the New Urbanism* also covers issues such as historic preservation, safe streets, green building, and the re-development of brownfield land. The ten Principles of Intelligent Urbanism also phrase guidelines for new urbanist approaches.

Architecturally, new urbanist developments are often accompanied by New Classical, postmodern, or vernacular styles, although that is not always the case.

Background

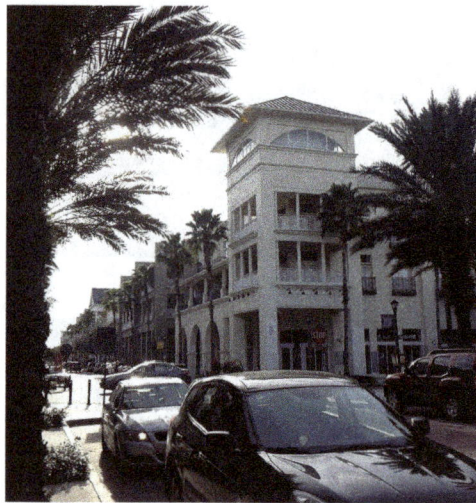

New Broad Street, Baldwin Park, Florida

Until the mid 20th century, cities were generally organized into and developed around mixed-use walkable neighborhoods. For most of human history this meant a city that was entirely walkable, although with the development of mass transit the reach of the city extended outward along transit lines, allowing for the growth of new pedestrian communities such as streetcar suburbs. But with the advent of cheap automobiles and favorable government policies, attention began to shift away from cities and towards ways of growth more focused on the needs of the car. Specifically, after World War II urban planning largely centered around the use of municipal zoning ordinances to segregate residential from commercial and industrial development, and focused on the construction of low-density single-family detached houses as the preferred housing format for the growing middle class. The physical separation of where people live from where they work, shop and frequently spend their recreational time, together with low housing density, which often drastically reduced population density relative to historical norms, made automobiles indispensable for practical transportation and contributed to the emergence of a culture of automobile dependency.

Beach Drive, St. Petersburg, Florida

This new system of development, with its rigorous separation of uses, arose after World War II and became known as "conventional suburban development" or pejoratively as urban sprawl. The majority of U.S. citizens now live in suburban communities built in the last fifty years, and automobile use per capita has soared.

Celebration, FL Post Office, designed by architect Michael Graves

Although New Urbanism as an organized movement would only arise later, a number of activists and thinkers soon began to criticize the modernist planning techniques being put into practice. Social philosopher and historian Lewis Mumford criticized the "anti-urban" development of postwar America. *The Death and Life of Great American Cities,* written by Jane Jacobs in the early 1960s, called for planners to reconsider the single-use housing projects, large car-dependent thoroughfares, and segregated commercial centers that had become the "norm." The French architect François Spoerry has developed in the 60' the concept of "soft architecture" that he applied to Port Grimaud, a new marina in south of France. The success of this project will have a considerable influence and will lead to many new projects of soft architecture like Port Liberté in New Jersey or Le Plessis Robisson in France.

Rooted in these early dissenters, the ideas behind New Urbanism began to solidify in the 1970s and 80s with the urban visions and theoretical models for the reconstruction of the "European" city proposed by architect Leon Krier, and the pattern language theories of Christopher Alexander. The term "new urbanism" itself started being used in this context in the mid-1980s, but it wasn't until the early 1990s that it was commonly written as a proper noun capitalized.

In 1991, the Local Government Commission, a private nonprofit group in Sacramento, California, invited architects Peter Calthorpe, Michael Corbett, Andrés Duany, Elizabeth Moule, Elizabeth Plater-Zyberk, Stefanos Polyzoides, and Daniel Solomon to develop a set of community principles for land use planning. Named the *Ahwahnee Principles* (after Yosemite National Park's Ahwahnee Hotel), the commission presented the principles to about one hundred government officials in the fall of 1991, at its first Yosemite Conference for Local Elected Officials.

Calthorpe, Duany, Moule, Plater-Zyberk, Polyzoides, and Solomon founded the Chicago-based Congress for the New Urbanism in 1993. The CNU has grown to more than three thousand members, and is the leading international organization promoting New Urbanist design principles. It holds annual Congresses in various U.S. cities.

In 2009, co-founders Elizabeth Moule, Hank Dittmar, and Stefanos Polyzoides authored the Canons of Sustainable Architecture and Urbanism to clarify and detail the relationship between New Urbanism and sustainability. The Canons are "a set of operating principles for human settlement that reestablish the relationship between the art of building, the making of community, and the conservation of our natural world." They promote the use of passive heating and cooling solutions, the use of locally obtained materials, and in general, a "culture of permanence."

New Urbanism is a broad movement that spans a number of different disciplines and geographic scales. And while the conventional approach to growth remains dominant, New Urbanist principles have become increasingly influential in the fields of planning, architecture, and public policy.

Defining Elements

Prospect New Town in Longmont, Colorado, showing a mix of aggregate housing and traditional detached homes

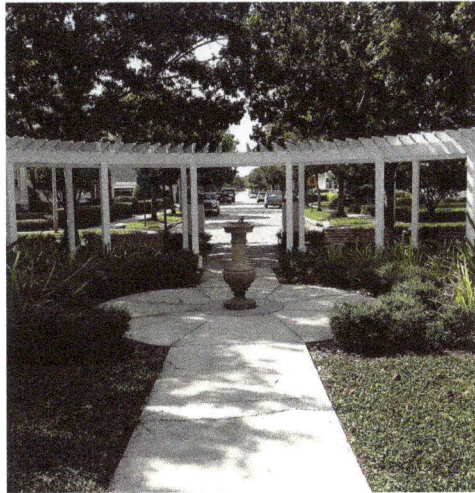

A park in Celebration, Florida

Andrés Duany and Elizabeth Plater-Zyberk, two of the founders of the Congress for the New Urbanism, observed mixed-use streetscapes with corner shops, front porches, and a diversity of well-crafted housing while living in one of the Victorian neighborhoods of New Haven, Connecticut. They and their colleagues observed patterns including the following:

- The neighborhood has a discernible center. This is often a square or a green and sometimes a busy or memorable street corner. A transit stop would be located at this center.

- Most of the dwellings are within a five-minute walk of the center, an average of roughly 0.25 miles (0.40 km).

- There are a variety of dwelling types — usually houses, rowhouses, and apartments — so that younger and older people, singles and families, the poor and the wealthy may find places to live.

- At the edge of the neighborhood, there are shops and offices of sufficiently varied types to supply the weekly needs of a household.

- A small ancillary building or garage apartment is permitted within the backyard of each house. It may be used as a rental unit or place to work (for example, an office or craft workshop).

- An elementary school is close enough so that most children can walk from their home.

- There are small playgrounds accessible to every dwelling — not more than a tenth of a mile away.

- Streets within the neighborhood form a connected network, which disperses traffic by providing a variety of pedestrian and vehicular routes to any destination.

- The streets are relatively narrow and shaded by rows of trees. This slows traffic, creating an environment suitable for pedestrians and bicycles.

- Buildings in the neighborhood center are placed close to the street, creating a well-defined outdoor room.

- Parking lots and garage doors rarely front the street. Parking is relegated to the rear of buildings, usually accessed by alleys.

- Certain prominent sites at the termination of street vistas or in the neighborhood center are reserved for civic buildings. These provide sites for community meetings, education, and religious or cultural activities.

Terminology

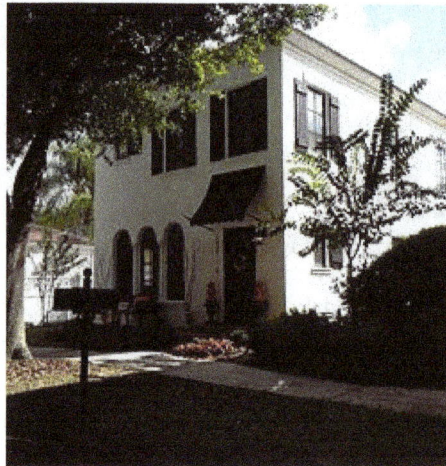

A Mediterranean Revival house in Celebration, Florida

A Key West style house in Baldwin Park, Florida

Several terms are viewed either as synonymous, included in, or overlapping with the New Urbanism. The terms Neotraditional Development or Traditional Neighborhood Development are often associated with the New Urbanism. These terms generally refer to complete New Towns or new neighborhoods, often built in traditional architectural styles, as opposed to smaller infill and redevelopment projects. The term Traditional Urbanism has also been used to describe the New Urbanism by those who object to the "new" moniker. The term "Walkable Urbanism" was proposed as an alternative term by developer and professor Christopher Leinberger. Many debate whether Smart Growth and the New Urbanism are the same or whether substantive differences exist between the two; overlap exists in membership and content between the two movements. Placemaking is another term that is often used to signify New Urbanist efforts or those of like-minded groups. The term Transit-Oriented Development is sometimes cited as being coined by prominent New Urbanist Peter Calthorpe and is heavily promoted by New Urbanists. The term Sustainable development is sometimes associated with the New Urbanism as there has been an increasing focus on the environmental benefits of New Urbanism associated with the rise of the term sustainability in the 2000s, however, this has caused some confusion as the term is also used by the United Nations and Agenda 21 to include human development issues (e.g., developing country) that exceed the scope of land development intended to be addressed by the New Urbanism or Sustainable Urbanism. The term livability or livable communities has also become popular under U.S. President Barack Obama's administration though it dates back at least to the mid-1990s when the term was used by the Local Government Commission.

Planning magazine discussed the proliferation of "urbanisms" in an article in 2011 titled "A Short Guide to 60 of the Newest Urbanisms." Several New Urbanists have popularized terminology under the umbrella of the New Urbanism including Sustainable Urbanism and Tactical Urbanism (of which Guerrilla Urbanism can be viewed as a subset). The term Tactical Urbanism was coined by Frenchman Michel de Certau in 1968 and revived in 2011 by New Urbanist Mike Lydon and the co-authors of the Tactical Urbanism Guide. In 2011 Andres Duany authored a book that used the term Agrarian Urbanism to describe an agriculturally-focused subset of New Urbanist town design. In 2013 a group of New Urbanists led by CNU co-founder Andres Duany began a research project under the banner of Lean Urbanism which purported to provide a bridge between Tactical Urbanism and the New Urbanism.

Other terms have surfaced in reaction to the New Urbanism intended to provide a contrast, alternative to, or a refinement of the New Urbanism. Some of these terms include Everyday Urbanism by Harvard Professor Margaret Crawford, John Chase, and John Kaliski, Ecological Urbanism, and True Urbanism by architect Bernard Zyscovich. Landscape urbanism was popularized by Charles Waldheim who explicitly defined it as in opposition to the New Urbanism in his lectures at Harvard University. A book called Landscape Urbanism and its Discontents, edited by Andres Duany and Emily Talen, specifically addressed the tension between these two views of urbanism. Michael E. Arth promotes what he describes as a variant of the New Urbanism called the New Pedestrianism, which is intended to be a more pedestrian-oriented and traces its origins to a 1929 planned community in Radburn, New Jersey.

Organizations

New urbanist *Sankt Eriksområdet* quarter in Stockholm, Sweden, built in the 1990s. (More photos)

Mixed use pedestrian-friendly street in Bitola, Republic of Macedonia.

The primary organization promoting the New Urbanism in the United States is the Congress for the New Urbanism (CNU). The Congress for the New Urbanism (CNU) is the leading organization promoting walkable, mixed-use neighborhood development, sustainable communities and healthier living conditions. CNU members promote the principles of CNU's Charter and the hallmarks of New Urbanism, including:

- Livable streets arranged in compact, walkable blocks.

- A range of housing choices to serve people of diverse ages and income levels.

- Schools, stores and other nearby destinations reachable by walking, bicycling or transit service.

- An affirming, human-scaled public realm where appropriately designed buildings define and enliven streets and other public spaces.

The CNU has met annually since 1993 when they held their first general meeting in Alexandria, Virginia, with approximately one hundred attendees. By 2008 the Congress was drawing two to three thousand attendees to the annual meetings.

The CNU began forming local and regional chapters circa 2004 with the founding of the New England and Florida Chapters. By 2011 there were 16 official chapters and interest groups for 7 more. As of 2013, Canada hosts two full CNU Chapters, one in Ontario (CNU Ontario), and one in British Columbia (Cascadia) which also includes a portion of the north-west US states.

While the CNU has international participation in Canada, sister organizations have been formed in other areas of the world including the Council for European Urbanism (CEU), the Movement for Israeli Urbanism (MIU) and the Australian Council for the New Urbanism.

By 2002 chapters of Students for the New Urbanism began appearing at universities including the Savannah College of Art and Design, University of Georgia, University of Notre Dame, and the University of Miami. In 2003, a group of younger professionals and students met at the 11th Congress in Washington, D.C. and began developing a "Manifesto of the Next Generation of New Urbanists". The Next Generation of New Urbanists held their first major session the following year at the 12th meeting of the CNU in Chicago in 2004. The group has continued meeting annually as of 2014 with a focus on young professionals, students, new member issues, and ensuring the flow of fresh ideas and diverse viewpoints within the New Urbanism and the CNU. Spinoff projects of the Next Generation of the New Urbanists include the Living Urbanism publication first published in 2008 and the first Tactical Urbanism Guide.

The CNU has spawned publications and research groups. Publications include the *New Urban News* and the *New Town Paper*. Research groups have formed independent nonprofits to research individual topics such as the Form-Based Codes Institute, The National Charrette Institute and the Center for Applied Transect Studies.

In the United Kingdom New Urbanist and European urbanism principles are practised and taught by the The Prince's Foundation for the Built Environment. Other organisations promote New Urbanism as part of their remit, such as INTBAU, A Vision of Europe, and others.

The CNU and other national organizations have also formed partnerships with like-minded groups. Organizations under the banner of Smart Growth also often work with the Congress for the New Urbanism. In addition the CNU has formed partnerships on specific projects such as working with the United States Green Building Council and the Natural Resources Defense Council to develop the LEED for Neighborhood Development standards, and with the Institute of Transportation Engineers to develop a Context Sensitive Solutions (CSS) Design manual.

Film

The New Urbanism Film Festival was held in 2013 and 2014 in Los Angeles to highlight films and short films about the New Urbanism and related topics. The 2011 film *Urbanized* by Gary Hustwit featured then CNU Board Chair Ellen Dunham-Jones and other urban thinkers on the international story of urbanization including the New Urbanist efforts in the United States.

The 2004 documentary *The End of Suburbia: Oil Depletion and the Collapse of the American Dream* argues that the depletion of oil will result in the demise of the sprawl-type development. *New Urban Cowboy: Toward a New Pedestrianism*, a feature length 2008 documentary about urban designer Michael E. Arth, explains the principles of his New Pedestrianism, a more ecological and pedestrian-oriented version of New Urbanism. The film also gives a brief history of New Urbanism, and chronicles the rebuilding of an inner city slum into a model of New Pedestrianism and New Urbanism called The Garden District.

Criticism

New Urbanism has drawn both praise and criticism from all parts of the political spectrum. It has been criticized both for being a social engineering scheme and for failing to address social equity and for both restricting private enterprise and for being a deregulatory force in support of private sector developers.

In an interview in *Reason*, a right-libertarian magazine, professor Peter Gordon, a professor of Urban Planning from University of Southern California, spoke in favor of suburbanization, stating that the New Urbanism ignores consumer preference the free market and that cities have moved towards car-oriented development because that is what people want.

On the other hand, journalist Alex Marshall has decried New Urbanism as essentially a marketing scheme that repackages conventional suburban sprawl behind a façade of nostalgic imagery and empty, aspirational slogans. In a 1996 article in Metropolis Magazine, Marshall denounced New Urbanism as "a grand fraud". The attack continued in numerous articles, including an opinion column in the Washington *Post* in September of the same year, and in Marshall's first book, *How Cities Work: Suburbs, Sprawl, and the Roads Not Taken*

Critics have asserted that the effectiveness claimed for the New Urbanist solution of mixed income developments lacks statistical evidence. Independent studies have supported the idea of addressing poverty through mixed-income developments, but the argument that New Urbanism produces such diversity has been challenged from findings from one community in Canada.

Some parties have criticized the New Urbanism for being too accommodating of motor vehicles and not going far enough to promote walking, cycling, and public transport. The Charter of the New Urbanism states that "communities should be designed for the pedestrian and transit as well as the car". Some critics suggest that communities should exclude the car altogether in favor of car-free developments. Michael E Arth proposes new pedestrianism as a way to further elevate the status of pedestrians by focusing on pedestrian-only paths. Steve Melia proposes the idea of "filtered permeability" which increases the connectivity of the pedestrian and cycling network resulting in a time and convenience advantage over drivers while still limited the connectivity of the vehicular network and thus maintaining the safety benefits of cul de sacs and horseshoe loops in resistance to property crime.

In response to critiques of a lack of evidence for the New Urbanism's claimed environmental benefits, a rating system for neighborhood environmental design, LEED-ND, was developed by the U.S. Green Building Council, Natural Resources Defense Council, and the Congress for the New Urbanism, to quantify the sustainability of New Urbanist neighborhood design. New Urbanist and board member of CNU, Doug Farr has taken a step further and coined Sustainable Urbanism, which combines New Urbanism and LEED-ND to create walkable, transit-served urbanism with high performance buildings and infrastructure.

New Urbanism has been criticized for being a form of centrally planned, large-scale development, "instead of allowing the initiative for construction to be taken by the final users themselves". It has been criticized for asserting universal principles of design instead of attending to local conditions.

United States

New Urbanism is having a growing influence on how and where metropolitan regions choose to grow. At least fourteen large-scale planning initiatives are based on the principles of linking transportation and land-use policies, and using the neighborhood as the fundamental building block of a region. Miami, Florida, has adopted the most ambitious New Urbanist-based zoning code reform yet undertaken by a major U.S. city.

More than six hundred new towns, villages, and neighborhoods, following New Urbanist principles, have been planned or are currently under construction in the U.S. Hundreds of new, small-scale, urban and suburban infill projects are under way to reestablish walkable streets and blocks. In Maryland and several other states, New Urbanist principles are an integral part of *smart growth* legislation.

In the mid-1990s, the U.S. Department of Housing and Urban Development (HUD) adopted the principles of the New Urbanism in its multibillion-dollar program to rebuild public housing projects nationwide. New Urbanists have planned and developed hundreds of projects in infill locations. Most were driven by the private sector, but many, including HUD projects, used public money.

University Place in Memphis

In 2010 University Place in Memphis became the second only U.S. Green Building Council (USGBC) LEED certified neighborhood. LEED ND (neighborhood development) standards integrates principles of smart growth, urbanism and green building and were developed through a collaboration between USGBC, Congress for the New Urbanism, and the Natural Resources Defense Council. University Place, developed by McCormack Baron Salazar, is a 405-unit, 30-acre, mixed-income, mixed use, multigenerational, HOPE VI grant community that revitalized the severely distressed Lamar Terrace public housing site.

The Cotton District

The Cotton District in Starkville, Mississippi, was the first New Urbanist development, begun in 1968 long before the New Urbanism movement was organized. The District borders Mississippi State University, and consists mostly of residential rental units for college students along with restaurants, bars and retail. The Cotton District got its name because it is built in the vicinity of an old cotton mill.

Seaside

Seaside, Florida, the first fully New Urbanist town, began development in 1981 on eighty acres (324,000 m²) of Florida Panhandle coastline. It was featured on the cover of the Atlantic Monthly in 1988, when only a few streets were completed, and has become internationally famous for its architecture, and the quality of its streets and public spaces.

Seaside is now a tourist destination and appeared in the 1998 movie *The Truman Show*. Lots sold for $15,000 in the early 1980s, and slightly over a decade later, the price had escalated to about $200,000. Today, most lots sell for more than a million dollars, and some houses top $5 million.

Mueller Community

The Mueller Community is located on the 700 acre site of the former Robert Mueller Municipal Airport in Austin, Texas, which closed in 1999. Per the developer, the value of the Mueller development upon completion will be $1.3 billion, and will comprise 4.2 million square feet of non-residential development, 650,000 square feet of retail space, 4,600 homes, and 140 acres of open space. An estimated 10,000 permanent jobs within the development will have been created by the time it is complete. The Mueller Community also has more electric cars per capita than any other neighborhood in the United States - a fact partially attributable to an incentive program.

Stapleton

The site of the former Stapleton International Airport in Denver, Colorado, closed in 1995, is now being redeveloped by Forest City Enterprises. Stapleton is expected to be home to at least 30,000 residents, six schools and 2 million square feet (180,000 m²) of retail. Construction began in 2001. Northfield Stapleton, one of the development's major retail centers, recently opened.

San Antonio

In 1997 San Antonio, Texas, as part of a new master plan, created new regulations called the Unified Development Code (UDC), largely influenced by New Urbanism. One feature of the UDC is six unique land development patterns that can be applied to certain districts: Conservation Development, Commercial Center Development, Office or Institutional Campus Development, Commercial Retrofit Development, Tradition Neighborhood Development, Transit Oriented Development. Each district has specific standards and design regulation. The six development patterns were created to reflect existing development patterns.

Mountain House

Mountain House, one of the latest New Urbanist projects in the United States, is a new town located near Tracy, California. Construction started in 2001. Mountain House will consist of 12 villages, each with its own elementary school, park, and commercial area. In addition, a future train station, transit center and bus system are planned for Mountain House.

Mesa Del Sol

Mesa del Sol, New Mexico—the largest New Urbanist project in the United States—was designed by architect Peter Calthorpe, and is being developed by Forest City Enterprises. Mesa del Sol may take five decades to reach full build-out, at which time it should have 38,000 residential units, housing a population of 100,000; a 1,400-acre (5.7 km²) industrial office park; four town centers; an urban center; and a downtown that would provide a twin city within Albuquerque.

I'On

Located in Mount Pleasant, South Carolina, I'On is a traditional neighborhood development, mixed with a new urbanism styled architecture, reflecting on the building designs of the nearby downtown areas of Charleston, South Carolina. Founded on April 30, 1995, I'On was designed by the town planning firms of Dover, Kohl & Partners and Duany Plater-Zyberk & Company, and currently holds over 750 single family homes. Features of the community include extensive sidewalks, shared public greens and parks, trails and a grid of narrow, traffic calming streets. Most homes are required to have a front porch of not less than eight feet (2.46 m) in depth. Floor heights of 10 feet (3.1 m), raised foundations and smaller lot sizes give the community a dense, vertical feel.

Haile Plantation

Haile Plantation, Florida, is a 2,600 household (1,700 acres (6.9 km²)) development of regional impact southwest of the city of Gainesville, within Alachua County. Haile Village Center is a traditional neighborhood center within the development. It was originally started in 1978 and completed in 2007. In addition to the 2,600 homes the neighborhood consists of two merchant centers (one a New England narrow street village and the other a chain grocery strip mall). There are also two public elementary schools and an 18-hole golf course.

Celebration, Florida

In June 1996, the Walt Disney Company unveiled its 5,000 acre (20 km²) town of Celebration, near Orlando, Florida. Celebration opened its downtown in October 1996, relying heavily on the experiences of Seaside, whose downtown was nearly complete. Disney shuns the label New Urbanism, calling Celebration simply a "town."

Celebration's Downtown has become one of the area's most popular tourist destinations making the community a showcase for New Urbanism as a prime example of the creation of a "sense of place".

Jersey City

The construction of the Hudson Bergen Light Rail in Hudson County, New Jersey has spurred transit-oriented development. In Jersey City, two projects are planned to transform brownfield sites, both of which have required remediation of toxic waste by previous owners. Bayfront, once site of a Honeywell plant is a 100 acres (0.40 km²) site on the Hackensack River, and is nearby the planned West Campus of New Jersey City University. Canal Crossing, named for the former Morris Canal, was once partially owned by PPG Industries, and is a 117 acres (0.47 km²) site west of Liberty State Park.

Old York Village, Chesterfield Township, New Jersey

The sparsely developed agricultural Township of Chesterfield in New Jersey covers approximately 21.61 square miles (56.0 km²) and has made farmland preservation a priority since the 1970s. Chesterfield has permanently preserved more than 7,000 acres (28 km²) of farmland through state and county programs and a township-wide transfer of development credits program that directs future growth to a designated "receiving area" known as Old York Village. Old York Village is a neo-traditional, new urbanism town on 560 acres (2.3 km²) incorporating a variety of housing types, neighborhood commercial facilities, a new elementary school, civic uses, and active and passive open space areas with preserved agricultural land surrounding the planned village. Construction began in the early 2000s and a significant percentage of the community is now complete. Old York Village was the winner of the American Planning Association National Outstanding Planning Award in 2004.

Civita

Civita is a sustainable, transit-oriented 230-acre master-planned village under development in the Mission Valley area of San Diego, California, United States. Located on a former quarry site, the urban- style village is organized around a 19-acre community park that cascades down the terraced property.

Civita development plans call for 60 to 70 acres of parks and open space, 4,780 residences (including approximately 478 affordable units), an approximately 480,000-square-foot retail center, and 420,000 square feet for an office/business campus.

Del Mar Station

Del Mar (Los Angeles Metro station), which won a Congress for the New Urbanism Charter Award in 2003, is a transit-oriented development surrounding a prominent Metro Rail stop on the Gold Line, which connects Los Angeles and Pasadena. Located at the southern edge of downtown Pasadena, it serves as a gateway to the city with 347 apartments, out of which 15% are affordable units. Approximately 20,000 square feet of retail is linked with a network of public plazas, paseos and private courtyards. The 3.4-acre, $77 million project sits above a 1,200-car multi-level subterranean parking garage, with 600 spaces dedicated to transit. The light rail right of way, detailed as a public street, bisects the site. It was designed by Moule & Polyzoides.

Other Countries

New Urbanism is closely related to the Urban village movement in Europe. They both occurred at similar times and share many of the same principles although urban villages has an emphasis on traditional city planning. In Europe many brown-field sites have been redeveloped since the 1980s following the models of the traditional city neighbourhoods rather than Modernist models. One well-publicized example is Poundbury in England, a suburban extension to the town of Dorchester, which was built on land owned by the Duchy of Cornwall under the overview of Prince Charles. The original masterplan was designed by Leon Krier. A report carried out after the first phase of construction found a high degree of satisfaction by residents, although the aspirations to reduce car dependency had not been successful. Rising house prices and a perceived premium have made the open market housing unaffordable for many local people.

The Council for European Urbanism (CEU), formed in 2003, shares many of the same aims as the U.S.'s New Urbanists. CEU's Charter is a development of the Congress for the New Urbanism Charter revised and reorganised to relate better to European conditions. An Australian organisation, Australian Council for New Urbanism has since 2001 run conferences and events to promote New Urbanism in that country. A New Zealand Urban Design Protocol was created by the Ministry for the Environment in 2005.

There are many developments around the world that follow New Urbanist principles to a greater or lesser extent:

Europe

Example of Neo-Traditionalism at Le Plessis Robinson

The new marketplace of Le Plessis-Robinson

- Le Plessis-Robinson, a 21st-century example of neo-traditionalism, in the south-west of Paris. This city is in the process of transforming itself, destroying old modern blocklike buildings and replacing them with traditional buildings and houses in one of the biggest worldwide projects with Val d'Europe. In 2008 the city was nominated best architectural project of the European Union.

Jakriborg, started in the late 1990s near Malmö

- Val d'Europe, east of Paris, France. Developed by Disneyland Resort Paris, this town is a kind of European counterpart to Walt Disney World Celebration City.

- Jakriborg, in Southern Sweden, is a recent example of the New Urbanist movement.

- Brandevoort, in Helmond, in the Netherlands, is a new example of the New Urbanist movement.

- *Sankt Eriksområdet* quarter in Stockholm, Sweden, built in the 1990s.

- Other developments can be found at Heulebrug, part of Knokke-Heist, in Belgium, and Fonti di Matilde in San Bartolomeo (outside of Reggio Emilia), Italy.

Americas

- Orchid Bay, Belize, is one of the largest New Urbanist projects in Central America and the Caribbean.

- Las Catalinas, Costa Rica, is a coastal town in the Guanacaste Province of Northwest Costa Rica. Envisioned as a compact, walkable beach town, Las Catalinas was founded in 2006 by Charles Brewer and incorporates many of the principles of New Urbanism.

- McKenzie Towne is a New Urbanist development which commenced in 1995 by Carma Developers LP in Calgary.

- Cornell, within the city of Markham, Ontario, was designed with walkable neighborhoods, density to support public transit, a variety of housing types and retail.

- New Amherst is a new urbanist development in the town of Cobourg, Ontario.

- UniverCity, beside the Simon Fraser University campus on Burnaby Mountain in Burnaby, British Columbia, is an award-winning sustainable community that is designed to be walkable, dense, and well connected to public transit networks.

Asia

- The structure plan for Thimphu, Bhutan, follows Principles of Intelligent Urbanism, which share underlying axioms with the New Urbanism.

Africa

There are several such developments in South Africa. The most notable is Melrose Arch in Johannesburg. Triple Point is a comparable mixed-use development in East London, in Eastern Cape province. The development, announced in 2007, comprises 30 hectares. It is made up of three apartment complexes together with over 30 residential sites as well as 20,000 sq m of residential and office space. The development is valued at over R2 billion ($250 million).

Australia

Most new developments on the edges of Australia's major cities are master planned, often guided expressly by the principles of New Urbanism. The relationship between housing, activity centres, the transport network and key social infrastructure (sporting facilities, libraries, community centres etc.) is defined at structure planning stage.

- Tullimbar Village, NSW Australia, is a new development which follows the principles of New Urbanism.

Ecological Footprint

The natural resources of the Earth are finite.

An ecological footprint is a measure of human impact on Earth's ecosystems. It's typically measured in area of wilderness or amount of natural capital consumed each year. A common way of estimating footprint is, the area of wilderness of both land and sea needed to supply resources to a human population; This includes the area of wilderness needed to assimilate human waste.

At a global scale, it is used to estimate how rapidly we are depleting natural capital. The Global Footprint Network calculates the global ecological footprint from UN and other data. They estimate that as of 2007 our planet has been using natural capital 1.5 times as fast as nature can renew it.

Footprint Measurements

In 2007, the Global Footprint Network estimated the global ecological footprint as 1.6 planet Earths; that is, they judged that ecological services were being used 1.6 times as quickly as they were being renewed.

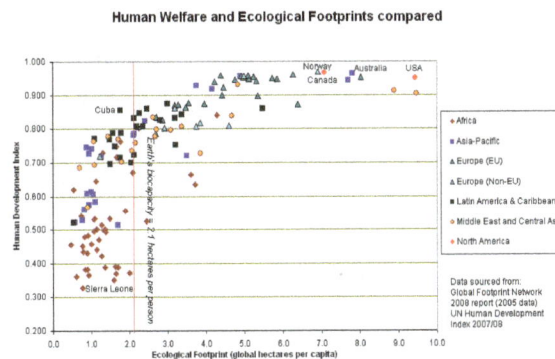

Ecological footprint for different nations compared to their Human Development Index.

Ecological footprints can be calculated at any scale: for an activity, a person, a community, a city, a region, a nation or humanity as a whole. Cities, due to population concentration, have large ecological footprints and have become ground zero for footprint reduction.

Global Footprints: Currently there is no fixed way to measure global footprints, and any attempts to describe the capacity of an ecosystem in a single number is a massive simplification of thousands of key renewable resources, which are not used or replenished at the same rate. However, there has been some convergence of metrics and standards since 2006.

City Ecological Footprints: are being measured. There are two types of measurements in use. The first measures ecosystem displacement which is defined as City Area minus remaining green spaces. This is an area measurement that does not include human or other biological activity. The Second attempts to quantify surviving ecosystem health. Specifically, it attempts to quantify both area and biological health of ecosystems surviving inside city areas such as nature reserves, parks, other green spaces. City footprints are being calculated and ranked with city ecological indexes.

Overview

The first academic publication about ecological footprints was by William Rees in 1992. The ecological footprint concept and calculation method was developed as the PhD dissertation of Mathis Wackernagel, under Rees' supervision at the University of British Columbia in Vancouver, Canada, from 1990–1994. Originally, Wackernagel and Rees called the concept "appropriated carrying capacity". To make the idea more accessible, Rees came up with the term "ecological footprint", inspired by a computer technician who praised his new computer's "small footprint on the desk". In early 1996, Wackernagel and Rees published the book *Our Ecological Footprint: Reducing Human Impact on the Earth* with illustrations by Phil Testemale.

Footprint values at the end of a survey are categorized for Carbon, Food, Housing, and Goods and Services as well as the total footprint number of Earths needed to sustain the world's population at that level of consumption. This approach can also be applied to an activity such as the manufacturing of a product or driving of a car. This resource accounting is similar to life-cycle analysis wherein the consumption of energy, biomass (food, fiber), building material, water and other resources are converted into a normalized measure of land area called global hectares (gha).

Per capita ecological footprint (EF), or ecological footprint analysis (EFA), is a means of comparing consumption and lifestyles, and checking this against nature's ability to provide for this consumption. The tool can inform policy by examining to what extent a nation uses more (or less) than is available within its territory, or to what extent the nation's lifestyle would be replicable worldwide. The footprint can also be a useful tool to educate people about carrying capacity and overconsumption, with the aim of altering personal behavior. Ecological footprints may be used to argue that many current lifestyles are not sustainable. Such a global comparison also clearly shows the inequalities of resource use on this planet at the beginning of the twenty-first century.

In 2007, the average biologically productive area per person worldwide was approximately 1.8 global hectares (gha) per capita. The U.S. footprint per capita was 9.0 gha, and that of Switzerland was 5.6 gha, while China's was 1.8 gha. The WWF claims that the human footprint has exceeded the biocapacity (the available supply of natural resources) of the planet by 20%. Wackernagel and Rees originally estimated that the available biological capacity for the 6 billion people on Earth at that time was about 1.3 hectares per person, which is smaller than the 1.8 global hectares published for 2006, because the initial studies neither used global hectares nor included bioproductive marine areas.

A number of NGOs offer ecological footprint calculators.

Ecological footprint analysis is now widely used around the Earth as an indicator of environmental sustainability. It can be used to measure and manage the use of resources throughout the economy. It can be used to explore the sustainability of individual lifestyles, goods and services, organizations, industry sectors, neighborhoods, cities, regions and nations. Since 2006, a first set of ecological footprint standards exist that detail both communication and calculation procedures.

Methodology

The ecological footprint accounting method at the national level is described in the l Footprint Atlas 2010 or in greater detail in the Calculation Methodology for the National Footprint Accounts. The National Accounts Review Committee has also published a research agenda on how the method will be improved.

In 2003, Jason Venetoulis, Carl Mas, Christopher Gaudet, Dahlia Chazan, and John Talberth developed Footprint 2., which offers a series of theoretical and methodological improvements to the standard footprint approach. The four primary improvements were that they included the entire surface of the Earth in biocapacity estimates, allocated space for other (i.e., non-human) species, updated the basis of equivalence factors from agricultural land to net primary productivity (NPP), and refined the carbon component of the footprint based on the latest global carbon models.

Studies in the United Kingdom

The UK's average ecological footprint is 5.45 global hectares per capita (gha) with variations between regions ranging from 4.80 gha (Wales) to 5.56 gha (East England).

Two recent studies have examined relatively low-impact small communities. BedZED, a 96-home mixed-income housing development in South London, was designed by Bill Dunster Architects and sustainability consultants BioRegional for the Peabody Trust. Despite being populated by relatively "mainstream" home-buyers, BedZED was found to have a footprint of 3.20 gha due to on-site renewable energy production, energy-efficient architecture, and an extensive green lifestyles program that included on-site London's first carsharing club. The report did not measure the added footprint of the 15,000 visitors who have toured BedZED since its completion in 2002. Findhorn Ecovillage, a rural intentional community in Moray, Scotland, had a total footprint of 2.56 gha, including both the many guests and visitors who travel to the community to undertake residential courses there and the nearby campus of Cluny Hill College. However, the residents alone have a footprint of 2.71 gha, a little over half the UK national average and one of the lowest ecological footprints of any community measured so far in the industrialized world. Keveral Farm, an organic farming community in Cornwall, was found to have a footprint of 2.4 gha, though with substantial differences in footprints among community members.

Critiques

Early criticism was published by van den Bergh and Verbruggen in 1999, which was updated in 2014. Another criticism was published in 2008. A more complete review commissioned by the Directorate-General for the Environment (European Commission) was published in June 2008.

A recent critique of the concept is due to Blomqvist et al., 2013a, with a reply from Rees and Wackernagel, 2013, and a rejoinder by Blomqvist et al., 2013b.

An additional strand of critique is due to Giampietro and Saltelli (2014a), with a reply from Goldfinger et al., 2014, a rejoinder by Giampietro and Saltelli (2014a), and additional comments from van den Bergh and Grazi (2015).

A number of countries have engaged in research collaborations to test the validity of the method. This includes Switzerland, Germany, United Arab Emirates, and Belgium.

Grazi et al. (2007) have performed a systematic comparison of the ecological footprint method with spatial welfare analysis that includes environmental externalities, agglomeration effects and trade advantages. They find that the two methods can lead to very distinct, and even opposite, rankings of different spatial patterns of economic activity. However this should not be surprising, since the two methods address different research questions.

Calculating the ecological footprint for densely populated areas, such as a city or small country with a comparatively large population — e.g. New York and Singapore respectively — may lead to the perception of these populations as "parasitic". This is because these communities have little intrinsic biocapacity, and instead must rely upon large *hinterlands*. Critics argue that this is a dubious characterization since mechanized rural farmers in developed nations may easily consume more resources than urban inhabitants, due to transportation requirements and the unavailability of economies of scale. Furthermore, such moral conclusions seem to be an argument for autarky. Some even take this train of thought a step further, claiming that the Footprint denies the benefits of trade. Therefore, the critics argue that the Footprint can only be applied globally.

The method seems to reward the replacement of original ecosystems with high-productivity agricultural monocultures by assigning a higher biocapacity to such regions. For example, replacing ancient woodlands or tropical forests with monoculture forests or plantations may improve the ecological footprint. Similarly, if organic farming yields were lower than those of conventional methods, this could result in the former being "penalized" with a larger ecological footprint. Of course, this insight, while valid, stems from the idea of using the footprint as one's only metric. If the use of ecological footprints are complemented with other indicators, such as one for biodiversity, the problem could maybe be solved. Indeed, WWF's Living Planet Report complements the biennial Footprint calculations with the Living Planet Index of biodiversity. Manfred Lenzen and Shauna Murray have created a modified Ecological Footprint that takes biodiversity into account for use in Australia.

Although the ecological footprint model prior to 2008 treated nuclear power in the same manner as coal power, the actual real world effects of the two are radically different. A life cycle analysis centered on the Swedish Forsmark Nuclear Power Plant estimated carbon dioxide emissions at 3.10 g/kWh and 5.05 g/kWh in 2002 for the Torness Nuclear Power Station. This compares to 11 g/kWh for hydroelectric power, 950 g/kWh for installed coal, 900 g/kWh for oil and 600 g/kWh for natural gas generation in the United States in 1999. Figures released by Mark Hertsgaard, however, show that because of the delays in building nuclear plants and the costs involved, investments in energy efficiency and renewable energies have seven times the return on investment of investments in nuclear energy.

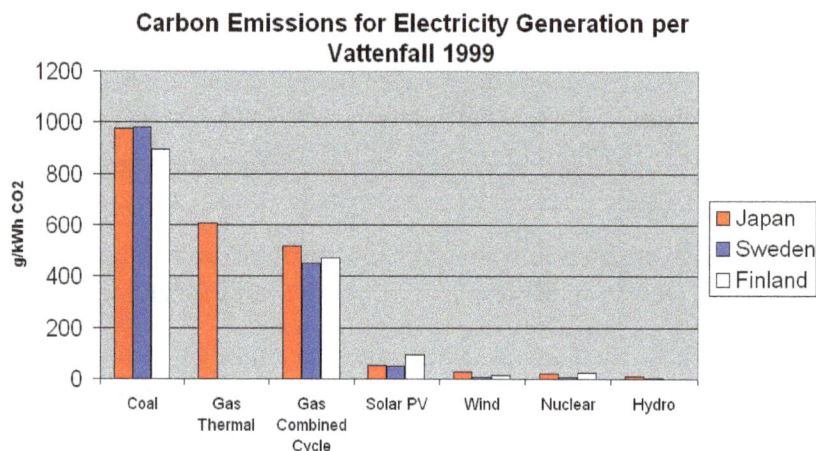

Carbon Emissions for Electricity Generation per Vattenfall 1999

The Swedish utility Vattenfall did a study of full life cycle emissions of Nuclear, Hydro, Coal, Gas, Solar Cell, Peat and Wind which the utility uses to produce electricity. The net result of the study was that nuclear power produced 3.3 grams of carbon dioxide per KW-Hr of produced power. This compares to 400 for natural gas and 700 for coal (according to this study). The study also concluded that nuclear power produced the smallest amount of CO_2 of any of their electricity sources.

Claims exist that the problems of nuclear waste do not come anywhere close to approaching the problems of fossil fuel waste. A 2004 article from the BBC states: "The World Health Organization (WHO) says 3 million people are killed worldwide by outdoor air pollution annually from vehicles and industrial emissions, and 1.6 million indoors through using solid fuel." In the U.S. alone, fossil fuel waste kills 20,000 people each year. A coal power plant releases 100 times as much radiation as a nuclear power plant of the same wattage. It is estimated that during 1982, US coal burning released 155 times as much radioactivity into the atmosphere as the Three Mile Island incident. In addition, fossil fuel waste causes global warming, which leads to increased deaths from hurricanes, flooding, and other weather events. The World Nuclear Association provides a comparison of deaths due to accidents among different forms of energy production. In their comparison, deaths per TW-yr of electricity produced (in UK and USA) from 1970 to 1992 are quoted as 885 for hydropower, 342 for coal, 85 for natural gas, and 8 for nuclear.

The Western Australian government State of the Environment Report included an Ecological Footprint measure for the average Western Australian seven times the average footprint per person on the planet in 2007, a total of about 15 hectares.

Footprint by Country

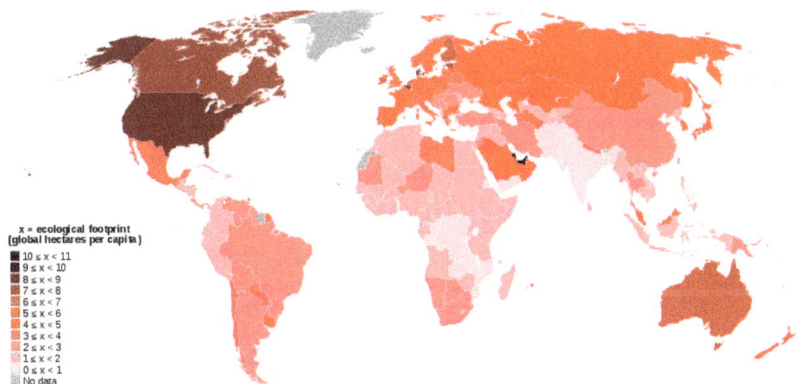

World map of countries by ecological footprint (data of 2007).

The world-average ecological footprint in 2012 was 1.8 global hectares per person. The average per country ranges from over 10 to under 1 hectares per person. There is also a high variation within countries, based on individual lifestyle and economic situation.

The GHG footprint differs from the ecological footprint in that the former is expressed in units of GHG warming potential (GGWP) and is generated by products or services, whereas the latter is expressed in units of land area and is generated by whole societies.

Implications

. . . the average world citizen has an eco-footprint of about 2.7 global average hectares while there are only 2.1 global hectare of bioproductive land and water per capita on earth. This means that humanity has already overshot global biocapacity by 30% and now lives unsustainabily by depleting stocks of "natural capital"

Ecological Indicator

Ecological indicators are used to communicate information about ecosystems and the impact human activity has on ecosystems to groups such as the public or government policy makers. Ecosystems are complex and ecological indicators can help describe them in simpler terms that can be understood and used by non-scientists to make management decisions. For example, the number of different beetle taxa found in a field can be used as an indicator of biodiversity.

Many different types of indicators have been developed. They can be used to reflect a variety of aspects of ecosystems, including biological, chemical and physical. Due to this variety, the development and selection of ecological indicators is a complex process.

Using ecological indicators is a pragmatic approach since direct documentation of changes in ecosystems as related to management measures, is cost and time intensive. For example, it would be expensive and time-consuming to count every bird, plant and animal in a newly restored wetland to see if the restoration was a success. Instead a few indicator species can be monitored to determine success of the restoration.

> *"It is difficult and often even impossible to characterize the functioning of a complex system, such as an eco-agrosystem, by means of direct measurements. The size of the system, the complexity of the interactions involved, or the difficulty and cost of the measurements needed are often crippling"*

The terms ecological indicator and environmental indicator are often used interchangeably. However, ecological indicators are actually a sub-set of environmental indicators. Generally, environmental indicators provide information on pressures on the environment, environmental conditions and societal responses. Ecological indicators refer only to ecological processes.

Policy Evaluation

Ecological indicators play an important role in evaluating policy regarding the environment.

Contributions

Indicators contribute to evaluation of policy development by:

- Providing decision-makers and the general public with relevant information on the current state and trends in the environment.

- Helping decision-makers better understand cause and effect relationships between the choices and practices of businesses and policy-makers versus the environment.

- Assisting to monitor and assess the effectiveness of measures taken to increase and enhance ecological goods and services.

Based on the United Nations convention to combat desertification and convention for biodiversity, indicators are planned to be built in order to evaluate the evolution of the factors. For instance, for the CCD, the Unesco-funded Observatoire du Sahara et du Sahel (OSS) has created the Réseau d'Observatoires du Sahara et du Sahel (ROSELT) (website) as a network of cross-Saharan observatories to establish ecological indicators.

Limitations

There are limitations and challenges to using indicators for evaluating policy programs.

For indicators to be useful for policy analysis, it is necessary to be able to use and compare indicator results on different scales (local, regional, national and international). Currently, indicators face the following spatial limitations and challenges:

1. Variable availability of data and information on local, regional and national scales.

2. Lack of methodological standards on an international scale.

3. Different ranking of indicators on an international scale which can result in different legal treatment.

4. Averaged values across a national level may hide regional and local trends.

5. When compiled, local indicators may be too diverse to provide a national result.

Indicators also face other limitations and challenges, such as:

1. Lack of reference levels, therefore it is unknown if trends in environmental change are strong or weak.

2. Indicator measures can overlap, causing over estimation of single parameters.

3. Long-term monitoring is necessary to identify long-term environmental changes.

4. Attention to more easily handled measurable indicators distracts from indicators less quantifiable such as aesthetics, ethics or cultural values.

Ecological Design

Ecological design is defined by Sim Van der Ryn and Stuart Cowan as "any form of design that minimizes environmentally destructive impacts by integrating itself with living processes." Ecological design is an integrative ecologically responsible design discipline.

It helps connect scattered efforts in green architecture, sustainable agriculture, ecological engineering, ecological restoration and other fields. The "eco" prefix was used to ninety sciences including eco-city, eco-management, eco-technique, eco-tecture. It was used by John Button in 1998 at the first time. The inchoate developing nature of ecological design was referred to the "adding in "of environmental factor to the design process, but later it was focused on the details of eco-design practice such as product system or individual product or industry as a whole. By including life cycle models through energy and materials flow, ecological design was related to the new interdisciplinary subject of industrial ecology. Industrial ecology meant a conceptual tool emulating models derived from natural ecosystem and a frame work for conceptualizing environmental and technical issues.

Living organisms exist in various systems of balanced symbiotic relationships. The ecological movement of the late twentieth-century is based on understanding that disruptions in these relationships has led to serious breakdown of natural ecosystems. In human history, technological means have resulted in growth of human populations through fire, implements and weapons. This dramatic increase in explosive population contributed the introduction of mechanical energies in machine production and there have been improvements in mechanized agriculture, manufactured chemical fertilizers and general health measures. Although the earlier invention inclined energy adjusting the ecological balance, population growth following the industrial revolution led to abnormal ecological change.

Ecological Design Issues and the Role of Designers

Since the Industrial Revolution, many propositions in the design field were raised with unsustainable design principles. The architect-designer Victor Papanek suggested that industrial design has murdered by creating new species of permanent garbage and by choosing materials and processes that pollute the air. For these issues, R. Buckminster Fuller, who was invited as University Professor at Southern Illinois University in Carbondale in 1960s, demonstrated how design could play a central role in identifying major world problems between 1965 and 1975. That included following contents:

- Review and analysis of world energy resources

- Defining more efficient uses of natural resources such as metals

- Integrating machine tools into efficient systems of industrial production

In the 1992 conference, 'The Agenda 21: The Earth Summit Strategy to Save Our Planet", a proposition was put forward that our world is on a path of energy production and consumption that cannot be sustained. The report drew attention to Individuals and groups around the world who have a set of principles to develop strategies for change that might be effective in world economics and trade policies, and the design professions will play a role in it. Namely, those meant that design profession becomes not what new products to make, but how to reinvent design culture likely to be realized. He noted designers firstly have to realize that design has historically been a dependent, contingent practice rather than one based on necessity. The design theorist, Clive Dilnot noted design becomes once again a means of ordering the world rather than merely of shaping products. As a broader approach, the conference of 'Agenda 21: The Earth Summit Strategy to Save Our Planet' 1992, emphasized that designers should challenge for facing human problems. These problems were mentioned to six themes: quality of life, efficient use of natural resources, protecting the global commons, managing human settlements, the use of chemicals and the management of human industrial waste, and fostering sustainable economic growth on a global scale.

History

- 1971 Ian McHarg, in his book "Design with Nature", popularized a system of analyzing the layers of a site in order to compile a complete understanding of the qualitative attributes of a place. McHarg gave every qualitative aspect of the site a layer, such as the history, hydrology, topography, vegetation, etc. This system became the foundation of today's Geographic Information Systems (GIS), a ubiquitous tool used in the practice of ecological landscape design.

- 1978 Permaculture. Bill Mollison and David Holmgren coin the phrase for a system of designing regenerative human ecosystems. (Founded in the work of Fukuoka, Yeoman, Smith, etc..

- 1994 David Orr, in his book "Earth in Mind: On Education, Environment, and the Human Prospect", compiled a series of essays on "ecolgocial design intelligence" and its power to create healthy, durable, resilient, just, and prosperous communities.

- 1994 Canadian biologists John Todd (biologist) and Nancy Jack Todd, in their book "From Eco-Cities to Living Machines" describe the precepts of ecological design.

- 2000 Ecosa Institute begins offering an Ecological Design Certificate, teaching designers to design with nature.

- 2004 Fritjof Capra, in his book "The Hidden Connections: A Science for Sustainable Living", wrote this primer on the science of living systems and considers the application of new thinking by life scientists to our understanding of social organization.

- 2004 K. Ausebel compiled compelling personal stories of the world's most innovative ecological designers in "Nature's Operating Instructions."

Influence

There are some clothing companies that are using several ecological design methods to change the future of the textile industry into a more environmentally friendly one. Recycling used clothing to minimize the use of resources, using biodegradable textile materials to reduce the impact on the environment, and using plant dyes instead of poisonous chemicals to improve the appearance of fabric.

Green Theory

Green Theory in international relations is a sub-field of international relations theory which concerns international environmental cooperation.

Liberal Institutionalism

The majority of scholarly literature in international relations approaches environmental problems from a liberal institutionalist perspective focusing on international environmental regimes. There is a relationship between Globalization and Environment which is among the forces behind the birth of green theory. However a unified theory is missed. Controversy of the human species as world managers where "conservation" is "right use" of nature and "preservation" is "right non-use" of nature and humanity as protecting nature against itself with little allowance for ecological dynamics is part of the cause for some "essentially contested concepts" where even "sustainable development" is sometimes contested.

The "collective action problem" is central to cooperation. It is discussed by Michael Laver (1997) in "Private Desires, Political Action" with examples of "The Prisoner's dilemma" and the "Tragedy of the commons." The disconnect between individual goals and group goals suggests a role for leadership and more than simple management. Green theory has championed consensus decision making as best it can be done. Empowering the disempowered is also a strand in Green theory.

Interactions are important to Green theory. From Stephen W. Littlejohn's (1983) book "Theories of Human Communication" (2nd ed.) we find a discussion of macronetworks with members and links. The five properties of links are: (1) symmetry (how much who relates to who or how equal the communication is); (2) strength (how often who relates to who); (3) reciprocity (agreement between members of links); (4) content (what the area or context of communication is); and, (5) mode (what the means or context of communication is). [Littlejohn is actually referring to Richard V. Farace, Peter R. Monge, and Hamish Russell from "Communicating and Organizing" (Reading, MA: Addison-Wesley, 1977)] Network communication is a bridge to appreciating the interdependence of ecosystems.

"International Regimes" is a classic IR work edited by Stephen D. Krasner (1983) which discusses emergent norms in complex (international) systems. International regimes are deemed to be a collective solution to problems of turbulence and unpredictability. On the micro level, Marshall Rosenberg's "Nonviolent Communication" (2nd ed., 2003) suggests people must agree on the description of the situation, agree on the stakeholders' feelings, agree on the stakeholders' needs, and then agree on the stakeholders' requests in order for healthy negotiation to be possible. In veridical conflict, when it occurs, there may not be a mutually satisfactory resolution possible.

Green IPE

This field studies the impact of IPE (international political economy) and it has been widely accepted as an area within IR theory. The strongest protesters of such irregular emigration movements are the ecofeminists who tend to gather once a month to hold non-peaceful and noisy demonstrations.

J. K. Galbraith said in "The Age of Uncertainty" that economics entails understanding the relationship of people with land. Green theory uses case studies of people living on land to better understand economy. Later, the idea of "ecological footprint" developed.

Green leaders use suasion, persuasion, exemplification and all the techniques of public relations and propaganda to shift the publics' tastes towards green decisions both in markets and in other areas where decisions, goals, or choices are being made. As Fraser P. Seitel (1989) in "The Practice of Public Relations" says there are many "publics" and many choices. Places or contexts of choices matter too. Green theory rewrites the rules for consumers. David R. Boyd & David T. Suzuki's (2008) "Green guide" comes to mind.

The public as workers is critical to Green theory. The greening of the labour market, workplace, and industry is important. One wonders, as democracy is valued in Green theory, what the appropriate attitude towards "economic democracy" would be. Robert A. Dahl (1985) in his "A Preface to Economic Democracy" argues weakly in favour of workplace democracy. If the factors of production are land, labour, capital, and entrepreneurship, it is unclear how these can be democratically related and what kind of property rights might survive.

Joseph Heath (2009), in "Filthy Lucre," describes "capitalism" as a "nexus of relationships" (as between suppliers, producers, consumers, marketers, regulators, for example). Green theory is complex in its management of policy networks. The New Age idea of "segmented, polycentric, integrated networks" (SPINs) suggests a possible complex replacement for capitalism. In public administration the idea of "governance" systems addresses some of the complexity.

Pierre Teilhard de Chardin's idea of "noosphere" as a connection and link thick environment where multilateral group, system, organization, and network interrelations can instantiate human wisdom is also possible for Green theory.

"Globalization" is commonly understood to involve cross-border flows. This can be transportation of material or people, transmission of information or ideas, transfer of capital or ownership, transactions altering status, transmissions of disease and disease causing organisms, trade in goods and services, transfers of technology or industrial arts and products, and may be considered a stage of "modernization" and/or "development." Global flows are to be monitored and controlled through many local actions. For this 'globalization' to work, local decisions must successfully restrict the perverse effects of global flows. A margin for error must be maintained.

Free trade appears to be focused upon free markets. Yet, even Raymond Vernon (1977) in "Storm Over the Multinationals" suggests that a multinational corporation could serve as a non-tariff barrier to free trade. You could go further and suggest that concentrations of purchasing power and control over property rights might serve to restrict "free trade" even if we do not wish to have the total equality of no concentrations of wealth. Free markets are not free if those who have wealth can control flows and restrictions as well as the litigation over disputes.

Control over energy and power, likewise, serves to replace the slave trade. "Nature" is never clearly enough defined. Natural resources and natural flows of energy and nutrients are neither entirely free nor entirely artificial. No one, apparently, wants total energy entropy (insofar as it can be delayed). It seems that a mixed economy may be best suited to the "natural" as complete rational planning and complete free markets are no friends to nature. This assumption is, of course, quite contestable.

Normative and Cosmopolitan

One aspect of Green IR theory is normative theorising such as bioregionalism. The idea of a "land ethic" and the belief that people can "think globally and act locally" have given hope that norms can be directly or indirectly derived from nature.

Futurology and counterfactual reasoning, such as that promoted by Philip E. Tetlock & Aaron Belkin (1996) in "Counterfactual Thought Experiments in World Politics," may just as easily produce dystopias as utopias. Environmental security, while still questionable, is at least more basic than human security. Likewise, "ecology" (the study of households or habitats) has priority over "economics" (the applied laws of households or habitats).

The fit between people and their environments brings up the topic of positive and negative eugenics which has been a background challenge in Green theory. Healthy and unhealthy have seemed to replace the theology of good and evil. Holistic health is a popular part of green theories combined with green living. Public health with prevention and health promotion are more consistent with Green theory too.

The non-violent thread in Green theory has led to an anti-hierarchical standard which can seem to be anarchical in this theory. Eco-feminists may hold anti-hierarchical views. Living wages and more equality may be emphasized. Ernest Callenbach's "Ecotopia" seems to almost be a matriarchal totalitarianism. The choices and liberties can be shifted in Green theory.

Turn taking among Green leaders means sometimes the responsibilities of power and sometimes being the follower. All this within groups attempting consensus with group skills added to communication skills which can be overpowering.

A behavioural space with contingencies, classical and operant conditioning, plus systems of semiotic systems with underlying structures, even for individual cells, organisms, groups, and all conceivable units of analysis could certainly be overpowering. It would seem to be a case of Michel Foucault's "governmentality" where both individuals and populations are simultaneously controlled. And the governmentality would grow and evolve.

Growth of any particular species can be in numbers, qualities, adaptations (fitting in), and adjustments (changes to environment) which makes for complex practical syllogisms. As Thomas Szasz said in "Ideology and Insanity": "stars move, machines function, animals behave, and people conduct themselves." It is an idea of "ecopsychology" that contact with nature promotes mental health and wellness.

According to Morris Berman there is a "Shadow Side of Systems" from the Journal of Humanistic Psychology (Winter 1996). Just as democracy can make for a tyranny of the majority so can systems thinking provide for authoritarianism and/or paternalism. A case study of Green theorists' dealing with the issue of "abortion" can illustrate this matter. Life and choice are both important to Green theorists. Disconnecting the conception decision from the birth decision can have eugenic consequences. Perhaps females could take charge of reproductive technology but then they would be in charge of eugenics too.

The compromises of living where you like and liking where you live; of doing what you like and liking what you do; and, of having the child you like and liking the child you have -- these do not make choices easier. Do we love the people we love and love the people we love -- both? Green theory does not always distinguish public space from private space. As for political ideologies and Green theory, Timothy O'Riordan (1990) in "Major Projects and the Environment" in Geographical Journal, indicates there to be "dry greens" (perhaps conservative and market centered), "shallow greens" (perhaps liberal and sustainable development centered), and "deep greens" (perhaps radical and ecosystem or earth centered). This, of course, may be an oversimplification. "Dry Greens" may be the least appreciated environmentalists and might not even be given that title. However, a very good economic treatment is given in "The Plundered Planet" by Paul Collier (2010). Collier may put the plight of the bottom billion (poorest and worst off people) on the planet above environmentalism. He believes protecting the viability of the planet and the bottom billion must both be done. Green theorists, of course, emphasize equality such that the bottom billion are equal to the top billion. How that works out is challenging. If the planet does need a billion less people, then there is no agreed upon way of choosing. Market solution would include rising food prices and only those with ability to pay survive. Each culture, religion, and academic discipline has developed and produced different solutions to the "Who survives?" conundrum.

Security Studies

Green IR approaches have challenged traditional approaches to security in international relations. This has included the concept of environmental security which has involved the 'securitization' of environmental threats.

While Green theory embraces non-violence and condemns the toxicity of much military materials, civilian defence and protest have long been used as tactics. Appropriate technology, even for military use, is becoming more preferred. Green theorists may dispense with much strategic security planning as letting the laws of nature take their course.

John H. Storer in "The Web of Life" (1953, 1956, pp. 76 – 77), mentions some of the laws of nature as: adaptation, succession, multiplication, and, control, such that a species occupies a niche with a carrying capacity and limiting factors. This is the long view of strategy. Strategy is assumed to be phylogenetic and tactics more ontogenetic (as may be morals and ethics respectively). Families and nations may be similar and different. "The Advent of Netwar" by John Arquilla & David Ronfeldt (RAND, 1996) suggests that fractal thinking may highlight fractal warfare. Swords and spears, the tools of regular warfare, may be changed to plows and pruning-hooks, the tools of eugenic management.

As for the short view, private security or acting locally may be of interest. Harvey Burstein (1994) in "Introduction to Security" gives five items that security staff must control: crime, waste, accidents, errors, and unethical practices. Such is the short view of tactics—albeit from an environmental perspective. Even the best "securitization" cannot eliminate error. The items that security controls also have their carrying capacities and limiting factors. Control may interact with higher level controls and counter-controls like a flow chart.

Agenda 21, that optimistic document for the beginning of the Twenty-first Century, seeks to control all accidents so that human beings can live in safety. It failed, but it can provide a guide for those who think globally and act locally. The move from human centered to ecosystem centered thinking and feeling may, ironically, be an improvement to humanity.

An area of challenge to all IR theories is secrecy and surveillance along with control of information. Monitoring is necessary for protection and control (security) but sometimes secrecy is necessary too. Realism, capitalism, and socialism could not handle secrecy well and it remains to be seen whether Green theory's handling of secrecy could not create worse disasters. Transparency and democracy may not be appropriate political technologies for all decisions.

Further discussion of power, secrecy vs. espionage, persuasion vs. evaluation, and information control is in "Policing Politics: Security Intelligence and the Liberal Democratic State" by Peter Gill (1994). Citizen policing of politics goes with empowering the disempowered and consensus. Perhaps "subsidiarity" may be needed with a secret "black box" at the appropriate level of decision making which is always the lowest possible.

It would be a good exercise in Green Theory to consider Derrick Jensen's "Twenty Premises" from his books "Endgame: Volumes I & II" (2006) with "Agenda 21" and its efforts to control humanity and the environment to produce security. Is a graceful transition from the present world to a greener future world possible? How possible?

A critical point in the history of Green Theory is 1992 when, of course, the Rio Conference happened. Also Biodiversity and Global Warming became more important. Emphasis may have moved from local ecosystems to the global ecosystem. One book published at this time in which the author seeks perhaps to rescue Greens from themselves and provide a revisionist Green policy is "Green Political Theory" by Robert E. Goodin (1992). It was a timely criticism. He compares a capitalist theory of value based upon consumer satisfaction; a socialist theory of value based upon labour; and, a Green theory of value based upon natural resources.

He also discusses the weakness for international organization of a highly decentralized political system. Perhaps in answer, "Leaders lead; but none too much," might provide a solution to Michel's 'iron law of oligarchy' where leaders can do job rotation with teaching environmental education, group decision-making, policy-making, as well as with other leaders, taking turns with administrative and diplomatic tasks. Much leaders make for much participation and many types of participation. Therefore diversity is increased.

Goodin (1992) also discusses norms or values by focusing on "goods" and "the good" as well as "agency" which is more in the direction of "the right." Such a critical analysis of Green Theory at what is also a critical point in history is invaluable.

An obvious modern point about environment and security is distribution of risk and even global flows of risk. If the rich are not to be protected too much, then it also might follow that the poor are not to be protected too much as well. Proper balance is not the mere matter of comforting the afflicted and afflicting the comfortable. Ulrich Beck and others have written on the "risk society" which seems to enter in to the discussions both of economy and security. Subsidiarity may include levels of government which share authority, for example, national government, religious or organizational government, family government, and individual government. There would not seem to be a monopoly over use of force. There may be a tendency for eco-anarchism here.

Just as we may use computer software to determine such things as financial crime and corruption, we might also use actuarial tables to determine unjust distribution of risk. We would have to find ways to prevent "fait accompli" acts of the powerful to the less powerful. This goes far beyond environmental impact statements or even studies. It is a whole fabric with the socio-economic structures.

Solving the problem of a gentle transition to a green future can be difficult. Between Derrick Jensen's "Endgame 1 & 2" and Donald R. Liddick's (2006) "Eco-Terrorism", it can seem like opposite sides must be taken. Liddick's book was published by Praeger Publishers which may have a reputation as being influenced by Intelligence agencies. Of course, there is not necessarily anything wrong with patriotism and good intelligence writing is often better than standard academic publishing. "Eco-Terrorism" could, however, be a bit of a propaganda piece. It might knock down weak forms of green arguments to establish those arguments of a more capitalist nature supporting free markets and corporations. Praeger Publishers may also have published "Crimes Against Nature" also by Donald R. Liddick (2011) and "The Rise of the Anti-Corporate Movement" by Evan Osborne (2007). If Green theorists are to take consensus decision-making quite seriously, then bringing the environmental skeptics on board to the gentle transition becomes very important to a Green future.

The question of whether (national) government alone must have a monopoly over the legitimate use of force has been contested for some time. Should national government also have a monopoly of the use of fear, violence, and terror and would this ever be legitimate? Can dissent, protest, and civil disobedience and holding the viewpoint of subversion of the current system be justified? Is the "right to rebel" really an euphemism for the "right to revolt"? It seems one other book: "Crime Wars: The Global Intersection of Crime, Political Violence, and International Law" by Paul Battersby, Joseph M. Siracusa, and Sasho Ripiloski (2011) might be worth reading. But, books like "Folks, this ain't normal" by Joel Salatin (2011) and "Green is the New Red" by Will Potter (2011) may also be worth reading.

Green theorists sometimes shift the meanings of "rights" and "freedoms" but a book to be published this year by Praeger Publishers about "Eco-psychology" (in two volumes) by Darlyne G. Nemeth, Robert B. Hamilton, and Judy Kuriansky, according to their catalogue would seem to shift some of the standard Green vocabulary. Even "eco-terrorism" is a use of a label with often pejorative connotations. It seems there is a battle going on over the future of Green discourse. We may have to rethink definitions for "genocide" and "ecocide" with both broad and narrow definitions for each and how much powers and authorities whether those of people and families, organizations (religious or otherwise), governments (at all levels), or businesses can be criminalized and especially where and how prevention can best be practiced. The Encyclopedia of Earth website is apparently neutral and non-controversial. Green theory has many essentially contested concepts. Encyclopedia of Earth may also be in the process of updating its ideas of biosecurity and bioterrorism. Antonio Gramsci's ideas of "war of manoeuvre" (real war in physical space) and "war of position" (war of words or what in Michel Foucault's terms could be a discussion and shaping of a discourse space) are pertinent to what is currently happening. John Stuart Mill's "On Liberty" applies to discourse rather well for a standard 'liberal' approach which, itself, is not neutral. It is supposed that Green thinkers clearly realize that much discussion need not lead to consensus and there can still remain veridical conflicts. Whether the 'soft energy path' and a 'steady state economy' can actually be made to work require further theory and practice too. The making of a fait accompli by mass ecocide or mass genocide (whether these be sudden or gradual) can hopefully be avoided.

I = PAT

I = PAT is the lettering of a formula put forward to describe the impact of human activity on the environment.

$$I = P \times A \times T$$

In words:

> Human Impact on the environment equals the product of Population, Affluence, and Technology. This shows how the population, affluence and technology produce an impact

The equation was developed in the 1970s during the course of a debate between Barry Commoner, Paul R. Ehrlich and John Holdren. Commoner argued that environmental impacts in the United States were caused primarily by changes in its production technology following World War II, while Ehrlich and Holdren argued that all three factors were important and emphasized in particular the role of human population growth.

The equation can aid in understanding some of the factors affecting human impacts on the environment, but it has also been cited as one of the primary factors underlying many of the dire environmental predictions of the 1970s by Paul Ehrlich, George Wald, Denis Hayes, Lester Brown, René Dubos, and Sidney Ripley that did not come to pass. Neal Koblitz classified equations of this type as "mathematical propaganda" and criticized Ehrlich's use of them in the media (e.g. on The Tonight Show) to sway the general public.

The Kaya identity is closely related to the I = PAT equation. The I = PAT equation is more general, describing an abstract "impact". The Kaya identity describes more clearly the impact of human activity on CO_2 emissions.

Population

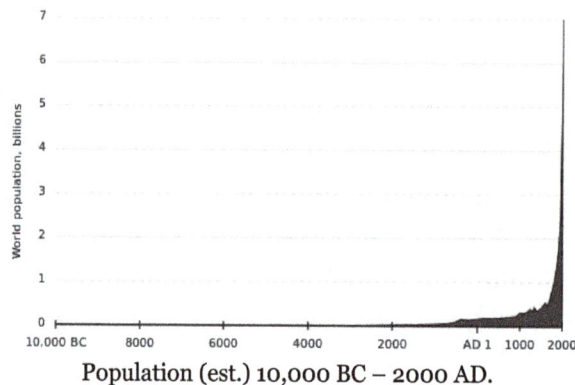

Population (est.) 10,000 BC – 2000 AD.

In the I=PAT equation, the variable P represents the population of an area, such as the world. Since the rise of industrial societies, human population has been increasing exponentially. This has caused Thomas Malthus and many others to postulate that this growth would continue until checked by widespread hunger and famine.

The United Nations and the US Census Bureau project that world population will increase from 7.4 billion today in 2016 (up from 7.0 billion forecast in 2005) to about 9.7 billion by 2050 (up from 9.2 billion). These projections take into consideration that population growth has slowed in recent years as women are having fewer children. This phenomenon is the result of demographic transition all over the world. The UN projects that human population might stabilize around 11.2 billion by 2100 (up from 9.2 billion). However, since the world population is set to keep rising for the next few decades, this factor of the I=PAT equation will likely keep increasing human impact on the environment for the near future.

Environmental Impacts

Increased population increases humans' environmental impact in many ways, which include but are not limited to:

- Increased land use - Results in habitat loss for other species.

- Increased resource use - Results in changes in land cover

- Increased pollution - Can cause sickness and damages ecosystems.

Affluence

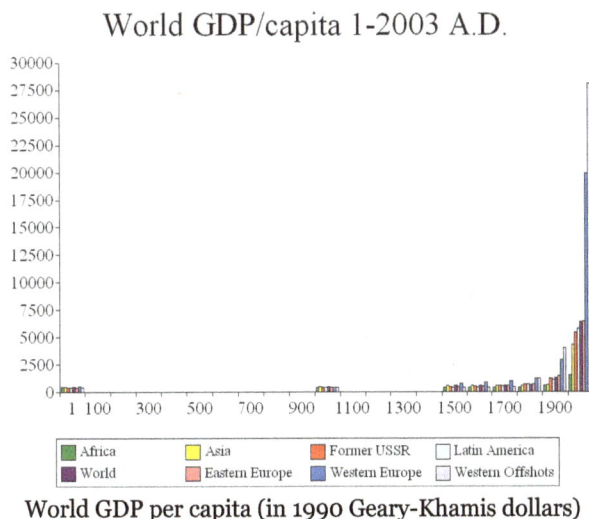

World GDP/capita 1-2003 A.D.

World GDP per capita (in 1990 Geary-Khamis dollars)

The variable **A**, in the I=PAT equation stands for affluence. It represents the average consumption of each person in the population. As the consumption of each person increases, the total environmental impact increases as well. A common proxy for measuring consumption is through GDP per capita. While GDP per capita measures production, it is often assumed that consumption increases when production increases. GDP per capita has been rising steadily over the last few centuries and is driving up human impact in the I=PAT equation.

Environmental Impacts

Increased consumption significantly increases human environmental impact. This is because each product consumed has wide ranging effects on the environment. For example, if the construction of a car had the following environmental impacts among others:

- 605,664 gallons of water for parts and tires

- 682 lbs. of pollution at a mine for the lead battery.

- 2178 lbs. of discharge into water supply for the 22 lbs. of copper contained in the car.

then the more cars per capita, the greater the impact. Since the ecological impacts of each product are far reaching, increases in consumption quickly result in large impacts on the environment.

Technology

The T variable in the I=PAT equation represents how resource intensive the production of affluence is; how much environmental impact is involved in creating, transporting and disposing of the goods, services and amenities used. Improvements in efficiency can reduce resource intensiveness, reducing the T multiplier. Since technology can affect environmental impact in many different ways, the unit for T is often tailored for the situation I=PAT is being applied to. For example, for a situation where the human impact on climate change is being measured, an appropriate unit for T might be greenhouse gas emissions per unit of GDP.

Environmental Impacts

Increases in efficiency can reduce overall environmental impact. However, since **P** has increased exponentially, and **A** has also increased drastically, the overall environmental impact, **I**, has still increased.

Reception

The I=PAT equation has been criticized for being too simplistic by assuming that P, A, and T are independent of each other. In reality, at least 7 interdependencies between P, A, and T could exist, indicating that it is more correct to rewrite the equation as I = f(P,A,T). For example, a doubling of technological efficiency, or equivalently a reduction of the T-factor by 50%, does not necessarily reduce the environmental impact (I) by 50% if efficiency induced price reductions stimulate additional consumption of the resource that was supposed to be conserved, a phenomenon called the rebound effect (conservation) or Jevons Paradox. As was shown by Alcott, despite significant improvements in the carbon intensity of GDP (i.e., the efficiency in carbon use) since 1980, world fossil energy consumption has increased in line with economic and population growth. Similarly, an extensive historical analysis of technological efficiency improvements has conclusively shown that energy and materials use efficiency improvements were almost always outpaced by economic growth, resulting in a net increase in resource use and associated pollution.

As a result of the interdependencies between P, A, and T and potential rebound effects, policies aimed at decreasing environmental impacts through reductions in P, A, and T may not only be very difficult to implement (i.e., population control and material sufficiency and degrowth movements have been very controversial) but also are likely to be rather ineffective compared to

References

- Jordan, Carl (1995). Replacing Quantity With Quality As a Goal for Global Management. Wiley. ISBN 0-471-59515-2.

- Meinig, Donald (1986). The Shaping of America: A Geographical Perspective on 500 Years of History, Volume 2. Yale University Press. p. 255. ISBN 978-0-300-17394-9. Retrieved 2014-09-06.

- Duany, Andres (2011). Garden Cities: Theory & Practice of Agrarian Urbanism. The Prince's Foundation for the Built Environment. ISBN 1-906384-04-5.

- Arth, Michael E. (2010). Democracy and the Common Wealth: Breaking the Stranglehold of the Special Interests Golden Apples Media, ISBN 978-0-912467-12-2. pp. 120–139, 363–386

- Wear, Andrew (16 February 2016). "Planning, Funding and Delivering Social Infrastructure in Australia's Outer Suburban Growth Areas". Urban Policy and Research. doi:10.1080/08111146.2015.1099523.

- Vincent, Roger and Groves, Martha (October 18, 2003) "L.A.'s Urban Model" Los Angeles Times. Retrieved 12 October 2016

- Sharifi, Ayyoob (September 2015). "From Garden City to Eco-urbanism: The quest for sustainable neighborhood development". Sustainable Cities and Society. doi:10.1016/j.scs.2015.09.002.

- Boeing; et al. (2014). "LEED-ND and Livability Revisited". Berkeley Planning Journal. 27: 31–55. Retrieved 2015-04-15.

- Art, Screen. "Tullimbar Village :: Wollongong NSW :: Contemporary Lifestyle Community". www.tullimbarvillage.com.au. Retrieved 2015-05-29.

- Van; den Bergh, J.; Grazi, Fabio (2015). "Reply to the first systematic response by the Global Footprint Network to criticism: A real debate finally?". Ecological Indicators. 58: 458–463. doi:10.1016/j.ecolind.2015.05.007.

- Reid, Barton (1985). The New Urbanism as a Way of Life: The Relationship Between Inner City Revitalization in Canada and the Rise of the New Middle Class. Retrieved 2014-09-06.

- Urban Design Update: Newsletter of the Institute for Urban Design, Volumes 7-15. Institute for Urban Design. 1991. Retrieved 2014-09-06.

- Barnett, Jonathan (April 2014). "A Short Guide to 60 of the Newest Urbanisms". 77 (4): 19–21. Retrieved 23 October 2014.

- Goldfinger; Wackernagel, S. M.; Galli, A.; Lazarus, E.; Lin, D. (2014). "Footprint facts and fallacies: A response to Giampietro and Saltelli (2014) "Footprints to Nowhere"". Ecological Indicators. 46: 622–632. doi:10.1016/j.ecolind.2014.04.025.

- Giampietro, M.; Saltelli, A.; et al. (2014b). "Footworking in circles: Reply to Goldfinger et al. (2014) "Footprint Facts and Fallacies: A Response to Giampietro and Saltelli (2014) Footprints to nowhere"". Ecological Indicators. 46: 260–263.

3

Understanding Natural Resources

Natural resources are resources that exist without the help of humankind, such as sunlight, atmosphere, land, water and animal life. Natural resource management, natural capital, natural capital accounting, non-renewable resource, exploitation of natural resources etc. are some of the topics included in this section. This text serves as a source to understand the concept of natural resources.

Natural Resource

Natural resources are resources that exist without actions of humankind. This includes all valued characteristics such as magnetic, gravitational, and electrical properties and forces. On earth it includes; sunlight, atmosphere, water, land (includes all minerals) along with all vegetation and animal life that naturally subsists upon or within the heretofore identified characteristics and substances.

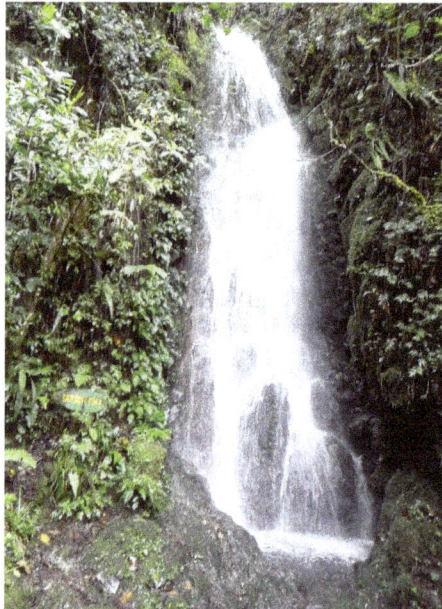

The Carson Fall in Mount Kinabalu, Malaysia is an example of undisturbed natural resource. Waterfalls provide spring water for humans, animals and plants for survival and also habitat for marine organisms. The water current can be used to turn turbines for hydroelectric generation.

Particular areas such as the rainforest in Fatu-Hiva are often characterized by the biodiversity and geodiversity existent in their ecosystems. Natural resources may be further classified in different ways. Natural resources are materials and components (something that can be used) that can be found within the environment. Every man-made product is composed of natural resources (at its fundamental level). A natural resource may exist as a separate entity such as fresh water, and air, as well as a living organism such as a fish, or it may exist in an alternate form which must be processed to obtain the resource such as metal ores, petroleum, and most forms of energy.

The rainforest in Fatu-Hiva, in the Marquesas Islands, is an example of an undisturbed natural resource. Forest provides timber for humans, food and shelter for the flora and fauna. The nutrient cycle between organisms form food chains and biodiversity of species.

The ocean is an example of a natural resource. Ocean waves can be used to generate wave power which is a renewable energy. Ocean water is important for salt production, desalination, and providing habitat for deep water fishes. There are biodiversity of marine species in the sea where nutrient cycles are common.

A picture of the Udachnaya pipe, an open-pit diamond mine in Siberia. An example of a non-renewable natural resource.

There is much debate worldwide over natural resource allocations, this is particularly true during periods of increasing scarcity and shortages (depletion and overconsumption of resources) but also because the exportation of natural resources is the basis for many economies (particularly for developed countries).

Some natural resources such as sunlight and air can be found everywhere, and are known as ubiquitous resources. However, most resources only occur in small sporadic areas, and are referred to as localized resources. There are very few resources that are considered inexhaustible (will not run out in foreseeable future) – these are solar radiation, geothermal energy, and air (though access to clean air may not be). The vast majority of resources are theoretically exhaustible, which means they have a finite quantity and can be depleted if managed improperly.

Classification

There are various methods of categorizing natural resources, these include source of origin, stage of development, and by their renewability. These classifications are described below. On the basis of origin, natural resources may be divided into:

- *Biotic* — Biotic resources are obtained from the biosphere (living and organic material), such as forests and animals, and the materials that can be obtained from them. Fossil fuels such as coal and petroleum are also included in this category because they are formed from decayed organic matter.

- *Abiotic* – Abiotic resources are those that come from non-living, non-organic material. Examples of abiotic resources include land, fresh water, air and heavy metals including ores such as gold, iron, copper, silver, etc.

Considering their stage of development, natural resources may be referred to in the following ways:

- *Potential resources* — Potential resources are those that exist in a region and may be used in the future. For example, petroleum occurs with sedimentary rocks in various regions, but until the time it is actually drilled out and put into use, it remains a potential resource.

- *Actual resources* — Actual resources are those that have been surveyed, their quantity and quality determined and are being used in present times. The development of an actual resource, such as wood processing depends upon the technology available and the cost involved.

- *Reserve resources* — The part of an actual resource which can be developed profitably in the future is called a reserve resource.

- *Stock resources* — Stock resources are those that have been surveyed but cannot be used by organisms due to lack of technology. For example: hydrogen.

Renewability is a very popular topic and many natural resources can be categorized as either renewable or non-renewable:

- *Renewable resources* — Renewable resources can be replenished naturally. Some of these resources, like sunlight, air, wind, water, etc., are continuously available and their quantity is not noticeably affected by human consumption. Though many renewable resources do not have such a rapid recovery rate, these resources are susceptible to depletion by overuse. Resources from a human use perspective are classified as renewable only so long as the rate of replenishment/recovery exceeds that of the rate of consumption.And they can replenish easily compared to Non-renewable resources.

- *Non-renewable resources* – Non-renewable resources either form slowly or do not naturally form in the environment. Minerals are the most common resource included in this category. By the human perspective, resources are non-renewable when their rate of consumption exceeds the rate of replenishment/recovery; a good example of this are fossil fuels, which are in this category because their rate of formation is extremely slow (potentially millions of years), meaning they are considered non-renewable. Some resources actually naturally deplete in amount without human interference, the most notable of these being radio-active elements such as uranium, which naturally decay into heavy metals. Of these, the metallic minerals can be re-used by recycling them, but coal and petroleum cannot be recycled.Once they are completely used they take millions of years to replenish.

Extraction

Resource extraction involves any activity that withdraws resources from nature. This can range in scale from the traditional use of preindustrial societies, to global industry. Extractive industries are, along with agriculture, the basis of the primary sector of the economy. Extraction produces raw material which is then processed to add value. Examples of extractive industries are hunting, trapping, mining, oil and gas drilling, and forestry. Natural resources can add substantial amounts to a country's wealth, however a sudden inflow of money caused by a resource boom can create social problems including inflation harming other industries ("Dutch disease") and corruption, leading to inequality and underdevelopment, this is known as the "resource curse".

Extractive industries represent a large growing activity in many less-developed countries but the wealth generated does not always lead to sustainable and inclusive growth. Extractive industry businesses often are assumed to be interested only in maximizing their short-term value, implying that less-developed countries are vulnerable to powerful corporations. Alternatively, host governments are often assumed to be only maximizing immediate revenue. Researchers argue there are areas of common interest where development goals and business cross. These present opportunities for international governmental agencies to engage with the private sector and host governments through revenue management and expenditure accountability, infrastructure development, employment creation, skills and enterprise development and impacts on children, especially girls and women.

Depletion

Wind is a natural resource that can be used to generate electricity, as with these 5MW wind turbines in Thorntonbank Wind Farm 28 km (17 mi) off the coast of Belgium

In recent years, the depletion of natural resources has become a major focus of governments and organizations such as the United Nations. This is evident in the UN's Agenda 21 Section Two, which outlines the necessary steps to be taken by countries to sustain their natural resources. The depletion of natural resources is considered to be a sustainable development issue. The term sustainable development has many interpretations, most notably the Brundtland Commission's 'to ensure that it meets the needs of the present without compromising the ability of future generations to meet their own needs', however in broad terms it is balancing the needs of the planet's people and species now and in the future. In regards to natural resources, depletion is of concern for sustainable development as it has the ability to degrade current environments and potential to impact the needs of future generations.

> The conservation of natural resources is the fundamental problem. Unless we solve that problem, it will avail us little to solve all others.
>
> *Theodore Roosevelt*

Depletion of natural resources is associated with social inequity. Considering most biodiversity are located in developing countries, depletion of this resource could result in losses of ecosystem services for these countries. Some view this depletion as a major source of social unrest and conflicts in developing nations.

At present, with it being the year of the forest, there is particular concern for rainforest regions which hold most of the Earth's biodiversity. According to Nelson deforestation and degradation affect 8.5% of the world's forests with 30% of the Earth's surface already cropped. If we consider that 80% of people rely on medicines obtained from plants and ¾ of the world's prescription medicines have ingredients taken from plants, loss of the world's rainforests could result in a loss of finding more potential life saving medicines.

The depletion of natural resources is caused by 'direct drivers of change' such as Mining, petroleum extraction, fishing and forestry as well as 'indirect drivers of change' such as demography, economy, society, politics and technology. The current practice of Agriculture is another factor causing depletion of natural resources. For example, the depletion of nutrients in the soil due to excessive use of nitrogen and desertification. The depletion of natural resources is a continuing concern for society. This is seen in the cited quote given by Theodore Roosevelt, a well-known conservationist and former United States president, who was opposed to unregulated natural resource extraction.

Protection

In 1982 the UN developed the World Charter for Nature, we have to protect it just like our mother, which recognized the need to protect nature from further depletion due to human activity. It states that measures need to be taken at all societal levels, from international to individual, to protect nature. It outlines the need for sustainable use of natural resources and suggests that the protection of resources should be incorporated into national and international systems of law. To look at the importance of protecting natural resources further, the World Ethic of Sustainability, developed by the IUCN, WWF and the UNEP in 1990, set out eight values for sustainability, including the need to protect natural resources from depletion. Since the development of these documents, many measures have been taken to protect natural resources including establishment of the scientific field and practice of conservation biology and habitat conservation, respectively.

Conservation biology is the scientific study of the nature and status of Earth's biodiversity with the aim of protecting species, their habitats, and ecosystems from excessive rates of extinction. It is an interdisciplinary subject drawing on science, economics and the practice of natural resource management. The term *conservation biology* was introduced as the title of a conference held at the University of California, San Diego, in La Jolla, California, in 1978, organized by biologists Bruce A. Wilcox and Michael E. Soulé.

Habitat conservation is a land management practice that seeks to conserve, protect and restore, habitat areas for wild plants and animals, especially conservation reliant species, and prevent their extinction, fragmentation or reduction in range.

Management

Natural resource management is a discipline in the management of natural resources such as land, water, soil, plants and animals, with a particular focus on how management affects the quality of life for both present and future generations.Hence sustainable development can be followed where there is a judicial use of resources which compromises the needs of the present generations as well as the future generations.

Management of natural resources involves identifying who has the right to use the resources and who does not for defining the boundaries of the resource. The resources are managed by the users according to the rules governing of when and how the resource is used depending on local condition.

A successful management of natural resources should engage the community because of the nature of the shared resources the individuals who are affected by the rules can participate in setting or changing them. The users have rights to devise their own management institutions and plans under the recognition by the government. The right to resources includes land, water, fisheries and pastoral rights. The users or parties accountable to the users have to actively monitor and ensure the utilisation of the resource compliance with the rules and to impose penalty on those peoples who violates the rules. These conflicts are resolved in a quick and low cost manner by the local institution according to the seriousness and context of the offence. The global science-based platform to discuss natural resources management is the World Resources Forum, based in Switzerland.

Natural Resource Management

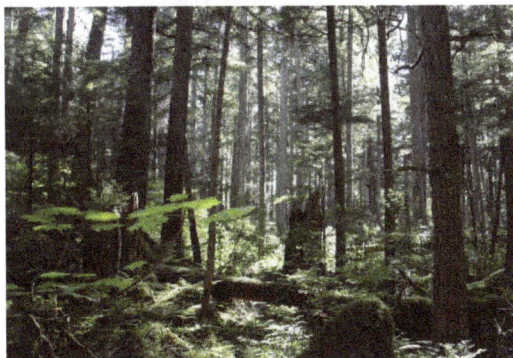

The Tongass National Forest in Alaska is managed by the United States Forest Service

Natural resource management refers to the management of natural resources such as land, water, soil, plants and animals, with a particular focus on how management affects the quality of life for both present and future generations (stewardship).

Natural resource management deals with managing the way in which people and natural landscapes interact. It brings together land use planning, water management, biodiversity conservation, and the future sustainability of industries like agriculture, mining, tourism, fisheries and forestry. It recognises that people and their livelihoods rely on the health and productivity of our landscapes, and their actions as stewards of the land play a critical role in maintaining this health and productivity.

Natural resource management specifically focuses on a scientific and technical understanding of resources and ecology and the life-supporting capacity of those resources. Environmental management is also similar to natural resource management. In academic contexts, the sociology of natural resources is closely related to, but distinct from, natural resource management.

History

The Bureau of Land Management in the United States manages America's public lands, totaling approximately 264 million acres (1,070,000 km2) or one-eighth of the landmass of the country.

The emphasis on sustainability can be traced back to early attempts to understand the ecological nature of North American rangelands in the late 19th century, and the resource conservation movement of the same time. This type of analysis coalesced in the 20th century with recognition that preservationist conservation strategies had not been effective in halting the decline of natural resources. A more integrated approach was implemented recognising the intertwined social, cultural, economic and political aspects of resource management. A more holistic, national and even global form evolved, from the Brundtland Commission and the advocacy of sustainable development.

In 2005 the government of New South Wales, established a *Standard for Quality Natural Resource Management*, to improve the consistency of practice, based on an adaptive management approach.

In the United States, the most active areas of natural resource management are wildlife management often associated with ecotourism and rangeland management. In Australia, water sharing, such as the Murray Darling Basin Plan and catchment management are also significant.

Ownership Regimes

Natural resource management approaches can be categorised according to the kind and right of stakeholders, natural resources:

- State property

- Private property
- Common property
- Non-property (open access)
- Hybrid

State Property Regime

Ownership and control over the use of resources is in hands of the state. Individuals or groups may be able to make use of the resources, but only at the permission of the state. National forest, National parks and military reservations are some US examples.

Private Property Regime

Any property owned by a defined individual or corporate entity. Both the benefit and duties to the resources fall to the owner(s). Private land is the most common example.

Common Property Regimes

It is a private property of a group. The group may vary in size, nature and internal structure e.g. indigenous neighbours of village. Some examples of common property are community forests.

Non-Property Regimes (Open Access)

There is no definite owner of these properties. Each potential user has equal ability to use it as they wish. These areas are the most exploited. It is said that "Everybody's property is nobody's property". An example is a lake fishery. Common land may exist without ownership, in which case in the UK it is vested in a local authority.

Hybrid Regimes

Many ownership regimes governing natural resources will contain parts of more than one of the regimes described above, so natural resource managers need to consider the impact of hybrid regimes. An example of such a hybrid is native vegetation management in NSW, Australia, where legislation recognises a public interest in the preservation of native vegetation, but where most native vegetation exists on private land.

Stakeholder Analysis

Stakeholder analysis originated from business management practices and has been incorporated into natural resource management in ever growing popularity. Stakeholder analysis in the context of natural resource management identifies distinctive interest groups affected in the utilisation and conservation of natural resources.

There is no definitive definition of a stakeholder as illustrated in the table below. Especially in natural resource management as it is difficult to determine who has a stake and this will differ according to each potential stakeholder.

Different Approaches to Who is a Stakeholder:

Source	Who is a stakeholder	Kind of research
Freeman.	"can affect or is affected by the achievement of the organization's objectives"	Business Management
Bowie	"without whose support the organization would cease to exist"	Business Management
Clarkson	"...persons or groups that have, or claim, ownership, rights, or interests in a corporation and its activities, past, present, or future."	Business Management
Grimble and Wellard	"...any group of people, organized or unorganized, who share a common interest or stake in a particular issue or system..."	Natural resource management
Gass et al.	"...any individual, group and institution who would potentially be affected, whether positively or negatively, by a specified event, process or change."	Natural resource management
Buanes et al	"...any group or individual who may directly or indirectly affect—or be affected—...planning to be at least potential stakeholders."	Natural resource management
Brugha and Varvasovszky	"...actors who have an interest in the issue under consideration, who are affected by the issue, or who—because of their position—have or could have an active or passive influence on the decision making and implementation process."	Health policy
ODA	"... persons, groups or institutions with interests in a project or programme."	Development

Therefore, it is dependent upon the circumstances of the stakeholders involved with natural resource as to which definition and subsequent theory is utilised.

Billgrena and Holme identified the aims of stakeholder analysis in natural resource management:

- Identify and categorise the stakeholders that may have influence
- Develop an understanding of why changes occur
- Establish who can make changes happen
- How to best manage natural resources

This gives transparency and clarity to policy making allowing stakeholders to recognise conflicts of interest and facilitate resolutions. There are numerous stakeholder theories such as Mitchell et al. however Grimble created a framework of stages for a Stakeholder Analysis in natural resource management. Grimble designed this framework to ensure that the analysis is specific to the essential aspects of natural resource management.

Stages in Stakeholder Analysis:

1. Clarify objectives of the analysis
2. Place issues in a systems context

3. Identify decision-makers and stakeholders

4. Investigate stakeholder interests and agendas

5. Investigate patterns of inter-action and dependence (e.g. conflicts and compatibilities, trade-offs and synergies)

Application:

Grimble and Wellard established that Stakeholder analysis in natural resource management is most relevant where issued can be characterised as;

- Cross-cutting systems and stakeholder interests

- Multiple uses and users of the resource.

- Market failure

- Subtractability and temporal trade-offs

- Unclear or open-access property rights

- Untraded products and services

- Poverty and under-representation

Case Studies:

In the case of the Bwindi Impenetrable National Park, a comprehensive stakeholder analysis would have been relevant and the Batwa people would have potentially been acknowledged as stakeholders preventing the loss of people's livelihoods and loss of life.

Nepal, Indonesia and Koreas' community forestry are successful examples of how stakeholder analysis can be incorporated into the management of natural resources. This allowed the stakeholders to identify their needs and level of involvement with the forests.

Criticisms:

- Natural resource management stakeholder analysis tends to include too many stakeholders which can create problems in of its self as suggested by Clarkson. "Stakeholder theory should not be used to weave a basket big enough to hold the world's misery."

- Starik proposed that nature needs to be represented as stakeholder. However this has been rejected by many scholars as it would be difficult to find appropriate representation and this representation could also be disputed by other stakeholders causing further issues.

- Stakeholder analysis can be used exploited and abused in order to marginalise other stakeholders.

- Identifying the relevant stakeholders for participatory processes is complex as certain stakeholder groups may have been excluded from previous decisions.

- On-going conflicts and lack of trust between stakeholders can prevent compromise and resolutions.

Alternatives/ Complementary Forms of Analysis:

- Social network analysis

- Common pool resource

Management of the Resources

Natural resource management issues are inherently complex they involve the ecological cycles, hydrological cycles, climate, animals, plants and geographetc. All these are dynamic and inter-related. A change in one of them may have far reaching and/or long term impacts which may even be irreversible. In addition to the natural systems, natural resource management also has to manage various stakeholders and their interests, policies, politics, geographical boundaries, economic implications and the list goes on. It is very difficult to satisfy all aspects at the same time. This results in conflicting situations.

After the United Nations Conference for the Environment and Development (UNCED) held in Rio de Janeiro in 1992, most nations subscribed to new principles for the integrated management of land, water, and forests. Although program names vary from nation to nation, all express similar aims.

The various approaches applied to natural resource management include:

- Top-down (command and control)

- Community-based natural resource management

- Adaptive management

- Precautionary approach

- Integrated natural resource management

Management Of The Resources

Natural resource management issues are inherently complex as they involve the ecological cycles, hydrological cycles, climate, animals, plants and geography etc. All these are dynamic and inter-related. A change in one of them may have far reaching and/or long term impacts which may even be irreversible. In addition to the natural systems, natural resource management also has to manage various stakeholders and their interests, policies, politics, geographical boundaries, economic implications and the list goes on. It is very difficult to satisfy all aspects at the same time. This results in conflicting situations.

After the United Nations Conference for the Environment and Development (UNCED) held in Rio de Janeiro in 1992, most nations subscribed to new principles for the integrated management of land, water, and forests. Although program names vary from nation to nation, all express similar aims.

The various approaches applied to natural resource management include:

- Top-down (command and control)

- Community-based natural resource management

- Adaptive management

- Precautionary approach

- Integrated natural resource management

Community-based natural resource management

The community-based natural resource management (CBNRM) approach combines conservation objectives with the generation of economic benefits for rural communities. The three key assumptions being that: locals are better placed to conserve natural resources, people will conserve a resource only if benefits exceed the costs of conservation, and people will conserve a resource that is linked directly to their quality of life. When a local people's quality of life is enhanced, their efforts and commitment to ensure the future well-being of the resource are also enhanced. Regional and community based natural resource management is also based on the principle of subsidiarity.

The United Nations advocates CBNRM in the Convention on Biodiversity and the Convention to Combat Desertification. Unless clearly defined, decentralised NRM can result an ambiguous socio-legal environment with local communities racing to exploit natural resources while they can e.g. forest communities in central Kalimantan (Indonesia).

A problem of CBNRM is the difficulty of reconciling and harmonising the objectives of socioeconomic development, biodiversity protection and sustainable resource utilisation. The concept and conflicting interests of CBNRM, show how the motives behind the participation are differentiated as either people-centred (active or participatory results that are truly empowering) or planner-centred (nominal and results in passive recipients). Understanding power relations is crucial to the success of community based NRM. Locals may be reluctant to challenge government recommendations for fear of losing promised benefits.

CBNRM is based particularly on advocacy by nongovernmental organizations working with local groups and communities, on the one hand, and national and transnational organizations, on the other, to build and extend new versions of environmental and social advocacy that link social justice and environmental management agendas with both direct and indirect benefits observed including a share of revenues, employment, diversification of livelihoods and increased pride and identity. CBNRM has raised new challenges, as concepts of community, territory, conservation, and indigenous are worked into politically varied plans and programs in disparate sites. Warner and Jones address strategies for effectively managing conflict in CBNRM.

The capacity of indigenous communities to conserve natural resources has been acknowledged by the Australian Government with the Caring for Country Program. Caring for our Country is an Australian Government initiative jointly administered by the Australian Government Department of Agriculture, Fisheries and Forestry and the Department of the Environment, Water, Heritage and the Arts. These Departments share responsibility for delivery of the Australian Government's environment and sustainable agriculture programs, which have traditionally been broadly referred to under the banner of 'natural resource management'.

These programs have been delivered regionally, through 56 State government bodies, successfully allowing regional communities to decide the natural resource priorities for their regions.

Governance is seen as a key consideration for delivering community-based or regional natural resource management. In the State of NSW, the 13 catchment management authorities (CMAs) are overseen by the Natural Resources Commission (NRC), responsible for undertaking audits of the effectiveness of regional natural resource management programs.

Adaptive management

The primary methodological approach adopted by catchment management authorities (CMAs) for regional natural resource management in Australia is adaptive management.

This approach includes recognition that adaption occurs through a process of 'plan-do-review-act'. It also recognises seven key components that should be considered for quality natural resource management practice:

- Determination of scale

- Collection and use of knowledge

- Information management

- Monitoring and evaluation

- Risk management

- Community engagement

- Opportunities for collaboration.

Integrated natural resource management

Integrated natural resource management (INRM) is a process of managing natural resources in a systematic way, which includes multiple aspects of natural resource use (biophysical, socio-political, and economic) meet production goals of producers and other direct users (e.g., food security, profitability, risk aversion) as well as goals of the wider community (e.g., poverty alleviation, welfare of future generations, environmental conservation). It focuses on sustainability and at the same time tries to incorporate all possible stakeholders from the planning level itself, reducing possible future conflicts. The conceptual basis of INRM has evolved in recent years through the convergence of research in diverse areas such as sustainable land use, participatory planning, integrated watershed management, and adaptive management. INRM is being used extensively and been successful in regional and community based natural management.

Adaptive management

The primary methodological approach adopted by catchment management authorities (CMAs) for regional natural resource management in Australia is adaptive management.

This approach includes recognition that adaption occurs through a process of 'plan-do-review-act'. It also recognises seven key components that should be considered for quality natural resource management practice:

- Determination of scale
- Collection and use of knowledge
- Information management
- Monitoring and evaluation
- Risk management
- Community engagement
- Opportunities for collaboration.

Integrated natural resource management

Integrated natural resource management (INRM) is a process of managing natural resources in a systematic way, which includes multiple aspects of natural resource use (biophysical, socio-political, and economic) meet production goals of producers and other direct users (e.g., food security, profitability, risk aversion) as well as goals of the wider community (e.g., poverty alleviation, welfare of future generations, environmental conservation). It focuses on sustainability and at the same time tries to incorporate all possible stakeholders from the planning level itself, reducing possible future conflicts. The conceptual basis of INRM has evolved in recent years through the convergence of research in diverse areas such as sustainable land use, participatory planning, integrated watershed management, and adaptive management. INRM is being used extensively and been successful in regional and community based natural management.

Frameworks and Modelling

There are various frameworks and computer models developed to assist natural resource management.

Geographic Information Systems (GIS)

GIS is a powerful analytical tool as it is capable of overlaying datasets to identify links. A bush regeneration scheme can be informed by the overlay of rainfall, cleared land and erosion. In Australia, Metadata Directories such as NDAR provide data on Australian natural resources such as vegetation, fisheries, soils and water. These are limited by the potential for subjective input and data manipulation.

Natural Resources Management Audit Frameworks

The NSW Government in Australia has published an audit framework for natural resource management, to assist the establishment of a performance audit role in the governance of regional natural resource management. This audit framework builds from other established audit methodologies, including performance audit, environmental audit and internal audit. Audits undertaken using this framework have provided confidence to stakeholders, identified areas for improvement and described policy expectations for the general public.

The Australian Government has established a framework for auditing greenhouse emissions and energy reporting, which closely follows Australian Standards for Assurance Engagements.

The Australian Government is also currently preparing an audit framework for auditing water management, focussing on the implementation of the Murray Darling Basin Plan.

Other Elements

Biodiversity Conservation

The issue of biodiversity conservation is regarded as an important element in natural resource management. What is biodiversity? Biodiversity is a comprehensive concept, which is a description of the extent of natural diversity. Gaston and Spicer (p. 3) point out that biodiversity is "the variety of life" and relate to different kinds of "biodiversity organization". According to Gray (p. 154), the first widespread use of the definition of biodiversity, was put forward by the United Nations in 1992, involving different aspects of biological diversity.

Precautionary Biodiversity Management

The "threats" wreaking havoc on biodiversity include; habitat fragmentation, putting a strain on the already stretched biological resources; forest deterioration and deforestation; the invasion of "alien species" and "climate change"(p. 2). Since these threats have received increasing attention from environmentalists and the public, the precautionary management of biodiversity becomes an important part of natural resources management. According to Cooney, there are material measures to carry out precautionary management of biodiversity in natural resource management.

Concrete "policy tools"

Cooney claims that the policy making is dependent on "evidences", relating to "high standard of proof", the forbidding of special "activities" and "information and monitoring requirements". Before making the policy of precaution, categorical evidence is needed. When the potential menace of "activities" is regarded as a critical and "irreversible" endangerment, these "activities" should be forbidden. For example, since explosives and toxicants will have serious consequences to endanger human and natural environment, the South Africa Marine Living Resources Act promulgated a series of policies on completely forbidding to "catch fish" by using explosives and toxicants.

Administration and guidelines

According to Cooney, there are 4 methods to manage the precaution of biodiversity in natural resources management;

1. "Ecosystem based Management" including "more risk-averse and precautionary management", where "given prevailing uncertainty regarding ecosystem structure, function, and inter-specific interactions, precaution demands an ecosystem rather than single-species approach to management".

2. "Adaptive management" is "a management approach that expressly tackles the uncertainty and dynamism of complex systems".

3. "Environmental impact assessment" and exposure ratings decrease the "uncertainties" of precaution, even though it has deficiencies, and

4. "Protectionist approaches", which "most frequently links to" biodiversity conservation in natural resources management.

Land management

In order to have a sustainable environment, understanding and using appropriate management strategies is important. In terms of understanding, Young emphasises some important points of land management:

- Comprehending the processes of nature including ecosystem, water, soils
- Using appropriate and adapting management systems in local situations
- Cooperation between scientists who have knowledge and resources and local people who have knowledge and skills

Dale et al. (2000) study has shown that there are five fundamental and helpful ecological principles for the land manager and people who need them. The ecological principles relate to time, place, species, disturbance and the landscape and they interact in many ways.It is suggested that land managers could follow these guidelines:

- Examine impacts of local decisions in a regional context, and the effects on natural resources.
- Plan for long-term change and unexpected events.
- Preserve rare landscape elements and associated species.
- Avoid land uses that deplete natural resources.
- Retain large contiguous or connected areas that contain critical habitats.
- Minimize the introduction and spread of non-native species.
- Avoid or compensate for the effects of development on ecological processes.
- Implement land-use and land-management practices that are compatible with the natural potential of the area.

Natural Capital

Natural capital is the world's stock of natural resources, which includes geology, soils, air, water and all living organisms. Natural capital assets provide people with a wide range of free goods and services, often called ecosystem services, which underpin our economy and society and some of which even make human life possible.

Mangrove swamp at Iriomote Island, Japan, providing beneficial services of sediment accumulation, coastal protection, nursery and fish-spawning grounds which may in turn support coastal fishing communities. At least 35% of the world's stock of mangrove swamps has been destroyed in just 20 years

Aerial view of the Amazon Rainforest. Looked at as a natural capital asset, rainforests provide air and water regulation services, potential sources of new medicines and natural carbon sequestration.

It is an extension of the economic notion of capital (resources which enable the production of more resources) to goods and services provided by the natural environment. For example, a well-maintained forest or river may provide an indefinitely sustainable flow of new trees or fish, whereas over-use of those resources may lead to a permanent decline in timber availability or fish stocks. Natural capital also provides people with essential services, like water catchment, erosion control and crop pollination by insects, which in turn ensure the long-term viability of other natural resources. Since the continuous supply of services from the available natural capital assets is dependent upon a healthy, functioning environment, the structure and diversity of habitats and ecosystems are important components of natural capital. Methods, called 'natural capital asset checks', help decision-makers understand how changes in the current and future performance of natural capital assets will impact on human well-being and the economy.

History of the Concept

Natural capital is one approach to ecosystem valuation which revolves around the idea, in contrast to traditional economics, that non-human life produces essential resources. Thus, ecological health is essential to the sustainability of the economy. In *Natural Capitalism: Creating the Next Industrial Revolution* the author claims that the global economy is within a larger economy of natural resources and ecosystem services that sustain us. In order to continue to reap the benefits of our natural environment, we need to recognize the importance of natural capital within the economy. According to the authors, the "next industrial revolution" depends on the espousal of four central strategies: "the conservation of resources through more effective manufacturing processes, the reuse of materials as found in natural systems, a change in values from quantity to quality, and investing in natural capital, or restoring and sustaining natural resources."

In a traditional economic analysis of the factors of production, natural capital would usually be classified as "land" distinct from traditional "capital". The historical distinction between "land" and "capital" defined "land" as naturally occurring with a fixed supply, whereas "capital", as originally defined referred only to man-made goods. (e.g., Georgism) It is however, misleading to view "land" as if its productive capacity is fixed, because natural capital can be improved or degraded by the actions of man over time. Moreover, natural capital yields benefits and goods, such as timber or food, which can be harvested by humans. These benefits are similar to those realized by owners of infrastructural capital which yields more goods, such as a factory which produces automobiles just as an apple tree produces apples.

The term 'natural capital' was first used in 1973 by E.F. Schumacher in his book *Small Is Beautiful* and is closely identified with Herman Daly, Robert Costanza, the Biosphere 2 project, and the Natural Capitalism economic model of Paul Hawken, Amory Lovins, and Hunter Lovins. Recently, it has begun to be used by politicians, notably Ralph Nader, Paul Martin Jr., and agencies of the UK government, including its Natural Capital Committee and the London Health Observatory. All users of the term currently differentiate natural from man-made or infrastructural capital in some way. Indicators adopted by United Nations Environment Programme's World Conservation Monitoring Centre and the Organisation for Economic Co-operation and Development (OECD) to measure natural biodiversity use the term in a slightly more specific way. According to the OECD, natural capital is "natural assets in their role of providing natural resource inputs and environmental services for economic production" and is "generally considered to comprise three principal categories: natural resources stocks, land, and ecosystems."

Within the international community the basic principle is not controversial, although much uncertainty exists over how best to value different aspects of ecological health, natural capital and ecosystem services. Full cost accounting, triple bottom line, measuring well-being and other proposals for accounting reform often include suggestions to measure an "ecological deficit" or "natural deficit" alongside a social and financial deficit. It is difficult to measure such a deficit without some agreement on methods of valuation and auditing of at least the global forms of natural capital (e.g. value of air, water, soil).

Ecologists are teaming up with economists to measure and express values of the wealth of ecosystems as a way of finding solutions to the biodiversity crisis. Some researchers have attempted to place a dollar figure on ecosystem services such as the value that the Canadian boreal forest's contribution to global ecosystem services. If ecologically intact, the boreal forest has an estimated value of US$3.7 trillion. The boreal forest ecosystem is one of the planet's great atmospheric regulators and it stores more carbon than any other biome on the planet. The annual value for ecological services of the Boreal Forest is estimated at US$93.2 billion, or 2.5 greater than the annual value of resource extraction. The economic value of 17 ecosystem services for the entire biosphere (calculated in 1997) has an estimated average value of US$33 trillion per year. These ecological economic values are not currently included in calculations of national income accounts, the GDP and they have no price attributes because they exist mostly outside of the global markets. The loss of natural capital continues to accelerate and goes undetected or ignored by mainstream monetary analysis.

Natural Capital Declaration

In June 2012 a 'national capital declaration' (NCD) was launched at the Rio+20 summit held in Brazil. An initiative of the global finance sector, it was signed by 40 CEOs to 'integrate natural capital considerations into loans, equity, fixed income and insurance products, as well as in accounting, disclosure and reporting frameworks.' They worked with supporting organisations to develop tools and metrics to integrate natural capital factors into existing business structures. In summary, its four key aims are to:

- Increase understanding of business dependency on natural capital assets;

- Support development of tools to integrate natural capital considerations into the decision-making process of all financial products and services;

- Help build a global consensus on integrating natural capital into private sector accounting and decision-making;

- Encourage a consensus on integrated reporting to include natural capital as one of the key components of an organisation's success.

Natural Capital Protocol

In July 2016, the Natural Capital Coalition released the Natural Capital Protocol. The Protocol provides a standardised framework for organisations to identify, measure and value their direct and indirect impacts and dependancies on natural capital. The Protocol harmonises existing tools and methodologies, and guides organisations towards the information they need to make strategic and operational decisions that include impacts and dependencies on natural capital.

The Protocol was developed in a unique collaboration between 38 organisations who signed voluntary, pre-competitive contracts.

The Protocol is available on a creative commons license and is free for organisations to apply.

Internationally Agreed Standard

Environmental-economic accounts provide the conceptual framework for integrated statistics on the environment and its relationship with the economy, including the impacts of the economy on the environment and the contribution of the environment to the economy. A coherent set of indicators and descriptive statistics can be derived from the accounts that inform a wide range of policies, including, but not limited to, green economy/green growth, natural resource management and sustainable development. The System of Environmental-Economic Accounting (SEEA) contains the internationally agreed standard concepts, definitions, classifications, accounting rules and tables for producing internationally comparable statistics on the environment and its relationship with the economy. The SEEA is a flexible system in the sense that its implementation can be adapted to countries' specific situations and priorities. Coordination of the implementation of the SEEA and on-going work on new methodological developments is managed and supervised by the UN Committee of Experts on Environmental-Economic Accounting (UNCEEA). The final, official version of the SEEA Central Framework was published in February 2014.

Criticism

Whilst measuring the components of natural capital in any region is a relatively straightforward process, both the task and the rationale of putting a monetary valuation on them, or on the value of the goods and services they freely give us, has proved more contentious. Within the UK, Guardian columnist, George Monbiot, has been critical of the work of the government's Natural Capital Committee and of other attempts to place any sort of monetary value on natural capital assets, or on the free ecosystem services they provide us with. In a speech referring to a report to government which suggested that better protection of the UK's freshwater ecosystems would yield an enhancement in aesthetic value of £700m, he derided attempts 'to compare things which cannot be directly compared'. He went on to say:

> These figures, ladies and gentlemen, are marmalade. They are finely shredded, boiled to a pulp, heavily sweetened ... and still indigestible. In other words they are total gibberish.
>
> —*G. Monbiot*

Others have defended efforts to integrate the valuation of natural capital into local and national economic decision-making, arguing that it puts the environment on a more balanced footing when weighed against other commercial pressures, and that 'valuation' of those assets is not the same as monetisation.

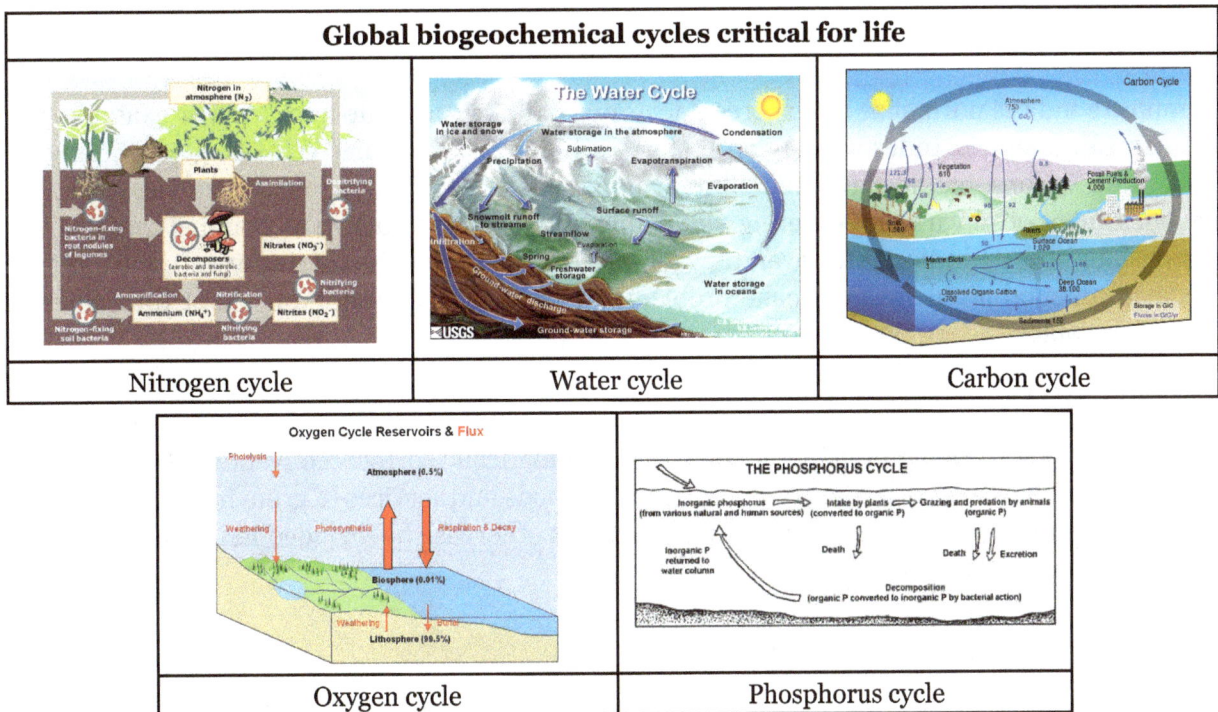

Global biogeochemical cycles critical for life

Nitrogen cycle | Water cycle | Carbon cycle

Oxygen cycle | Phosphorus cycle

Natural Capital Accounting

Natural capital accounting is the process of calculating the total stocks and flows of natural resources and services in a given ecosystem or region. Accounting for such goods may occur in physical or monetary terms. This process can subsequently inform government, corporate and consumer decision making as each relates to the use or consumption of natural resources and land, and sustainable behaviour.

Methods of Accounting

There are several methods of accounting which aim to address the issue of sustainability. These are: large and eclectic dashboards; composite indices; indices focusing on overconsumption; adjusted economic indicators.

Large and Eclectic Dashboards

These dashboards bring together a number of indicators that are directly and indirectly related to the durability of socio-economic progress. One example of this is the Eurostat Sustainable Development Indicators, which is a list of over 100 indicators used to monitor the EU Sustainable Development Strategy. The criticism associated with these dashboards is that a large number of indicators risks muddling a clear message about sustainability that resonates with policy makers or citizens. In response, there has been a greater tendency to select headline indicators that "track central elements of green growth And [Are] Representative Of A Broader Set Of Green Growth Issues."

Composite Indices

Composite indices normalize and aggregate various data into a single number. For example, the Human Development Index, Osberg and Sharpe's Index of Economic Well-Being, the Changing Wealth of Nations, or the Environmental Sustainability Index, which ranks countries based on an assessment of 76 variables covering 5 domains. It is often instructive to examine the separate dimensions of these indices. However, they may present a skewed view of countries' contributions to environmental problems and make problematic, normative assumptions about the values of certain variables.

Adjusted GDPs

Adjusted gross domestic product, or green GDP, systematically corrects conventional GDP by taking into account aspects of a country's production of goods and services (e.g. environmental degradation and natural resource depletion) that would not otherwise be included in the indicator, but are relevant to sustainability.

Indices Focusing on Overconsumption

Indicators that fall in this category conceive of sustainability with respect to consumption levels and investment in natural resources. Examples include adjusted net savings (ANS) and ecological footprint accounts. ANS is calculated as the change in total wealth over a given time period, while ecological footprint assessments determine how much of the regenerative capacity of the biosphere is required to maintain the consumption habits of a defined population. The explicit emphasis on sustainability makes these indices useful; however computing them by country fails to capture the global nature of sustainability.

Monetary or Physical Indicators

All sustainability indicators can be grouped broadly into two types. Specifically, they will be calculated in monetary terms, using one or more valuation techniques, or in physical terms. It is more likely for monetary indicators to be expressed as flows, and physical indicators as stocks.

Global Initiatives

General commitment by the international community to support the development of natural capital accounting was motivated early on by the Brundtland Report in 1987 and the 1992 Rio Summit. At the Summit in particular, Agenda 21 – in which Chapter 40 called for signatories to develop quantitative information regarding their activities – was adopted.

System of Environmental-economic Accounting

In September 1992, the Commission on the Environment of the Organization of American States (OAS) Permanent Council held a Seminar on Natural Resource and Environmental Accounts for Development Policy. Many of the country participants expressed interest in developing accounting capacities for natural resources. A proposal was made at that time to create a program to coordinate and strengthen the efforts of countries and institutions undertaking such initiatives.

The development of the first system of environmental-economic accounting (SEEA) in 1993 (SEEA-1993) was a major step towards establishing standards around integrating the environment into national accounts, and subsequently, environmentally-adjusting or "greening" macroeconomic indicators such as GDP. While the SEEA-2003 and subsequent revisions being undertaken for 2013 have expanded the range of analyses within the framework, the purpose of the SEEA has remained the same. It is an accounting framework that records the stocks and flows that are relevant to both the environment and the economy. Its Central Framework comprises three main accounts that can be integrated with the existing United Nations System of National Accounts (SNA), and each focuses on a different aspect of the interaction between the economy and the environment: physical flow accounts; functional accounts for environmental transactions; and asset accounts in physical and monetary terms.

The latest version of the SEEA (Q3, 2012) has two other parts, aside from the Central Framework: SEEA Experimental Ecosystem Accounts and SEEA Extensions. The Experimental Ecosystem Accounts, specifically, introduces an accounting framework for ecosystems, despite the fact that many of its relevant stocks and flows are centered on non-market assets. While some of the measurement concepts involved in the accounting process are still evolving, it is possible that the eventual valuation of ecosystems and their depletion could be included in the calculations of environmentally-adjusted macroeconomic indicators. This has implications for future policy, since the emphasis on certain projects or activities undertaken by governments will likely change, depending on how the above-mentioned measurements impact their respective accounts, and subsequent environmental adjustments to certain indicators.

The London Group on Environmental Accounting and the UN Committee of Experts on Environmental-Economic Accounting are two groups, created in 1994 and 2005 respectively, to assist in the development of the SEEA and its implementation. As well, the Working Group on Environmental Auditing, a subgroup of the International Organization of Supreme Audit Institutions, is working to improve auditing standards related to environmental issues.

Wealth Accounting and Ecosystem Partnership Services

The ability of developing countries to build their natural capital account capacities is being improved significantly through the Wealth Accounting and Ecosystem Partnership Services (WAVES), by encouraging the development of relevant measurement frameworks. WAVES is global partnership that was inaugurated in October 2010 by World Bank President Robert Zoellick at the Convention on Biological Diversity COP-10 meeting in Japan. It aims to promote sustainable development by encouraging the inclusion of natural capital measurements in national accounts. Several projects have been initiated in developing countries such as Botswana, Colombia and Madagascar with a view to improving their capacity to implement the SEEA, in collaboration with UNEP, the UNDP, the United Nations Statistical Commission, and the financial support of NGOs and the governments of Australia, Canada, France, Japan, Norway and the United States.

Ecological Footprint Accounts

Alternately, there have been many attempts to move away from integrated accounts, and towards novel sustainability indices or statistics. Ecological footprint accounts, developed by Monfreda et al. (2004) and since 2007, the Global Footprint Network, or the proposal for nine planetary boundaries within which humans can safely operate, by Rockström et al., are projects which advocate for new approaches to global sustainability.

Inclusive Wealth Index

The UN International Human Dimensions Programme has created an inclusive sustainability indicator, the Inclusive Wealth Index (IWI), which measures the productive bases of an economy: produced, natural and human capital, and based on these three assessments, calculates the trajectory of a country's wealth. The calculation of natural capital in the IWI is based on the shadow value of an economy's natural capital assets.

A similar conceptual direction was taken by the Commission on the Measurement of Economic Performance and Social Progress, under the direction of economists Joseph Stiglitz, Amartya Sen and Jean-Paul Fitoussi, at the behest of former French President Nicholas Sarkozy, in 2008. The authors concluded that a pragmatic measure of sustainability would combine an indicator based on the extended wealth approach, and a small dashboard of physical indicators.

While they have different theoretical underpinnings, what these approaches have in common with each other is a fundamental recognition of the limitations of traditional indicators in measuring economic performance and social progress, and the importance of sustainability in the long-run.

Beyond GDP

There have been several initiatives organized at the regional level that seek to move away from traditional GDP as the major indicator of wealth and well-being. The first of these is Beyond GDP, an initiative started by the EU in 2007 to develop highly aggregated environmental and social indicators and extend the national accounts to environmental and social issues. In May 2010, the Summit for Sustainability in Africa was held in Gaborone, Botswana. It also brought together regional leaders to discuss sustainable development planning, and in particular, to commit to a set of goals on fully accounting for natural capital, and integrating it into national planning, reporting and policies.

Economics of Ecosystems and Biodiversity

Four publications were presented at the CBD COP-10 by The Economics of Ecosystems and Biodiversity (TEEB) initiative: TEEB Ecological and Economic Foundations, TEEB in in National and International Policy Making, TEEB in Local and Regional Policy, and TEEB in Business and Enterprise. These, along with an Interim Report released at the CBD COP-9 in Bonn, Germany, represent a comprehensive analysis of the economic value of biodiversity, and the consequences it holds for different levels of public and private policy analysis. TEEB also seeks to systematize the role of corporations, under the TEEB for Business Coalition (2012), by formulating standards and assessing externalities. Since natural capital accounting requires the identification of replenishment activities as well as environmental degradation, the inclusion of corporations into the valuation process is key.

Natural Capital Declaration

In June 2012, the Rio+20 conference "marked a watershed in the world wide interest on Natural Capital Accounting." The Natural Capital Declaration (NCD), a commitment by CEOs in the financial sector to embed ESG considerations in management and investment activities, was revealed prior to the conference. As well, the World Bank started the WAVES 50:50 Initiative to analyze the progress and next steps required for improving efforts to account for natural capital and enhance countries' sustainable decision-making capabilities. At the time of the conference, 62 countries, 90 corporations, and 17 civil society members had signed on to the campaign.

Country Initiatives

Many countries are undertaking projects to develop environmental accounts, to integrate them with national accounts or to create environmentally-adjusted macroeconomic indicators, including Green GDP. Early adopters of integrated environmental-economic accounts include the Netherlands, France, and the Philippines. This section documents the initiatives undertaken by, or concerning the G20 countries.

Argentina

A team of researchers at the University of Buenos Aires and CONICET organized the ARKLEMS+LAND project. Their research, based on the KLEMS database framework (Capital, Labour, Energy, Material and Service Inputs), measures and analyses the sources of economic growth, productivity and competitiveness in the Argentinian economy. The database includes the contributions of "Natural Resource as Land and Subsoil Assets" services to GDP growth.

Professor George Santopietro, at Radford University in Virginia, examined several alternative methods for estimating resource rent and relatedly, depletion costs of natural capital. The methods he engages with are: net price, El Serafy's depletion cost, sustainable price, transaction value and replacement cost. He derives data from the privatization of Yacimientos Petroliferos Fiscales (YPF), Argentina's state-owned oil enterprise, and using each method, generates estimated resource rent and depletion cost values for Argentina's reserves of petroleum. The results show that the net price and transaction value methods overvalue the resource rent of petroleum reserves. He concludes that rent should be derived using the value of a firm's stock.

Australia

Several Australian organizations produce national environmental accounts. The Australian Bureau of Statistics produces Water Account Australia and a pilot Land Account for the Great Barrier Reef, both of which are based on the SEEA framework. The Bureau of Meteorology produces a complementary National Water Account on the water available to be used. The Department of Climate Change and Energy Efficiency reports on greenhouse gas (GHG) emissions. The Wentworth Group of Concerned Scientists has created a trial, environmental accounting model that can be applied to regions of any size. These regional accounts use a common unit of measurement, which allows comparisons to be made between different natural capital assets. The Victorian Department of Sustainability and Environment is in the process of developing experimental ecosystem accounts.

The energy and water accounts found on the National Balance Sheet are produced annually. As well, subsoil assets, timber for producing logs, and land are valued monetarily, and included on the non-produced assets in the National Balance Sheet. The Australian Bureau of Statistics (ABS) is currently working with several other national departments on the National Plan for Environmental Information (NPEI), the State of the Environment Report, and planning for national environmental accounts.

In 2012, following the adoption of the SEEA as an international statistical standard, the Australian Bureau of Statistics (ABS) also published a report titled "Completing the Picture – Environmental Accounting in Practice," which develops a number of accounting tables rooted in SEEA formatting, and explores how these could influence policy decisions.

Britain

The UK produces Environmental Accounts, consistent with the SEEA framework, that are separated into three dimensions: natural resource accounts, physical flows and monetary accounts. The majority of the data is measured in physical units; monetary units are used where relevant, and if the necessary data is available.

The current environmental asset accounts produced by the Office for National Statistics (ONS) are: oil and gas reserves, forestry and land cover. In 2012, the ONS planned to run a study on producing a full ecosystem account, with high priority on physical and monetary forestry accounts. The Office also set out a series of goals to include land use and cover accounts in the 2013 Environmental Accounts.

The UK National Ecosystem Assessment (NEA) was a two-year initiative from 2009 to 2011 that assessed the benefits of the natural environment to society and the economy.

The Government has pledged to develop full UK Environmental Accounts by 2020. This work is being led by the Office for National Statistics (ONS) and the Department for Environment, Food and Rural Affairs (Defra). The UK Natural Capital Committee is providing advice to help inform this project. In December 2012, the ONS published a 'roadmap' that set out the work-plan for the project. The incorporation of natural capital into the national accounts, if done properly, would enable a high level picture to be obtained of trends in the state of the nation's natural assets through time. This in turn would help demonstrate the implications of actions impacting on the natural environment and may, therefore, encourage policy-makers to ensure natural assets are in future used sustainably through the appropriate use of policy levers.

The Natural Capital Committee is also developing a methodology for corporate natural capital accounting. It is undertaking a pilot project throughout 2014 with a range of organisations to test and refine this methodology and is encouraging organisations that own or manage land to develop corporate natural capital accounts.

Brazil

Despite reports that Brazil would produce a Green GDP in the lead-up to the Rio+20, there is no indication that this process was undertaken.

Canada

Statistics Canada does not redefine or supplement existing national accounts aggregates with environmental information. However, it has developed the Canadian System of Environmental and Resource Accounts (CSERA), which "provides some information for those who may wish to calculate such "green aggregates."" It also includes estimates of land in the "tangible non-produced assets" section of the Canadian balance sheet.

The CSERA has three dimensions: Natural Resource Stock Accounts; Material and Energy Flow Accounts; and Environmental Protection Expenditure Accounts. The Natural Resource Stock Accounts are measured in physical and monetary units. These, in turn, comprise natural resource wealth estimates that are included on the National Balance Sheet Accounts.

In 2005, the Pembina Institute published a report on the Canada Boreal Initiative, which sought to quantify the full economic value of goods and services provided by Canada's Boreal region. The Boreal Ecosystem Wealth Accounting System (BEWAS) was constructed for this purpose. It considered the physical conditions in the boreal region by using physical inventory and spatial data. The two-year study approximated the market value, for 2002, of the region's forestry, mining, oil and gas, and hydroelectric generation sectors at $48.9 billion. For the same time period, it also estimated the net market value of natural capital extraction at $37.8 billion and the non-market value of ecosystem services at $93.2 billion.

Professor Nancy Olewiler, at Simon Fraser University, conducted several case studies to value natural capital in settled areas of Canada. Her agricultural lands case studies found the total net value of conservation efforts to be approximately $195/ha/yr in the Ontario Grand River Watershed, $65/ha/yr in the Upper Assiniboine River Basin and $126/ha/yr in the Prince Edward Island Mill River Watershed.

China

In 1997, Beijing authorities carried out a project assessing the city's Green GDP. This set a series of precedents for environmentally adjusted indicators in the country. Over the next decade, several regional pilot projects were undertaken by local authorities. Between 2001 and 2004, Chinese authorities worked with Statistics Norway to carry out a Green GDP assessment of Chongqing.

Then, in 2004, the Green GDP Accounting Research Project was launched by the State Environmental Protection Administration of China (SEPA) and the National Bureau of Statistics (NBS). The findings, released in the China Green National Accounting Study Report 2004 in 2006, reported that environmental pollution cost the economy 511.8 billion yuan or 3.5% of GDP in 2004. A breakdown of the figure shows that water pollution, air pollution, and solid waste and accidents cost 286.28 billion yuan, 219.8 billion yuan, and 5.74 billion yuan, respectively. According to the report, only about ten of the items the project intended to distinguish were accounted for. The costs of resource depletion and ecological damage were not included in the calculations because of methodological difficulties, limited technological capabilities and a lack of relevant data.

At the same time as the report was released, SEPA Vice Minister Zhu Guangya issued an independent report stating that each year, environmental damage in the country cost approximately 10% of GDP. This estimate was consistent with the one that scientists, economists and the World Bank had expected, of 8-12% of GDP.

The government withdrew its official support for Green GDP in 2007, after early results showed the reduction of growth rates in some provinces to nearly zero. It also did not release an official Green GDP report for 2005, which had been scheduled for publication in March 2007.

Reports have periodically surfaced about re-calculating China's green GDP. In 2007, the All-China Environment Federation (ACEF) called on the National Development and Reform Commission (NDRC) to carry out a national accounting system. "China Daily" reported that the Ministry of Environmental Protection valued the cost of pollution to the economy at 1.4 trillion yuan in 2009.

Starting in 1998, the NBS developed rudimentary environmental accounts in forestry and energy. Since then, the NBS has expanded its environmental accounts to include pollution treatment, water, and minerals, along with the development of a comprehensive SEEA for China.

European Union

The interaction between Eurostat and national statistical offices of EU Member States was formalized in 2011, by the adoption of Regulation No 691 on European Environmental Economic Accounts. It requires Member States to report data and accounts on air emissions, taxes related to the environment, and material flows from 2012. Eurostat is also constructing environmental accounts expressed in physical and monetary terms, and asset accounts, as a step towards developing a regional SEEA.

The creation of inclusive wealth indicators is also a recognized priority of the EU. The proposed 7th Environment Action Programme (EAP) of the EC explicitly identifies this issue, by calling for further development and integration of economic and environmental indicators. The Shared Environmental Information System (SEIS) proposes to streamline the collection of data required for designing environmental indicators. The EU also committed to the Aichi Accord at the CBD COP-10, in which Parties agree to integrate biodiversity into their national accounts. This is on top of the experimental ecosystem accounting framework that was launched in 2009. The European Environment Agency (EEA) proposed that given the compatibility of the design with the SNA, it would be possible to use one particular indicator, Consumption of Ecosystem Capital (CEC) to adjust National Accounts aggregates to create, in particular, CEC Adjusted Net Domestic Product and CEC Adjusted Net National Income.

The EC Communication Roadmap to a Resource Efficient Europe sets 2020 as the year by which businesses, along with public authorities, will properly account for natural capital and ecosystem services.

France

The French Ministry of Sustainable Development produces satellite, environmental-economic accounts each year. However, it is planning on fully expanding the accounts to correspond with the SEEA (2012). The National Institute of Statistics and Economics also includes estimates of land, subsoil assets, and non-cultivated bio and water resources in the "tangible non-produced assets" section of the French balance sheet.

From 2008 to 2010, an exploratory study on ecosystem valuation was carried out. In accordance with the adoption of the Aichi Accord at the CBD COP-10, the French National Biodiversity Strategy Target 7 was set to "include preservation of biodiversity in economic decisions." The Ministry of Sustainable Development is presently consulting with stakeholders and refining methodological options in order to begin assembling physical ecosystem assessments. Annual expert workshops on monetary valuation and economic instruments are also organized by the Ministry.

Germany

The German Environmental Economic Accounts (GEEA) follow the SEEA framework. The data from these accounts, along with the German Socio-economic Accounts, is used to calculate an indicator set, as outlined in the National Strategy for Sustainable Development (2002). While no adjusted macroeconomic aggregates are estimated in the GEEA, two of the GEEA indicators are embedded into the National Accounts: productivity of energy and raw materials, and transport intensity and share of the railways in providing transport.

India

An Expert Group, led by IHDP Scientific Committee chairman Professor Partha Dasgupta, is in the process of developing a system to "green" India's national accounts. It plans to adjust GDP to account for environmental costs and impacts by 2015. Some of the issues the Group will address include establishing coordination mechanisms within the country and with international partners, and standardizing data collection and valuation methodologies.

The Green Accounting for Indian States Project (GAISP) was the first initiative of the Green Indian States Trust (GIST), an NGO started in 2004. The Project used data from Indian national databases to measure sustainable development and create green accounts at the state-level that were consistent with SEEA-2003 guidelines. Following this, GIST calculated a Green GDP total which adjusted the traditional indicator for "all major externalities."

The South Asian Network for Development and Environmental Economics (SANDEE) and the Indian Society for Ecological Economics operate in India, although SANDEE is based in Nepal. Both are networks that conduct research on and analyses of environmental accounting, which is instructive for teachers, researchers and members of the policy community.

Indonesia

Members of the World Resources Institute, led by David Repetto, integrated environmental effects into Indonesia's national accounts in 1990. He subtracted net natural resource depreciation for the petroleum, timber, and soils sectors from GDP to estimate the environmentally-adjusted NDP. They found that estimates of net income and growth of net income were overstated when calculated using the conventional accounts.

In 1997, the United Nations University-Institute of Advanced Studies (UNU-IAS) measured the environmental impacts of industrialization and trade in Indonesia (along with China and Japan). In their study, the UNU-IAS constructed an international environmental input-output model for the Asia Pacific region, and from this, was able to compile a preliminary SEEA and Green GDP approximation. In 1990, environmental costs were equal to 4.9% of NDP for Indonesia. In particular, changes in land use and oil exploitation were major costs.

Since 2002, Statistics Indonesia (BPS) has produced several Indicators of Sustainable Development reports, based on the 134 UN Commission on Sustainable Development Indicators developed by the UN Department of Economic and Social Affairs. BPS also regularly generates environmental statistics and statistics of marine and coastal resources.

Italy

The National Institute of Statistics (ISTAT) currently constructs three types of environmental accounting modules: material flows (MFA), the environmental account matrix integrated with national economic account (NAMEA), and the expenditure estimate for environmental protection (SERIEE-EPEA).

In 1999, Silvia Teizzi at the University of Sienna applied a method of monetary valuation to externalities arising from agricultural production in Italy. She separately estimated a shadow price and a quantity for the externalities and calculated their values between for each year between 1961 and 1991. She subtracted these figures from the value added of the agricultural sector "as a first step towards the correction of national accounting aggregates to take environmental degradation into account."

In their case study on Trento, Italy, Professors Andrea Francesconi and Paolo Penasa analyze the accounting schemes for environmental expenditures developed by local governments in Italy. Under the local Agenda 21 process, the City and Local Environmental Accounting and Reporting (CLEAR) project and the CONTAROMA project were undertaken by Italian municipalities. The CLEAR project was started in 2001 to pair the financial statements of municipalities with an environmental report. The CLEAR method reclassified the aforementioned financial statements such that expenditures for environmental purposes could be identified, and compiled into separate monetary accounts. Similarly, the CONTAROMA project developed an environmental accounting system for the municipal budget of Rome.

Japan

In 1995, the Japanese Economic Planning Agency made initial estimates of the SEEA and Green GDP from 1985 to 1990. In 1998, they extended their estimates from 1970 to 1995. According to these calculations, environmental costs in Japan fell from 8% of NDP in 1970 to nearly 1% in 1995.

The E-10 are the official economic-environmental accounts produced by the Japanese Ministry of the Environment. They are based on data compiled by 10 government ministries, and describe the environmental burden of particular natural resources in physical units. There are three tables which make up the E-10: a Basic Transaction Table; resource and environmental burden tables; and supplementary tables.

The Statistics Bureau of Japan currently includes estimates of land in the "tangible non-produced assets" section of the Japanese balance sheet.

Republic of Korea

In 2001, the Korean Ministry of the Environment planned to introduce an environmentally-adjusted GDP indicator. It began developing a SEEA (KORSEEA) in 2002, and the development of Economy-wide Material Flow Accounts, and NAMEA took place in the following years. Several government institutions and departments provide economic and environmental data required for the KORSEEA. Statistics Korea also includes estimates of land, subsoil assets, and non-cultivated bio and water resources in the "tangible non-produced assets" section of the Koreah balance sheet.

The Korea Rural Economic Institute calculated the Green GDP of the Korean agricultural sector in the period 1980 to 1997 using pilot economic-environmental accounts. Environmental adjustment in this study is limited to the subtraction of "degradation costs of natural resources" from NDP. They report that the ratio of environmentally adjusted domestic product to NDP decreases from 100.6% to 99.5% in the agricultural sector during the specified period.

Mexico

In 1985, the UN, World Bank and Mexican government executed a joint pilot project using 1985 data to produce two environmentally-adjusted indicators: resource depletion subtracted from NDP and environmental degradation subtracted from NDP.

The Mexican System of Economic and Ecological Accounts (MSEEA) was started in 1988, and has been published annually since 1991. The National Institute of Statistics and Geography of Mexico (INEGI) have used the accounts to derive key information about environmental sustainability in the country. Yearly environmental costs are equal to 8.5% of GDP. Since 2003, the INEGI has calculated a yearly Green GDP aggregate. Between 2003 and 2009, the ratio of Green GDP to GDP increased from 90.4% to 92.1%. The INEGI reports that while the scope of action should be increased, this trend reflects the successful efforts of the government to reduce impacts that negatively affect the environment.

Russia

The Federal State Statistics Service (Rosstat) currently includes estimates of land and non-cultivated bio and water resources in the "tangible non-produced assets" section of the Russian balance sheet. Rosstat has outlined a series of priorities for developing a Russian SEEA. Its highest priority is asset accounts, and with respect to the environment, a particular record of environmental protection expenditures. After 2015, the basis for integration will appear with the creation of input-output tables. Physical flow accounts are a medium priority; further development will depend on the completion of the input-output tables and whether new surveys are introduced on time. While accounts for extended SNA aggregates, such as the depletion or degradation of natural capital are being given conceptual consideration, they are the lowest priority of Rosstat.

Saudi Arabia

Saudi Arabia has not undertaken any relevant initiatives.

South Africa

The environmental-economic accounts are developed in the National Accounts Division of Statistics South Africa (SA). Information for the accounts is sourced from various government departments, surveys and geographic and spatial data. According to a 2011 report to the United Nations, environmental statistics are not a mainstream priority for the country, and Statistics SA does not have adequate resources to collect environmental data. Additionally, a "lengthy institutionalization process" has prevented the publication of any official reports.

Turkey

The Turkish Statistical Institute produces a range of environmental statistics.

United States of America

In 1992, the Bureau of Economic Analysis within the Department of Commerce began intensively developing environmental accounting methodologies. However, Congress directed the Department to suspend all activities in this area in 1994, after the first U.S. Integrated Environmental and Economic Satellite Accounts were published. It also ordered an external review of environmental accounting. This review, Nature's Numbers: Expanding the National Economic Accounts to Include the Environment was concluded in 1999.

William Nordhaus (Nature's Numbers), along with Nicholas Muller and Robert Mendelsohn, co-authored a study on integrating environmental externalities into a SNA. Their model estimates gross damages from air pollution in each industry in the United States. It estimates that damages from activities such as combustion of waste, sewage treatment and firing power plants by oil or coal are larger than their value added. The largest externalities, for which damages range from 0.8 to 5.6 times value added, arise from coal-fired electric generation.

Renewable Resource

A 'renewable resource' is an organic natural resource which can replenish to overcome usage and consumption, either through biological reproduction or other naturally recurring processes in a finite amount of time. Renewable resources are a part of Earth's natural environment and the largest components of its ecosphere. A positive life cycle assessment is a key indicator of a resource's sustainability.

Definitions of renewable: resources may also include agricultural production, as in sustainable agriculture and to an extent water resources. In 1962 Paul Alfred Weiss defined Renewable Resources as: "*The total range of living organisms providing man with food, fibres, drugs, etc...*". Another type of renewable resources is renewable energy resources. Common sources of renewable energy include solar, geothermal and wind power, which are all categorised as renewable resources.

Air, Food, Water

Water Resources

Water can be considered a *renewable* material when carefully controlled usage, treatment, and release are followed. If not, it would become a non-renewable resource at that location. For example, groundwater is usually removed from an aquifer at a rate much greater than its very slow natural recharge, and so groundwater is considered non-renewable. Removal of water from the pore spaces may cause permanent compaction (subsidence) that cannot be renewed. 97.5% of the water on the Earth is salt water, and 3% is fresh water; slightly over two thirds of this is frozen in glaciers and polar ice caps. The remaining unfrozen freshwater is found mainly as groundwater, with only a small fraction (0.008%) present above ground or in the air.

Water pollution is one of the main concerns regarding water resources. It is estimated that 22% of worldwide water is used in industry. Major industrial users include hydroelectric dams, thermoelectric power plants, which use water for cooling, ore and oil refineries, which use water in chemical processes, and manufacturing plants, which use water as a solvent.

Panoramic of a natural wetland (Sinclair Wetlands, New Zealand)

Non Agricultural Food

Food is any substance consumed to provide nutritional support for the body. Most food has its origin in renewable resources. Food is obtained directly from plants and animals.

Wild berries and other fruits, mushrooms, plants, seeds and naturally growing edible resources, still represent a valuable source of nutrition in many countries, especially in rural areas. In fact many wild animals are dependent on wild plants and fruits as a source of food.

Alaska wild "berries" from the Innoko National Wildlife Refuge - Renewable Resources

Hunting may not be the first source of meat in the modernised world, but it is still an important and essential source for many rural and remote groups. It is also the sole source of feeding for wild carnivores.

Sustainable Agriculture

The phrase sustainable agriculture was coined by Australian agricultural scientist Gordon McClymont. It has been defined as "an integrated system of plant and animal production practices having a site-specific application that will last over the long term". Expansion of agricultural land reduces biodiversity and contributes to deforestation. The Food and Agriculture Organisation of the United Nations estimates that in coming decades, cropland will continue to be lost to industrial and urban development, along with reclamation of wetlands, and conversion of forest to cultivation, resulting in the loss of biodiversity and increased soil erosion.

Polyculture practices in Andhra Pradesh

Although air and sunlight are available everywhere on Earth, crops also depend on soil nutrients and the availability of water. Monoculture is a method of growing only one crop at a time in a given field, which can damage land and cause it to become either unusable or suffer from reduced yields. Monoculture can also cause the build-up of pathogens and pests that target one specific species. The Great Irish Famine (1845–1849) is a well-known example of the dangers of monoculture.

Crop rotation and long-term crop rotations confer the replenishment of nitrogen through the use of green manure in sequence with cereals and other crops, and can improve soil structure and fertility by alternating deep-rooted and shallow-rooted plants. Other methods to combat lost soil nutrients are returning to natural cycles that annually flood cultivated lands (returning lost nutrients indefinitely) such as the Flooding of the Nile, the long-term use of biochar, and use of crop and livestock landraces that are adapted to less than ideal conditions such as pests, drought, or lack of nutrients.

Agricultural practices are the single greatest contributor to the global increase in soil erosion rates. It is estimated that "more than a thousand million tonnes of southern Africa's soil are eroded every year. Experts predict that crop yields will be halved within thirty to fifty years if erosion continues at present rates." The Dust Bowl phenomenon in the 1930s was caused by severe drought combined with farming methods that did not include crop rotation, fallow fields, cover crops, soil terracing and wind-breaking trees to prevent wind erosion.

The tillage of agricultural lands is one of the primary contributing factors to erosion, due to mechanised agricultural equipment that allows for deep plowing, which severely increases the amount of soil that is available for transport by water erosion. The phenomenon called *Peak Soil* describes how large-scale factory farming techniques are jeopardizing humanity's ability to grow food in the present and in the future. Without efforts to improve soil management practices, the availability of arable soil will become increasingly problematic.

Methods to combat erosion include no-till farming, using a keyline design, growing wind breaks to hold the soil, and widespread use of compost. Chemical fertiliser and pesticides can also have an effect of soil erosion, which can contribute to soil salinity and prevent other species from growing. Phosphate is a primary component in the chemical fertiliser applied most commonly in modern agricultural production. However, scientists estimate that rock phosphate reserves will be depleted in 50–100 years and that *Peak Phosphate* will occur in about 2030.

Industrial processing and logistics also have an effect on agriculture's sustainability. The way and locations crops are sold requires energy for transportation, as well as the energy cost for materials, labour, and transport. Food sold at a local location, such a farmers' market, have reduced energy overheads.

Illegal slash and burn practice in Madagascar, 2010

Air

Air is a renewable resource. All living organisms need oxygen, nitrogen(directly or indirectly), carbon(directly or indirectly) and many other gases in small quantities for their survival.

Non-food Resources

An important renewable resource is wood provided by means of forestry, which is used for construction, housing and firewood since ancient times.   Plants provide the main sources for renewable resources, the main distinction is made between energy crops and Non-food crops. A large variety of lubricants, industrially used vegetable oils, textiles and fibre made e.g. of cotton, copra or hemp, paper derived from wood, rags or grasses, bioplastic are based on plant renewable resources. A large variety of chemical base products like latex, ethanol, resin, sugar and starch can be provided with plant renewables. Animal based renewables include fur, leather, technical fat and lubricants and further derived products, as e.g. animal glue, tendons, casings or in historical times ambra and baleen provided by whaling.

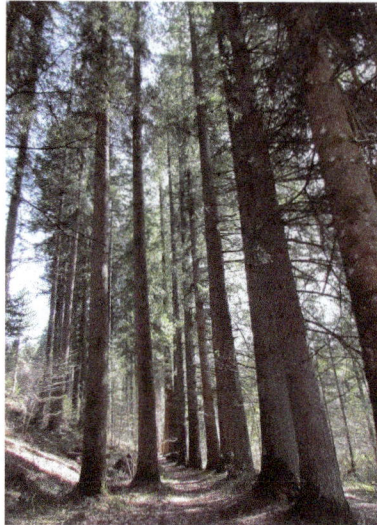

Douglas (Pseudotsuga menziesii) forest created in 1850, Meymac (Corrèze), France

With regard to pharmacy ingredients and legal and illegal drugs, plants are important sources, however e.g. venom of snakes, frogs and insects has been a valuable renewable source of pharmacological ingredients. Before GMO production set in, insulin and important hormones were based on animal sources. Feathers, an important byproduct of poultry farming for food, is still being used as filler and as base for keratin in general. Same applies for the chitin produced in farming Crustaceans which may be used as base of chitosan. The most important part of the human body used for non-medical purposes is human hair as for artificial hair integrations, which is being traded worldwide.

Historical role

Historically, renewable resources like firewood, latex, guano, charcoal, wood ash, plant colors as indigo, and whale products have been crucial for human needs but failed to supply demand in the beginning of the industrial era. Early modern times faced large problems with overuse of renewable resources as in deforestation, overgrazing or overfishing.

An adult and sub-adult Minke whale are dragged aboard the Nisshin Maru, a Japanese whaling vessel

Hemp insulation, a renewable resource used as building material

Besides fresh meat and milk, which is as a food item not topic of this section, livestock farmers and artisans used further animal ingredients as tendons, horn, bones, bladders. Complex technical constructions as the composite bow were based on combination of animal and plant based materials. The current distribution conflict between biofuel and food production is being described as Food vs. fuel. Conflicts between food needs and usage, as supposed by fief obligations were in so far common in historical times as well. However, a significant percentage of (middle European) farmers yields went into livestock, which provides as well organic fertiliser. Oxen and horses were important for transportation purposes, drove engines as e.g. in treadmills.

Other regions solved the transportation problem with terracing, urban and garden agriculture. Further conflicts as between forestry and herding, or (sheep) herders and cattle farmers led to various solutions. Some confined wool production and sheep to large state and nobility domains or outsourced to professional shepherds with larger wandering herds.

The British Agricultural Revolution was mainly based on a new system of crop rotation, the four-field rotation. British agriculturist Charles Townshend recognised the invention in Dutch Waasland and popularised it in the 18th century UK, George Washington Carver in the USA. The system used wheat, turnips and barley and introduced as well clover. Clover is able to fix nitrogen from air, a practically non exhaustive renewable resource, into fertilizing compounds to the soil and allowed to increase yields by large. Farmers opened up a fodder crop and grazing crop. Thus livestock could to be bred year-round and winter culling was avoided. The amount of manure rose and allowed more crops but to refrain from wood pasture.

Early modern times and the 19th century saw the previous resource base partially replaced respectively supplemented by large scale chemical synthesis and by the use of fossil and mineral resources respectively. Besides the still central role of wood, there is a sort of renaissance of renewable products based on modern agriculture, genetic research and extraction technology. Besides fears about an upcoming global shortage of fossil fuels, local shortages due to boycotts, war and blockades or just transportation problems in remote regions have contributed to different methods of replacing or substituting fossil resources based on renewables.

Challenges

The use of certain basically renewable products as in TCM endangers various species. Just the black market in rhinoceros horn reduced the world's rhino population by more than 90 percent over the past 40 years.

Renewables used for Autarky Approaches

In vitro-culture of Vitis (grapevine), Geisenheim Grape Breeding Institute

The success of the German chemical industry till World War I was based on the replacement of colonial products. The predecessors of IG Farben dominated the world market for synthetic dyes at the beginning of the 20th century and had an important role in artificial pharmaceuticals, photographic film, agricultural chemicals and electrochemicals.

However the former Plant breeding research institutes took a different approach. After the loss of the German colonial empire, important players in the field as Erwin Baur and Konrad Meyer switched to using local crops as base for economic autarky. Meyer as a key agricultural scientist and spatial planner of the Nazi era managed and lead Deutsche Forschungsgemeinschaft resources and focused about a third of the complete research grants in Nazi Germany on agricultural and genetic research and especially on resources needed in case of a further German war effort. A wide array of agrarian research institutes still existing today and having importance in the field was founded or enlarged in the time.

There were some major failures as trying to e.g. grow frost resistant olive species, but some success in the case of hemp, Linum, rapeseed, which are still of current importance. During the war, German scientists tried to systematically exploit foreign research results in occupied countries. Heinrich Himmler personally supported a research project using Russian Taraxacum (dandelion) species to manufacture natural rubber. The project was conducted using 150 female KZ prisoners and captured Russian scientists kept together as 'Kommando Pflanzenzucht' (Plant breeding command) in a subcamp (SS) of Konzentrationslager Auschwitz led by SS agrarian research officer Joachim Caesar. Rubber dandelions are still of interest, as scientists in the Fraunhofer Institute for Molecular Biology and Applied Ecology (IME) announced 2013 to have developed a cultivar that is suitable for commercial production of natural rubber.

Legal Situation and Subsidies

Several legal and economic means have been used to enhance the market share of renewables. The UK uses Non-Fossil Fuel Obligations (NFFO), a collection of orders requiring the electricity Distribution Network Operators in England and Wales to purchase electricity from the nuclear power and renewable energy sectors. Similar mechanisms operate in Scotland (the Scottish Renewable Orders under the Scottish Renewables Obligation) and Northern Ireland (the Northern Ireland Non-Fossil Fuel Obligation). In the USA, Renewable Energy Certificates (RECs), use a similar approach. German Energiewende is using fed-in tariffs. An unexpected outcome of the subsidies was the quick increase of pellet byfiring in conventional fossil fuel plants (compare Tilbury power stations) and cement works, making wood respectively biomass accounting for about half of Europe's renewable-energy consumption.

Examples of Industrial Use

Bioplastics

A packaging blister made from cellulose acetate, a bioplastic

Bioplastics are a form of plastics derived from renewable biomass sources, such as vegetable fats and oils, lignin, corn starch, pea starch or microbiota. The most common form of bioplastic is thermoplastic starch. Other forms include Cellulose bioplastics, biopolyester, Polylactic acid, and bio-derived polyethylene.

The production and use of bioplastics is generally regarded as a more sustainable activity when compared to plastic production from petroleum (petroplastic); however, manufacturing of bioplastic materials is often still reliant upon petroleum as an energy and materials source. Because of the fragmentation in the market and ambiguous definitions it is difficult to describe the total market size for bioplastics, but the global production capacity is estimated at 327,000 tonnes. In contrast, global consumption of all flexible packaging is estimated at around 12.3 million tonnes.

Bioasphalt

Bioasphalt is an asphalt alternative made from non-petroleum based renewable resources. Manufacturing sources of bioasphalt include sugar, molasses and rice, corn and potato starches, and vegetable oil based waste. Asphalt made with vegetable oil based binders was patented by Colas SA in France in 2004.

Renewable Energy

Renewable energy refers to the provision of energy via renewable resources which are naturally replenished fast enough as being used. It includes e.g. sunlight, wind, biomass, rain, tides, waves and geothermal heat. Renewable energy may replace or enhance fossil energy supply various distinct areas: electricity generation, hot water/space heating, motor fuels, and rural (off-grid) energy services.

Biomass

Biomass is referring to biological material from living, or recently living organisms, most often referring to plants or plant-derived materials.

A sugarcane plantation in Brazil (State of São Paulo). Cane is used for biomass energy.

Sustainable harvesting and use of renewable resources (i.e., maintaining a positive renewal rate) can reduce air pollution, soil contamination, habitat destruction and land degradation. Biomass energy is derived from six distinct energy sources: garbage, wood, plants, waste, landfill gases, and alcohol fuels. Historically, humans have harnessed biomass-derived energy since the advent of burning wood to make fire, and wood remains the largest biomass energy source today.

However, low tech use of biomass, which still amounts for more than 10% of world energy needs may induce indoor air pollution in developing nations and results in between 1.5 million and 2 million deaths in 2000.

The biomass used for electricity generation varies by region. Forest by-products, such as wood residues, are common in the United States. Agricultural waste is common in Mauritius (sugar cane residue) and Southeast Asia (rice husks). Animal husbandry residues, such as poultry litter, are common in the UK. The biomass power generating industry in the United States, which consists of approximately 11,000 MW of summer operating capacity actively supplying power to the grid, produces about 1.4 percent of the U.S. electricity supply.

Biofuel

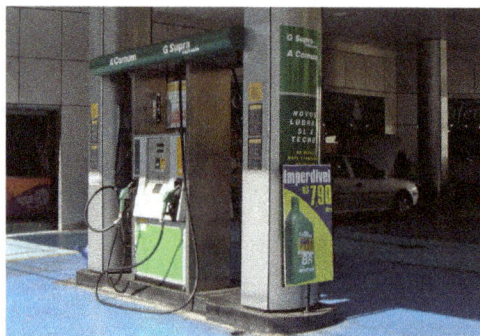

Brazil has bioethanol made from sugarcane available throughout the country. Shown a typical Petrobras gas station at São Paulo with dual fuel service, marked A for alcohol (ethanol) and G for gasoline.

A biofuel is a type of fuel whose energy is derived from biological carbon fixation. Biofuels include fuels derived from biomass conversion, as well as solid biomass, liquid fuels and various biogases.

Bioethanol is an alcohol made by fermentation, mostly from carbohydrates produced in sugar or starch crops such as corn, sugarcane or switchgrass.

Biodiesel is made from vegetable oils and animal fats. Biodiesel is produced from oils or fats using transesterification and is the most common biofuel in Europe.

Biogas is methane produced by the process of anaerobic digestion of organic material by anaerobes., etc. is also a renewable source of energy.

Biogas

Biogas typically refers to a mixture of gases produced by the breakdown of organic matter in the absence of oxygen. Biogas is produced by anaerobic digestion with anaerobic bacteria or fermentation of biodegradable materials such as manure, sewage, municipal waste, green waste, plant material, and crops. It is primarily methane (CH_4) and carbon dioxide (CO_2) and may have small amounts of hydrogen sulphide (H_2S), moisture and siloxanes.

Natural Fibre

Natural fibres are a class of hair-like materials that are continuous filaments or are in discrete elongated pieces, similar to pieces of thread. They can be used as a component of composite materials. They can also be matted into sheets to make products such as paper or felt. Fibres are of two types: natural fibre which consists of animal and plant fibres, and man made fibre which consists of synthetic fibres and regenerated fibres.

Threats to Renewable Resources

Renewable resources are endangered by non-regulated industrial developments and growth. They must be carefully managed to avoid exceeding the natural world's capacity to replenish them. A life cycle assessment provides a systematic means of evaluating renewability. This is a matter of sustainability in the natural environment.

Overfishing

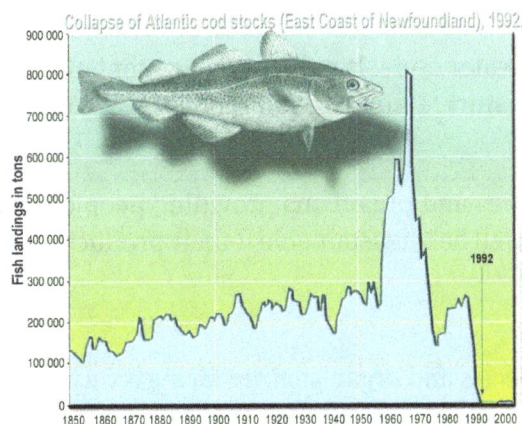

Atlantic cod stocks severely overfished leading to abrupt collapse

National Geographic has described ocean over fishing as "simply the taking of wildlife from the sea at rates too high for fished species to replace themselves."

Tuna meat is driving overfishing as to endanger some species like the bluefin tuna. The European Community and other organisations are trying to regulate fishery as to protect species and to prevent their extinctions.

Examples of overfishing exist in areas such as the North Sea of Europe, the Grand Banks of North America and the East China Sea of Asia.

The decline of penguin population is caused in part by overfishing, caused by human competition over the same renewable resources

Deforestation

Deforestation in Brazil 1996

Besides their role as a resource for fuel and building material, trees protect the environment by absorbing carbon dioxide and by creating oxygen. The destruction of rain forests is one of the critical causes of climate change. Deforestation causes carbon dioxide to linger in the atmosphere. As carbon dioxide accrues, it produces a layer in the atmosphere that traps radiation from the sun. The radiation converts to heat which causes global warming, which is better known as the greenhouse effect.

Deforestation also affects the water cycle. It reduces the content of water in the soil and groundwater as well as atmospheric moisture. Deforestation reduces soil cohesion, so that erosion, flooding and landslides ensue.

Rain forests house many species and organisms providing people with food and other commodities. In this way biofuels may well be unsustainable if their production contributes to deforestation.

Endangered Species

Some renewable resources, species and organisms are facing a very high risk of extinction caused by growing human population and over-consumption. It has been estimated that over 40% of all living species on Earth are at risk of going extinct. Many nations have laws to protect hunted species and to restrict the practice of hunting. Other conservation methods include restricting land development or creating preserves. The IUCN Red List of Threatened Species is the best-known worldwide conservation status listing and ranking system. Internationally, 199 countries have signed an accord agreeing to create Biodiversity Action Plans to protect endangered and other threatened species.

Over-hunting of American Bison.

Non-renewable Resource

A non-renewable resource (also called a finite resource) is a resource that does not renew itself at a sufficient rate for sustainable economic extraction in meaningful human time-frames. An example is carbon-based, organically-derived fuel. The original organic material, with the aid of heat and pressure, becomes a fuel such as oil or gas. Earth minerals and metal ores, fossil fuels (coal, petroleum, natural gas) and groundwater in certain aquifers are all considered non-renewable resources, though individual elements are almost always conserved.

A coal mine in Wyoming, United States. Coal, produced over millions of years, is a finite and non-renewable resource on a human time scale.

In contrast, resources such as timber (when harvested sustainably) and wind (used to power energy conversion systems) are considered renewable resources, largely because their localized replenishment can occur within time frames meaningful to humans.

Earth Minerals and Metal Ores

Earth minerals and metal ores are examples of non-renewable resources. The metals themselves are present in vast amounts in Earth's crust, and their extraction by humans only occurs where they are concentrated by natural geological processes (such as heat, pressure, organic activity, weathering and other processes) enough to become economically viable to extract. These processes generally take from tens of thousands to millions of years, through plate tectonics, tectonic subsidence and crustal recycling.

The localized deposits of metal ores near the surface which can be extracted economically by humans are non-renewable in human time-frames. There are certain rare earth minerals and elements that are more scarce and exhaustible than others. These are in high demand in manufacturing, particularly for the electronics industry.

Most metal ores are considered vastly greater in supply to fossil fuels, because metal ores are formed by crustal-scale processes which make up a much larger portion of the Earth's near-surface environment, than those that form fossil fuels which are limited to areas where carbon-based life forms flourish, die, and are quickly buried.

Fossil Fuels

Natural resources such as coal, petroleum (crude oil) and natural gas take thousands of years to form naturally and cannot be replaced as fast as they are being consumed. Eventually it is considered that fossil-based resources will become too costly to harvest and humanity will need to shift its reliance to other sources of energy such as solar or wind power.

An alternative hypothesis is that carbon based fuel is virtually inexhaustible in human terms, if one includes all sources of carbon-based energy such as methane hydrates on the sea floor, which are vastly greater than all other carbon based fossil fuel resources combined. These sources of carbon are also considered non-renewable, although their rate of formation/replenishment on the sea floor is not known. However their extraction at economically viable costs and rates has yet to be determined.

At present, the main energy source used by humans is non-renewable fossil fuels. Since the dawn of internal combustion engine technologies in the 17th century, petroleum and other fossil fuels have remained in continual demand. As a result, conventional infrastructure and transport systems, which are fitted to combustion engines, remain prominent throughout the globe. The continual use of fossil fuels at the current rate is believed to increase global warming and cause more severe climate change.

Nuclear Fuels

In 1987, the World Commission on Environment and Development (WCED) an organization set up by but independent from the United Nations classified fission reactors that produce more fissile nuclear fuel than they consume -i.e. breeder reactors, and when it is developed, fusion power, among conventional renewable energy sources, such as solar and falling water. The American Petroleum Institute likewise does not consider conventional nuclear fission as renewable, but that breeder reactor nuclear power fuel is considered renewable and sustainable, before explaining that radioactive waste from used spent fuel rods remains dangerous, and so has to be very carefully stored for up to a thousand years. With the careful monitoring of radioactive waste products also being required upon the use of other renewable energy sources, such as geothermal energy.

Rössing uranium mine is the longest-running and one of the largest open pit uranium mines in the world, every year it produces eight percent of global uranium needs.

The use of nuclear technology relying on fission requires Naturally occurring radioactive material as fuel. Uranium, the most common fission fuel, and is present in the ground at relatively low concentrations and mined in 19 countries. This mined uranium is used to fuel energy-generating nuclear reactors with fissionable uranium-235 which generates heat that is ultimately used to power turbines to generate electricity.

Nuclear power provides about 6% of the world's energy and 13–14% of the world's electricity. Nuclear energy production is associated with potentially dangerous radioactive contamination as it relies upon unstable elements. In particular, nuclear power facilities produce about 200,000 metric tons of low and intermediate level waste (LILW) and 10,000 metric tons of high level waste (HLW) (including spent fuel designated as waste) each year worldwide.

Issues entirely separate from the question of the sustainability of nuclear fuel, relate to the use of nuclear fuel and the high-level radioactive waste the nuclear industry generates that if not properly contained, is highly hazardous to people and wildlife. Radiocontaminants in the environment can enter the food chain and become bioaccumulated. Internal or external exposure can cause mutagenic DNA breakage producing teratogenic generational birth defects, cancers and other damage. The United Nations (UNSCEAR) estimated in 2008 that average annual human radiation exposure includes 0.01 millisievert (mSv) from the legacy of past atmospheric nuclear testing plus the Chernobyl disaster and the nuclear fuel cycle, along with 2.0 mSv from natural radioisotopes and 0.4 mSv from cosmic rays; all exposures vary by location. natural uranium radioisotopes in nuclear waste and naturally in the ground emits radiation for the prolonged period of 4.5 billion years or more, and storage has risks of containment. The storage of waste, health implications and dangers of radioactive fuel continue to be a topic of debate, resulting in a controversial and unresolved industry.

Renewable Resources

Natural resources, known as renewable resources, are replaced by natural processes and forces persistent in the natural environment. There are intermittent and reoccurring renewables, and recyclable materials, which are utilized during a cycle across a certain amount of time, and can be harnessed for any number of cycles.

The Three Gorges Dam, the largest renewable energy generating station in the world.

The production of goods and services by manufacturing products in economic systems creates many types of waste during production and after the consumer has made use of it. The material is then either incinerated, buried in a landfill or recycled for reuse. Recycling turns materials of value that would otherwise become waste into valuable resources again.

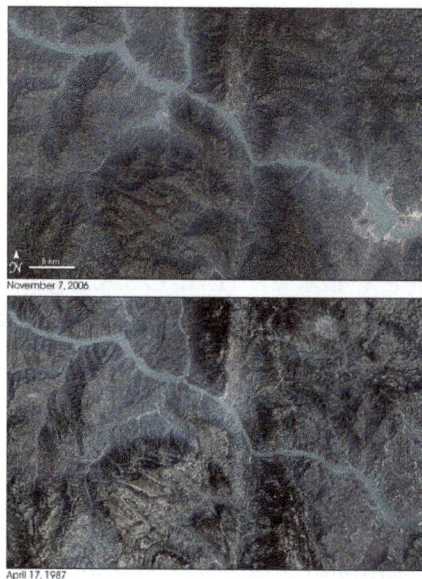

Satellite map showing areas flooded by the Three Gorges reservoir. Compare November 7, 2006 (above) with April 17, 1987 (below). The energy station required the flooding of archaeological and cultural sites and displaced some 1.3 million people, and is causing significant ecological changes, including an increased risk of landslides. The dam has been a controversial topic both domestically and abroad.

The natural environment, with soil, water, forests, plants and animals are all renewable resources, as long as they are adequately monitored, protected and conserved. Sustainable agriculture is the cultivation of plant and animal materials in a manner that preserves plant and animal ecosystems over the long term. The overfishing of the oceans is one example of where an industry practice or method can threaten an ecosystem, endanger species and possibly even determine whether or not a fishery is sustainable for use by humans. An unregulated industry practice or method can lead to a complete resource depletion.

The renewable energy from the sun, wind, wave, biomass and geothermal energies are based on renewable resources. Renewable resources such as the movement of water (hydropower, tidal power and wave power), wind and radiant energy from geothermal heat (used for geothermal power) and solar energy (used for solar power) are practically infinite and cannot be depleted, unlike their non-renewable counterparts, which are likely to run out if not used sparingly.

The potential wave energy on coastlines can provide 1/5 of world demand. Hydroelectric power can supply 1/3 of our total energy global needs. Geothermal energy can provide 1.5 more times the energy we need. There is enough wind to power the planet 30 times over, wind power could power all of humanity's needs alone. Solar currently supplies only 0.1% of our world energy needs, but there is enough out there to power humanity's needs 4,000 times over, the entire global projected energy demand by 2050.

Renewable energy and energy efficiency are no longer niche sectors that are promoted only by governments and environmentalists. The increasing levels of investment and that more of the capital is from conventional financial actors, both suggest that sustainable energy has become mainstream and the future of energy production, as non-renewable resources decline. This is reinforced by climate change concerns, nuclear dangers and accumulating radioactive waste, high oil prices, peak oil and increasing government support for renewable energy. These factors are commercializing renewable energy, enlarging the market and growing demand, the adoption of new products to replace obsolete technology and the conversion of existing infrastructure to a renewable standard.

Economic Models

In economics, a non-renewable resource is defined as goods, where greater consumption today implies less consumption tomorrow. David Ricardo in his early works analysed the pricing of exhaustible resources, where he argued that the price of a mineral resource should increase over time. He argued that the spot price is always determined by the mine with the highest cost of extraction, and mine owners with lower extraction costs benefit from a differential rent. The first model is defined by Hotelling's rule, which is a 1931 economic model of non-renewable resource management by Harold Hotelling. It shows that efficient exploitation of a nonrenewable and non-augmentable resource would, under otherwise stable conditions, lead to a depletion of the resource. The rule states that this would lead to a net price or "Hotelling rent" for it that rose annually at a rate equal to the rate of interest, reflecting the increasing scarcity of the resources. The Hartwick's rule provides an important result about the sustainability of welfare in an economy that uses non-renewable source.

However, nearly all metal prices have been declining over time in inflation adjusted terms, because of a number of false assumptions in the above. Firstly, metal resources are non-renewable, but on a world scale, largely inexhaustible. This is because they are present throughout the earth's crust on a vast scale, far exceeding human demand on all time scales. Metal ores however, are only extracted in those areas where nature has concentrated the metal in the crust to a level whereby it is locally economic to extract. This also depends on the available technology for both finding the metal ores as well as extracting them, which is constantly changing. If the technology or demand changes, vast amounts of metal previously ignored can become economically extractable. This is why Ricardo's simplistic notion that the price of a mineral resource should increase over time has in fact turned out to be the opposite, nearly all metal ores have decreased in inflation adjusted prices since well before the early 20th century. The main reason he was wrong is that he assumed that metals are exhaustible on a world scale, and he also misunderstood the effect of globally competing markets; in human terms the amount of metal in the earth's crust is essentially limitless. It is only in localized areas that metal ores can become depleted, as these local areas compete with extraction costs of resources elsewhere, which does have ramifications for the sustainability of local economies.

Exploitation of Natural Resources

The exploitation of natural resources is the use of natural resources for economic growth, sometimes with a negative connotation of accompanying environmental degradation. It started to emerge on an industrial scale in the 19th century as the extraction and processing of raw materials (such as in mining, steam power, and machinery) developed much further than it had in preindustrial eras. During the 20th century, energy consumption rapidly increased. Today, about 80% of the world's energy consumption is sustained by the extraction of fossil fuels, which consists of oil, coal and gas. Another non-renewable resource that is exploited by humans is subsoil minerals such as precious metals that are mainly used in the production of industrial commodities. Intensive agriculture is an example of a mode of production that hinders many aspects of the natural environment, for example the degradation of forests in a terrestrial ecosystem and water pollution in an aquatic ecosystem. As the world population rises and economic growth occurs, the depletion of natural resources influenced by the unsustainable extraction of raw materials becomes an increasing concern.

Timber

Why Resources are Under Pressure

- Increase in the sophistication of technology enabling natural resources to be extracted quickly and efficiently. E.g., in the past, it could take long hours just to cut down one tree only using saws. Due to increased technology, rates of deforestation have greatly increased

- A rapid increase in population that is now decreasing gradually. The current number of 7.132 billion humans consume many natural resources.

- Cultures of consumerism. Materialistic views lead to the mining of gold and diamonds to produce jewelry, unnecessary commodities for human life or advancement.

- Excessive demand often leads to conflicts due to intense competition. Organizations such as Global Witness and the United Nations have documented the connection.

- Non-equitable distribution of resources.

Problems Arising from the Exploitation of Natural Resources

Natural resources are not limitless, and the following consequences can arise from the careless and excessive consumption of these resources:

- Deforestation
- Desertification
- Extinction of species
- Forced migration
- Soil erosion
- Oil depletion
- Ozone depletion
- Greenhouse gas increase
- Extreme energy
- Water pollution
- Natural hazard/Natural disaster

Effects on Local Communities

The Global South

Human Resources Macon, Georgia, 1909

When a mining company enters a developing country to extract raw materials, advocating the advantages of the industry's presence and minimizing the potential negative effects gain cooperation of the local people. Advantageous factors are primarily in economic development so services that the government could not provide such as health centers, police departments and schools can be established. However, with economic development, money becomes a dominant subject of interest. This can bring about major conflicts that a local community in a developing country has never dealt with before. These conflicts emerge by a change to more egocentric views among the locals influenced by consumerist values.

The effects of the exploitation of natural resources in the local community of a developing country are exhibited in the impacts from the Ok Tedi Mine. After BHP, now BHP Billiton, entered into Papua New Guinea to exploit copper and gold, the economy of the indigenous peoples boomed. Although their quality of life has improved, initially disputes were common among the locals in terms of land rights and who should be getting the benefits from the mining project. The consequences of the Ok Tedi environmental disaster illustrate the potential negative effects from the exploitation of natural resources. The resulting mining pollution includes toxic contamination of the natural water supply for communities along the Ok Tedi River, causing widespread killing of aquatic life. When a mining company ends a project after extracting the raw materials from an area of a developing country, the local people are left to manage with the environmental damage done to their community and the long run sustainability of the economic benefits stimulated by the mining company's presence becomes a concern.

References

- Soule, Michael E. (1986). Conservation Biology: The Science of Scarcity and Diversity. Sinauer Associates. p. 584. ISBN 0-87893-795-1.

- Hunter, M. L. (1996). Fundamentals of Conservation Biology. Blackwell Science Inc., Cambridge, Massachusetts., ISBN 0-86542-371-7.

- Groom, M.J., Meffe, G.K. and Carroll, C.R. (2006) Principles of Conservation Biology (3rd ed.). Sinauer Associates, Sunderland, MA. ISBN 0-87893-518-5

- van Dyke, Fred (2008). Conservation Biology: Foundations, Concepts, Applications, 2nd ed. Springer Verlag. p. 478. ISBN 978-1-4020-6890-4.

- Committee on 21st Century Systems Agriculture (2010). Toward Sustainable Agricultural Systems in the 21st Century. National Academies Press. ISBN 978-0-309-14896-2.

- Blanco, Humberto & Lal, Rattan (2010). "Tillage erosion". Principles of Soil Conservation and Management. Springer. ISBN 978-90-481-8529-0.

- Lobb, D.A. (2009). "Soil movement by tillage and other agricultural activities". In Jorgenson, Sven E. Applications in Ecological Engineering. Academic Press. ISBN 978-0-444-53448-4.

- A short history of livestock production, J. Hartung, in Livestock housing, Modern management to ensure optimal health and welfare of farm animals, edited by: Andres Aland and Thomas Banhazi, © 2013 ISBN 978-90-8686-217-7

- Gustav Comberg, Die deutsche Tierzucht im 19. und 20. Jahrhundert, Ulmer, 1984, ISBN 3-8001-3061-0, (History of livestock breeding in Germany)

- Autarkie und Ostexpansion: Pflanzenzucht und Agrarforschung im Nationalsozialismus, (agrarian research during the NS regime) Susanne Heim, Wallstein, 2002, ISBN 389244496X

- Heim, Susanne (2002). Autarkie und Ostexpansion: Pflanzenzucht und Agrarforschung im Nationalsozialismus, (agrarian research during the NS regime). Wallstein. ISBN 389244496X.

- America's Climate Choices: Panel on Advancing the Science of Climate Change; National Research Council (2010). Advancing the Science of Climate Change. Washington, D.C.: The National Academies Press. ISBN 0-309-14588-0.

- Pedro, Antonio M.A. (2004). Mainstreaming Mineral Wealth in Growth and Poverty Reduction Strategies (PDF). Economic Commission for Africa. pp. 5–6. ISBN 9789211250978. Retrieved 20 March 2012.

- "Mangrove Forests: One of the World's Threatened Major Tropical Environments". Bioscience. 51 (10): 807–815. 2001. ISSN 1525-3244. Retrieved 5 January 2016.

- Conniff, Richard (18 October 2012). "What's Wrong with Putting a Price on Nature?". e360.yale.edu. Yale University. Retrieved 5 January 2016.

- Monbiot, George (24 July 2014). "Put a price on nature? We must stop this neoliberal road to ruin". The Guardian. Retrieved 5 January 2016.

- Paddison, Laura (1 August 2014). "Is natural capital a 'neoliberal road to ruin'?- experts discuss". The Guardian. Retrieved 5 January 2016.

- "Natural Capital (Does putting a value on natural capital mean it will be 'sold off'?)". www.naturalcapitalcommittee.org. Natural Capital Committee. Retrieved 5 January 2016.

4

Sustainability: An Integrated Study

Biological systems are diverse in nature and can be indefinitely productive. This ability of biological systems is known as sustainability. Aspects of sustainability include circles of sustainability, sustainable living, sustainable design, sustainable development, sustainability and environmental management and sustainability and systemic change resistance. The topics discussed in the section are of great importance to broaden the existing knowledge on sustainability.

Sustainability

In ecology, sustainability (from sustain and ability) is the property of biological systems to remain diverse and productive indefinitely. Long-lived and healthy wetlands and forests are examples of sustainable biological systems. In more general terms, sustainability is the endurance of systems and processes. The organizing principle for sustainability is sustainable development, which includes the four interconnected domains: ecology, economics, politics and culture. Sustainability science is the study of sustainable development and environmental science.

Achieving sustainability will enable the Earth to continue supporting human life.

Sustainability can also be defined as a socio-ecological process characterized by the pursuit of a common ideal.An ideal is by definition unattainable in a given time/space but endlessly approachable and it is this endless pursuit that forms a sustainable system in the process (ibid). Healthy ecosystems and environments are necessary to the survival of humans and other organisms. Ways of reducing negative human impact are environmentally-friendly chemical engineering, environmental resources management and environmental protection. Information is gained from green chemistry, earth science, environmental science and conservation biology. Ecological economics studies the fields of academic research that aim to address human economies and natural ecosystems.

Batad rice terraces, The Philippines —UNESCO World Heritage site

Moving towards sustainability is also a social challenge that entails international and national law, urban planning and transport, local and individual lifestyles and ethical consumerism. Ways of living more sustainably can take many forms from reorganizing living conditions (e.g., ecovillages, eco-municipalities and sustainable cities), reappraising economic sectors (permaculture, green building, sustainable agriculture), or work practices (sustainable architecture), using science to develop new technologies (green technologies, renewable energy and sustainable fission and fusion power), or designing systems in a flexible and reversible manner, and adjusting individual lifestyles that conserve natural resources.

"The term 'sustainability' should be viewed as humanity's target goal of human-ecosystem equilibrium (homeostasis), while 'sustainable development' refers to the holistic approach and temporal processes that lead us to the end point of sustainability." (305) Despite the increased popularity of the use of the term "sustainability", the possibility that human societies will achieve environmental sustainability has been, and continues to be, questioned—in light of environmental degradation, climate change, overconsumption, population growth and societies' pursuit of indefinite economic growth in a closed system.

Etymology

The name sustainability is derived from the Latin *sustinere* (*tenere*, to hold; *sub*, up). *Sustain* can mean "maintain", "support", or "endure". Since the 1980s *sustainability* has been used more in the sense of human sustainability on planet Earth and this has resulted in the most widely quoted definition of sustainability as a part of the concept *sustainable development*, that of the Brundtland Commission of the United Nations on March 20, 1987: "sustainable development is development that meets the needs of the present without compromising the ability of future generations to meet their own needs."

Components

Three Pillars of Sustainability

The 2005 World Summit on Social Development identified sustainable development goals, such as economic development, social development and environmental protection. This view has been expressed as an illustration using three overlapping ellipses indicating that the three pillars of sustainability are not mutually exclusive and can be mutually reinforcing. In fact, the three pillars are interdependent, and in the long run none can exist without the others. The three pillars have served as a common ground for numerous sustainability standards and certification systems in recent years, in particular in the food industry. Standards which today explicitly refer to the triple bottom line include Rainforest Alliance, Fairtrade and UTZ Certified. Some sustainability experts and practitioners have illustrated four pillars of sustainability, or a quadruple bottom line. One such pillar is future generations, which emphasizes the long-term thinking associated with sustainability.

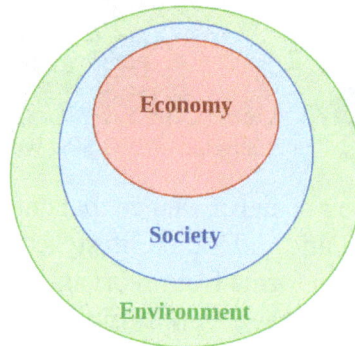

A diagram indicating the relationship between the "three pillars of sustainability", in which both economy and society are constrained by environmental limits

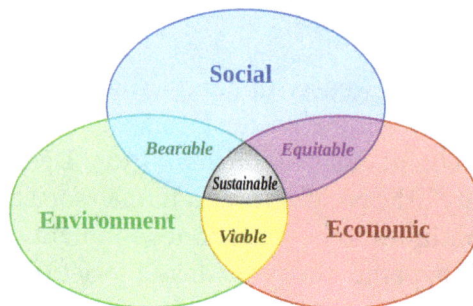

Venn diagram of sustainable development:at the confluence of three constituent parts

Sustainable development consists of balancing local and global efforts to meet basic human needs without destroying or degrading the natural environment. The question then becomes how to represent the relationship between those needs and the environment.

A study from 2005 pointed out that environmental justice is as important as is sustainable development. Ecological economist Herman Daly asked, "what use is a sawmill without a forest?" From this perspective, the economy is a subsystem of human society, which is itself a subsystem of the biosphere, and a gain in one sector is a loss from another. This perspective led to the nested circles figure of 'economics' inside 'society' inside the 'environment'.

The simple definition that sustainability is something that improves "the quality of human life while living within the carrying capacity of supporting eco-systems", though vague, conveys the idea of sustainability having quantifiable limits. But sustainability is also a call to action, a task in progress or "journey" and therefore a political process, so some definitions set out common goals and values. The Earth Charter speaks of "a sustainable global society founded on respect for nature, universal human rights, economic justice, and a culture of peace." This suggested a more complex figure of sustainability, which included the importance of the domain of 'politics'.

More than that, sustainability implies responsible and proactive decision-making and innovation that minimizes negative impact and maintains balance between ecological resilience, economic prosperity, political justice and cultural vibrancy to ensure a desirable planet for all species now and in the future. Specific types of sustainability include, sustainable agriculture, sustainable architecture or ecological economics. Understanding sustainable development is important but without clear targets an unfocused term like "liberty" or "justice". It has also been described as a "dialogue of values that challenge the sociology of development".

Circles of Sustainability

While the United Nations Millennium Declaration identified principles and treaties on sustainable development, including economic development, social development and environmental protection it continued using three domains: economics, environment and social sustainability. More recently, using a systematic domain model that responds to the debates over the last decade, the Circles of Sustainability approach distinguished four domains of economic, ecological, political and cultural sustainability. This in accord with the United Nations Agenda 21, which specifies **culture** as the fourth domain of sustainable development. The model is now being used by organizations such as the United Nations Cities Programme. and Metropolis

Shaping the Future

Integral elements of sustainability are research and innovation activities. A telling example is the European environmental research and innovation policy. It aims at defining and implementing a transformative agenda to greening the economy and the society as a whole so to make them sustainable. Research and innovation in Europe are financially supported by the programme Horizon 2020, which is also open to participation worldwide.

Resiliency

Resiliency in ecology is the capacity of an ecosystem to absorb disturbance and still retain its basic structure and viability. Resilience-thinking evolved from the need to manage interactions between human-constructed systems and natural ecosystems in a sustainable way despite the fact that to policymakers a definition remains elusive. Resilience-thinking addresses how much planetary ecological systems can withstand assault from human disturbances and still deliver the services current and future generations need from them. It is also concerned with commitment from geopolitical policymakers to promote and manage essential planetary ecological resources in order to promote resilience and achieve sustainability of these essential resources for benefit of future generations of life? The resiliency of an ecosystem, and thereby, its sustainability, can be reasonably measured at junctures or events where the combination of naturally occurring regenerative forces (solar energy, water, soil, atmosphere, vegetation, and biomass) interact with the energy released into the ecosystem from disturbances.

A practical view of sustainability is closed systems that maintain processes of productivity indefinitely by replacing resources used by actions of people with resources of equal or greater value by those same people without degrading or endangering natural biotic systems. In this way, sustainability can be concretely measured in human projects if there is a transparent accounting of the resources put back into the ecosystem to replace those displaced. In nature, the accounting occurs naturally through a process of adaptation as an ecosystem returns to viability from an external disturbance. The adaptation is a multi-stage process that begins with the disturbance event (earthquake, volcanic eruption, hurricane, tornado, flood, or thunderstorm), followed by absorption, utilization, or deflection of the energy or energies that the external forces created.

In analysing systems such as urban and national parks, dams, farms and gardens, theme parks, open-pit mines, water catchments, one way to look at the relationship between sustainability and resiliency is to view the former with a long-term vision and resiliency as the capacity of human engineers to respond to immediate environmental events.

History

The history of sustainability traces human-dominated ecological systems from the earliest civilizations to the present time. This history is characterized by the increased regional success of a particular society, followed by crises that were either resolved, producing sustainability, or not, leading to decline.

In early human history, the use of fire and desire for specific foods may have altered the natural composition of plant and animal communities. Between 8,000 and 10,000 years ago, agrarian communities emerged which depended largely on their environment and the creation of a "structure of permanence."

The Western industrial revolution of the 18th to 19th centuries tapped into the vast growth potential of the energy in fossil fuels. Coal was used to power ever more efficient engines and later to generate electricity. Modern sanitation systems and advances in medicine protected large populations from disease. In the mid-20th century, a gathering environmental movement pointed out that there were environmental costs associated with the many material benefits that were now being enjoyed. In the late 20th century, environmental problems became global in scale. The 1973 and 1979 energy crises demonstrated the extent to which the global community had become dependent on non-renewable energy resources.

In the 21st century, there is increasing global awareness of the threat posed by the human greenhouse effect, produced largely by forest clearing and the burning of fossil fuels.

Principles and Concepts

The philosophical and analytic framework of sustainability draws on and connects with many different disciplines and fields; in recent years an area that has come to be called sustainability science has emerged.

The United Nations Millennium Declaration identified principles and treaties on sustainable development, including economic development, social development and environmental protection. The Circles of Sustainability approach distinguishes the four domains of economic, ecological, political and cultural sustainability. This in accord with the United Nations Agenda 21, which specifies culture as the fourth domain of sustainable development.

Scale and context

Sustainability is studied and managed over many scales (levels or frames of reference) of time and space and in many contexts of environmental, social and economic organization. The focus ranges from the total carrying capacity (sustainability) of planet Earth to the sustainability of economic sectors, ecosystems, countries, municipalities, neighbourhoods, home gardens, individual lives, individual goods and services, occupations, lifestyles, behaviour patterns and so on. In short, it can entail the full compass of biological and human activity or any part of it. As Daniel Botkin, author and environmentalist, has stated: "We see a landscape that is always in flux, changing over many scales of time and space."

The sheer size and complexity of the planetary ecosystem has proved problematic for the design of practical measures to reach global sustainability. To shed light on the big picture, explorer and sustainability campaigner Jason Lewis has drawn parallels to other, more tangible closed systems. For example, he likens human existence on Earth — isolated as the planet is in space, whereby people cannot be evacuated to relieve population pressure and resources cannot be imported to prevent accelerated depletion of resources — to life at sea on a small boat isolated by water. In both cases, he argues, exercising the precautionary principle is a key factor in survival.

Consumption

A major driver of human impact on Earth systems is the destruction of biophysical resources, and especially, the Earth's ecosystems. The environmental impact of a community or of humankind as a whole depends both on population and impact per person, which in turn depends in complex ways on what resources are being used, whether or not those resources are renewable, and the scale of the human activity relative to the carrying capacity of the ecosystems involved. Careful resource management can be applied at many scales, from economic sectors like agriculture, manufacturing and industry, to work organizations, the consumption patterns of households and individuals and to the resource demands of individual goods and services.

One of the initial attempts to express human impact mathematically was developed in the 1970s and is called the I PAT formula. This formulation attempts to explain human consumption in terms of three components: population numbers, levels of consumption (which it terms "affluence", although the usage is different), and impact per unit of resource use (which is termed "technology", because this impact depends on the technology used). The equation is expressed:

$$I = P \times A \times T$$

Where: I = Environmental impact, P = Population, A = Affluence, T = Technology

Measurement

Sustainability measurement is a term that denotes the measurements used as the quantitative basis for the informed management of sustainability. The metrics used for the measurement of sustainability (involving the sustainability of environmental, social and economic domains, both individually and in various combinations) are evolving: they include indicators, benchmarks, audits, sustainability standards and certification systems like Fairtrade and Organic, indexes and accounting, as well as assessment, appraisal and other reporting systems. They are applied over a wide range of spatial and temporal scales.

Some of the best known and most widely used sustainability measures include corporate sustainability reporting, Triple Bottom Line accounting, World Sustainability Society, Circles of Sustainability, and estimates of the quality of sustainability governance for individual countries using the Environmental Sustainability Index and Environmental Performance Index.

Population

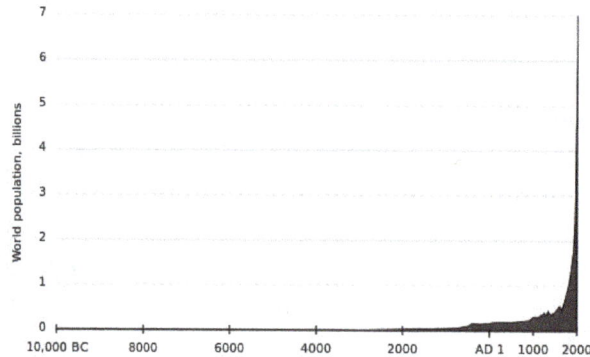

Graph showing human population growth from 10,000 BC – 2000 AD, illustrating current exponential growth

According to the 2008 Revision of the official United Nations population estimates and projections, the world population is projected to reach 7 billion early in 2012, up from the current 6.9 billion (May 2009), to exceed 9 billion people by 2050. Most of the increase will be in developing countries whose population is projected to rise from 5.6 billion in 2009 to 7.9 billion in 2050. This increase will be distributed among the population aged 15–59 (1.2 billion) and 60 or over (1.1 billion) because the number of children under age 15 in developing countries is predicted to decrease. In contrast, the population of the more developed regions is expected to undergo only slight increase from 1.23 billion to 1.28 billion, and this would have declined to 1.15 billion but for a projected net migration from developing to developed countries, which is expected to average 2.4 million persons annually from 2009 to 2050. Long-term estimates in 2004 of global population suggest a peak at around 2070 of nine to ten billion people, and then a slow decrease to 8.4 billion by 2100.

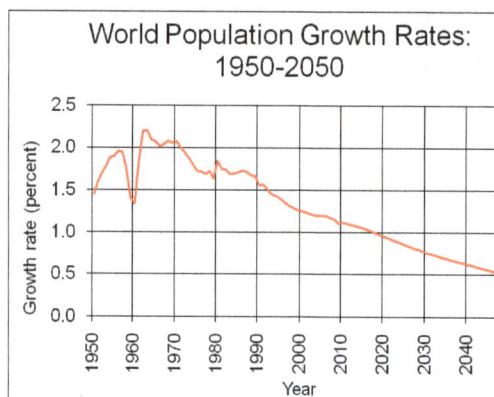

World population growth rate, 1950–2050, as estimated in 2011 by the U.S. Census Bureau, International Data Base

Emerging economies like those of China and India aspire to the living standards of the Western world as does the non-industrialized world in general. It is the combination of population increase in the developing world and unsustainable consumption levels in the developed world that poses a stark challenge to sustainability.

Carrying Capacity

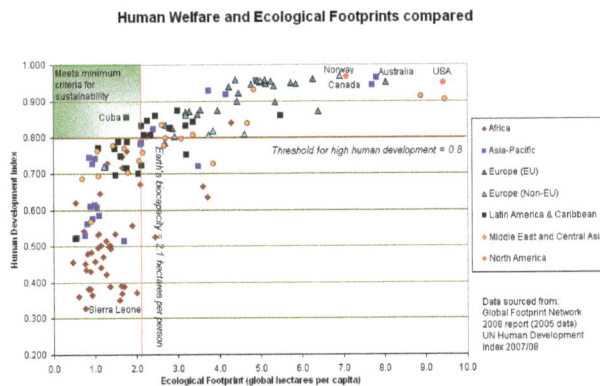

Human Welfare and Ecological Footprints compared

Ecological footprint for different nations compared to their Human Development Index (HDI)

At the global scale, scientific data now indicates that humans are living beyond the carrying capacity of planet Earth and that this cannot continue indefinitely. This scientific evidence comes from many sources but is presented in detail in the Millennium Ecosystem Assessment and the planetary boundaries framework. An early detailed examination of global limits was published in the 1972 book *Limits to Growth*, which has prompted follow-up commentary and analysis. A 2012 review in *Nature* by 22 international researchers expressed concerns that the Earth may be "approaching a state shift" in its biosphere.

The Ecological footprint measures human consumption in terms of the biologically productive land needed to provide the resources, and absorb the wastes of the average global citizen. In 2008 it required 2.7 global hectares per person, 30% more than the natural biological capacity of 2.1 global hectares (assuming no provision for other organisms). The resulting ecological deficit must be met from unsustainable *extra* sources and these are obtained in three ways: embedded in the goods and services of world trade; taken from the past (e.g. fossil fuels); or borrowed from the future as unsustainable resource usage (e.g. by over exploiting forests and fisheries).

The figure (right) examines sustainability at the scale of individual countries by contrasting their Ecological Footprint with their UN Human Development Index (a measure of standard of living). The graph shows what is necessary for countries to maintain an acceptable standard of living for their citizens while, at the same time, maintaining sustainable resource use. The general trend is for higher standards of living to become less sustainable. As always, population growth has a marked influence on levels of consumption and the efficiency of resource use. The sustainability goal is to raise the global standard of living without increasing the use of resources beyond globally sustainable levels; that is, to not exceed "one planet" consumption. Information generated by reports at the national, regional and city scales confirm the global trend towards societies that are becoming less sustainable over time.

Romanian American economist Nicholas Georgescu-Roegen, a progenitor in economics and a paradigm founder of ecological economics, has argued that the carrying capacity of Earth — that is, Earth's capacity to sustain human populations and consumption levels — is bound to decrease sometime in the future as Earth's finite stock of mineral resources is presently being extracted and put to use. Leading ecological economist and steady-state theorist Herman Daly, a student of Georgescu-Roegen, has propounded the same argument.

Global Human Impact on Biodiversity

At a fundamental level energy flow and biogeochemical cycling set an upper limit on the number and mass of organisms in any ecosystem. Human impacts on the Earth are demonstrated in a general way through detrimental changes in the global biogeochemical cycles of chemicals that are critical to life, most notably those of water, oxygen, carbon, nitrogen and phosphorus.

The *Millennium Ecosystem Assessment* is an international synthesis by over 1000 of the world's leading biological scientists that analyzes the state of the Earth's ecosystems and provides summaries and guidelines for decision-makers. It concludes that human activity is having a significant and escalating impact on the biodiversity of world ecosystems, reducing both their resilience and biocapacity. The report refers to natural systems as humanity's "life-support system", providing essential "ecosystem services". The assessment measures 24 ecosystem services concluding that only four have shown improvement over the last 50 years, 15 are in serious decline, and five are in a precarious condition.

Sustainable Development Goals

The Sustainable Development Goals (SDGs) are the current harmonized set of seventeen future international development targets.

The Official Agenda for Sustainable Development adopted on 25 September 2015 has 92 paragraphs, with the main paragraph (51) outlining the 17 Sustainable Development Goals and its associated 169 targets. This included the following seventeen goals:

1. Poverty – End poverty in all its forms everywhere

2. Food – End hunger, achieve food security and improved nutrition and promote sustainable agriculture

3. Health – Ensure healthy lives and promote well-being for all at all ages

4. Education – Ensure inclusive and equitable quality education and promote lifelong learning opportunities for all

5. Women – Achieve gender equality and empower all women and girls

6. Water – Ensure availability and sustainable management of water and sanitation for all

7. Energy – Ensure access to affordable, reliable, sustainable and modern energy for all

8. Economy – Promote sustained, inclusive and sustainable economic growth, full and productive employment and decent work for all

9. Infrastructure – Build resilient infrastructure, promote inclusive and sustainable industrialization and foster innovation

10. Inequality – Reduce inequality within and among countries

11. Habitation – Make cities and human settlements inclusive, safe, resilient and sustainable

12. Consumption – Ensure sustainable consumption and production patterns

13. Climate – Take urgent action to combat climate change and its impacts

14. Marine-ecosystems – Conserve and sustainably use the oceans, seas and marine resources for sustainable development

15. Ecosystems – Protect, restore and promote sustainable use of terrestrial ecosystems, sustainably manage forests, combat desertification, and halt and reverse land degradation and halt biodiversity loss

16. Institutions – Promote peaceful and inclusive societies for sustainable development, provide access to justice for all and build effective, accountable and inclusive institutions at all levels

17. Sustainability – Strengthen the means of implementation and revitalize the global partnership for sustainable development

As of August 2015, there were 169 proposed targets for these goals and 304 proposed indicators to show compliance.

The Sustainable Development Goals (SDGs) replace the eight Millennium Development Goals (MDGs), which expired at the end of 2015. The MDGs were established in 2000 following the Millennium Summit of the United Nations. Adopted by the 189 United Nations member states at the time and more than twenty international organizations, these goals were advanced to help achieve the following sustainable development standards by 2015.

1. To eradicate extreme poverty and hunger

2. To achieve universal primary education

3. To promote gender equality and empower women

4. To reduce child mortality

5. To improve maternal health

6. To combat HIV/AIDS, malaria, and other diseases

7. To ensure environmental sustainability (one of the targets in this goal focuses on increasing sustainable access to safe drinking water and basic sanitation)

8. To develop a global partnership for development

Sustainable Development

According to the data that member countries represented to the United Nations, Cuba was the only nation in the world in 2006 that met the World Wide Fund for Nature's definition of sustainable development, with an ecological footprint of less than 1.8 hectares per capita, 1.5, and a Human Development Index of over 0.8, 0.855.

Environmental Dimension

Healthy ecosystems provide vital goods and services to humans and other organisms. There are two major ways of reducing negative human impact and enhancing ecosystem services and the first of these is environmental management. This direct approach is based largely on information gained from earth science, environmental science and conservation biology. However, this is management at the end of a long series of indirect causal factors that are initiated by human consumption, so a second approach is through demand management of human resource use.

Management of human consumption of resources is an indirect approach based largely on information gained from economics. Herman Daly has suggested three broad criteria for ecological sustainability: renewable resources should provide a sustainable yield (the rate of harvest should not exceed the rate of regeneration); for non-renewable resources there should be equivalent development of renewable substitutes; waste generation should not exceed the assimilative capacity of the environment.

Environmental Management

At the global scale and in the broadest sense environmental management involves the oceans, freshwater systems, land and atmosphere, but following the sustainability principle of scale it can be equally applied to any ecosystem from a tropical rainforest to a home garden.

Atmosphere

At a March 2009 meeting of the Copenhagen Climate Council, 2,500 climate experts from 80 countries issued a keynote statement that there is now "no excuse" for failing to act on global warming and that without strong carbon reduction "abrupt or irreversible" shifts in climate may occur that "will be very difficult for contemporary societies to cope with". Management of the global atmosphere now involves assessment of all aspects of the carbon cycle to identify opportunities to address human-induced climate change and this has become a major focus of scientific research because of the potential catastrophic effects on biodiversity and human communities.

Other human impacts on the atmosphere include the air pollution in cities, the pollutants including toxic chemicals like nitrogen oxides, sulfur oxides, volatile organic compounds and airborne particulate matter that produce photochemical smog and acid rain, and the chlorofluorocarbons that degrade the ozone layer. Anthropogenic particulates such as sulfate aerosols in the atmosphere reduce the direct irradiance and reflectance (albedo) of the Earth's surface. Known as global dimming, the decrease is estimated to have been about 4% between 1960 and 1990 although the trend has subsequently reversed. Global dimming may have disturbed the global water cycle by reducing evaporation and rainfall in some areas. It also creates a cooling effect and this may have partially masked the effect of greenhouse gases on global warming.

Freshwater and Oceans

Water covers 71% of the Earth's surface. Of this, 97.5% is the salty water of the oceans and only 2.5% freshwater, most of which is locked up in the Antarctic ice sheet. The remaining freshwater is found in glaciers, lakes, rivers, wetlands, the soil, aquifers and atmosphere. Due to the water cycle, fresh water supply is continually replenished by precipitation, however there is still a limited amount necessitating management of this resource. Awareness of the global importance of preserving water for ecosystem services has only recently emerged as, during the 20th century, more than half the world's wetlands have been lost along with their valuable environmental services. Increasing urbanization pollutes clean water supplies and much of the world still does not have access to clean, safe water. Greater emphasis is now being placed on the improved management of blue (harvestable) and green (soil water available for plant use) water, and this applies at all scales of water management.

Ocean circulation patterns have a strong influence on climate and weather and, in turn, the food supply of both humans and other organisms. Scientists have warned of the possibility, under the influence of climate change, of a sudden alteration in circulation patterns of ocean currents that could drastically alter the climate in some regions of the globe. Ten per cent of the world's population – about 600 million people – live in low-lying areas vulnerable to sea level rise.

Land Use

A rice paddy in Bangladesh. Rice, wheat, corn and potatoes make up more than half the world's food supply.

Loss of biodiversity stems largely from the habitat loss and fragmentation produced by the human appropriation of land for development, forestry and agriculture as natural capital is progressively converted to man-made capital. Land use change is fundamental to the operations of the biosphere because alterations in the relative proportions of land dedicated to urbanisation, agriculture, forest, woodland, grassland and pasture have a marked effect on the global water, carbon and nitrogen biogeochemical cycles and this can impact negatively on both natural and human systems. At the local human scale, major sustainability benefits accrue from sustainable parks and gardens and green cities.

Since the Neolithic Revolution about 47% of the world's forests have been lost to human use. Present-day forests occupy about a quarter of the world's ice-free land with about half of these occurring in the tropics. In temperate and boreal regions forest area is gradually increasing (with the exception of Siberia), but deforestation in the tropics is of major concern.

Food is essential to life. Feeding more than seven billion human bodies takes a heavy toll on the Earth's resources. This begins with the appropriation of about 38% of the Earth's land surface and about 20% of its net primary productivity. Added to this are the resource-hungry activities of industrial agribusiness – everything from the crop need for irrigation water, synthetic fertilizers and pesticides to the resource costs of food packaging, transport (now a major part of global trade) and retail. Environmental problems associated with industrial agriculture and agribusiness are now being addressed through such movements as sustainable agriculture, organic farming and more sustainable business practices.

Management of Human Consumption

The Helix of Sustainability

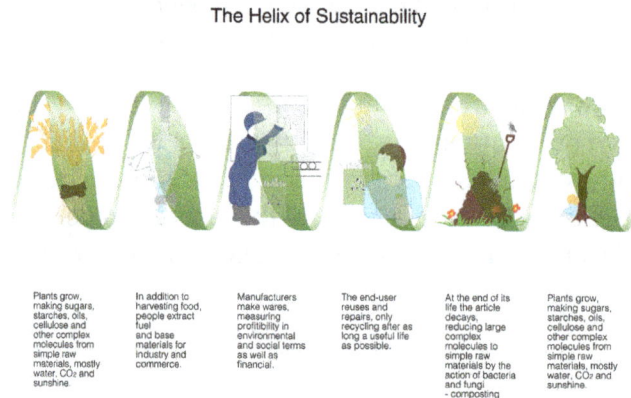

| Plants grow, making sugars, starches, oils, cellulose and other complex molecules from simple raw materials, mostly water, CO₂ and sunshine. | In addition to harvesting food, people extract fuel and base materials for industry and commerce. | Manufacturers make wares, measuring profitibility in environmental and social terms as well as financial. | The end-user reuses and repairs, only recycling after as long a useful life as possible. | At the end of its life the article decays, reducing large complex molecules to simple raw materials by the action of bacteria and fungi - composting | Plants grow, making sugars, starches, oils, cellulose and other complex molecules from simple raw materials, mostly water, CO₂ and sunshine. |

Helix of sustainability – the carbon cycle of manufacturing

The underlying driver of direct human impacts on the environment is human consumption. This impact is reduced by not only consuming less but by also making the full cycle of production, use and disposal more sustainable. Consumption of goods and services can be analysed and managed at all scales through the chain of consumption, starting with the effects of individual lifestyle choices and spending patterns, through to the resource demands of specific goods and services, the impacts of economic sectors, through national economies to the global economy. Analysis of consumption patterns relates resource use to the environmental, social and economic impacts at the scale or context under investigation. The ideas of embodied resource use (the total resources needed to produce a product or service), resource intensity, and resource productivity are important tools for understanding the impacts of consumption. Key resource categories relating to human needs are food, energy, materials and water.

In 2010, the International Resource Panel, hosted by the United Nations Environment Programme (UNEP), published the first global scientific assessment on the impacts of consumption and production and identified priority actions for developed and developing countries. The study found that the most critical impacts are related to ecosystem health, human health and resource depletion. From a production perspective, it found that fossil-fuel combustion processes, agriculture and fisheries have the most important impacts. Meanwhile, from a final consumption perspective, it found that household consumption related to mobility, shelter, food and energy-using products cause the majority of life-cycle impacts of consumption.

Energy

The Sun's energy, stored by plants (primary producers) during photosynthesis, passes through the food chain to other organisms to ultimately power all living processes. Since the industrial revolution the concentrated energy of the Sun stored in fossilized plants as fossil fuels has been a major driver of technology which, in turn, has been the source of both economic and political power. In 2007 climate scientists of the IPCC concluded that there was at least a 90% probability that atmospheric increase in CO_2 was human-induced, mostly as a result of fossil fuel emissions but, to a lesser extent from changes in land use. Stabilizing the world's climate will require high-income countries to reduce their emissions by 60–90% over 2006 levels by 2050 which should hold CO_2 levels at 450–650 ppm from current levels of about 380 ppm. Above this level, temperatures could rise by more than 2 °C to produce "catastrophic" climate change. Reduction of current CO_2 levels must be achieved against a background of global population increase and developing countries aspiring to energy-intensive high consumption Western lifestyles.

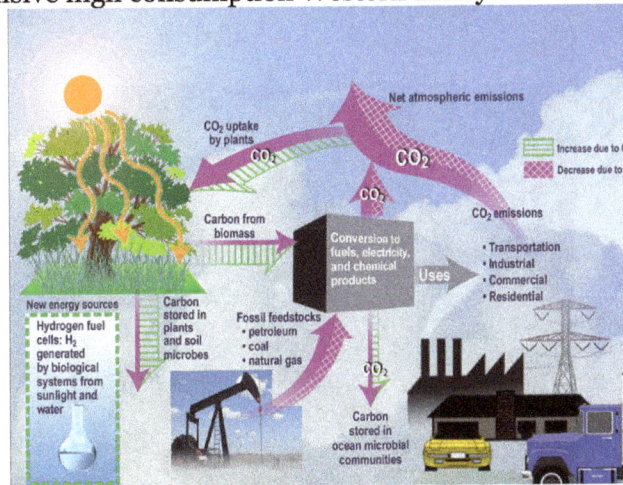

Flow of CO_2 in an ecosystem

Reducing greenhouse emissions, is being tackled at all scales, ranging from tracking the passage of carbon through the carbon cycle to the commercialization of renewable energy, developing less carbon-hungry technology and transport systems and attempts by individuals to lead carbon neutral lifestyles by monitoring the fossil fuel use embodied in all the goods and services they use. Engineering of emerging technologies such as carbon-neutral fuel and energy storage systems such as power to gas, compressed air energy storage, and pumped-storage hydroelectricity are necessary to store power from transient renewable energy sources including emerging renewables such as airborne wind turbines.

Water

Water security and food security are inextricably linked. In the decade 1951–60 human water withdrawals were four times greater than the previous decade. This rapid increase resulted from scientific and technological developments impacting through the economy – especially the increase in irrigated land, growth in industrial and power sectors, and intensive dam construction on all continents. This altered the water cycle of rivers and lakes, affected their water quality and had a significant impact on the global water cycle. Currently towards 35% of human water use is unsustainable, drawing on diminishing aquifers and reducing the flows of major rivers: this percentage is likely to increase if climate change impacts become more severe, populations increase, aquifers become progressively depleted and supplies become polluted and unsanitary. From 1961 to 2001 water demand doubled — agricultural use increased by 75%, industrial use by more than 200%, and domestic use more than 400%. In the 1990s it was estimated that humans were using 40–50% of the globally available freshwater in the approximate proportion of 70% for agriculture, 22% for industry, and 8% for domestic purposes with total use progressively increasing.

Water efficiency is being improved on a global scale by increased demand management, improved infrastructure, improved water productivity of agriculture, minimising the water intensity (embodied water) of goods and services, addressing shortages in the non-industrialized world, concentrating food production in areas of high productivity, and planning for climate change, such as through flexible system design. A promising direction towards sustainable development is to design systems that are flexible and reversible. At the local level, people are becoming more self-sufficient by harvesting rainwater and reducing use of mains water.

Food

Feijoada — A typical black bean food dish from Brazil

The American Public Health Association (APHA) defines a "sustainable food system" as "one that provides healthy food to meet current food needs while maintaining healthy ecosystems that can also provide food for generations to come with minimal negative impact to the environment. A sustainable food system also encourages local production and distribution infrastructures and makes nutritious food available, accessible, and affordable to all. Further, it is humane and just, protecting farmers and other workers, consumers, and communities." Concerns about the environmental impacts of agribusiness and the stark contrast between the obesity problems of the Western world and the poverty and food insecurity of the developing world have generated a strong movement towards healthy, sustainable eating as a major component of overall ethical consumerism. The environmental effects of different dietary patterns depend on many factors, including the proportion of animal and plant foods consumed and the method of food production. The World Health Organization has published a *Global Strategy on Diet, Physical Activity and Health* report which was endorsed by the May 2004 World Health Assembly. It recommends the Mediterranean diet which is associated with health and longevity and is low in meat, rich in fruits and vegetables, low in added sugar and limited salt, and low in saturated fatty acids; the traditional source of fat in the Mediterranean is olive oil, rich in monounsaturated fat. The healthy rice-based Japanese diet is also high in carbohydrates and low in fat. Both diets are low in meat and saturated fats and high in legumes and other vegetables; they are associated with a low incidence of ailments and low environmental impact.

At the global level the environmental impact of agribusiness is being addressed through sustainable agriculture and organic farming. At the local level there are various movements working towards local food production, more productive use of urban wastelands and domestic gardens including permaculture, urban horticulture, local food, slow food, sustainable gardening, and organic gardening.

Sustainable seafood is seafood from either fished or farmed sources that can maintain or increase production in the future without jeopardizing the ecosystems from which it was acquired. The sustainable seafood movement has gained momentum as more people become aware about both overfishing and environmentally destructive fishing methods.

Materials, toxic Substances, Waste

An electric wire reel reused as a center table in a Rio de Janeiro decoration fair. The reuse of materials is a sustainable practice that is rapidly growing among designers in Brazil.

As global population and affluence has increased, so has the use of various materials increased in volume, diversity and distance transported. Included here are raw materials, minerals, synthetic chemicals (including hazardous substances), manufactured products, food, living organisms and waste. By 2050, humanity could consume an estimated 140 billion tons of minerals, ores, fossil fuels and biomass per year (three times its current amount) unless the economic growth rate is decoupled from the rate of natural resource consumption. Developed countries' citizens consume an average of 16 tons of those four key resources per capita, ranging up to 40 or more tons per person in some developed countries with resource consumption levels far beyond what is likely sustainable.

Sustainable use of materials has targeted the idea of dematerialization, converting the linear path of materials (extraction, use, disposal in landfill) to a circular material flow that reuses materials as much as possible, much like the cycling and reuse of waste in nature. This approach is supported by product stewardship and the increasing use of material flow analysis at all levels, especially individual countries and the global economy. The use of sustainable biomaterials that come from renewable sources and that can be recycled is preferred to the use on non-renewables from a life cycle standpoint.

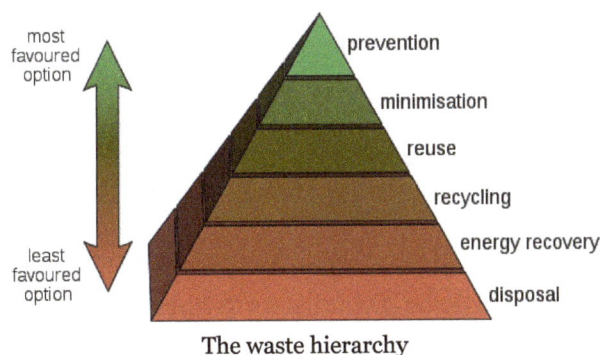

The waste hierarchy

Synthetic chemical production has escalated following the stimulus it received during the second World War. Chemical production includes everything from herbicides, pesticides and fertilizers to domestic chemicals and hazardous substances. Apart from the build-up of greenhouse gas emissions in the atmosphere, chemicals of particular concern include: heavy metals, nuclear waste, chlorofluorocarbons, persistent organic pollutants and all harmful chemicals capable of bioaccumulation. Although most synthetic chemicals are harmless there needs to be rigorous testing of new chemicals, in all countries, for adverse environmental and health effects. International legislation has been established to deal with the global distribution and management of dangerous goods. The effects of some chemical agents needed long-term measurements and a lot of legal battles to realize their danger to human health. The classification of the toxic carcinogenic agents is handle by the International Agency for Research on Cancer.

Every economic activity produces material that can be classified as waste. To reduce waste, industry, business and government are now mimicking nature by turning the waste produced by industrial metabolism into resource. Dematerialization is being encouraged through the ideas of industrial ecology, ecodesign and ecolabelling. In addition to the well-established "reduce, reuse and recycle," shoppers are using their purchasing power for ethical consumerism.

The European Union is expected to table by the end of 2015 an ambitious Circular Economy package which is expected to include concrete legislative proposals on waste management, ecodesign and limits on land fills.

Economic Dimension

On one account, sustainability "concerns the specification of a set of actions to be taken by present persons that will not diminish the prospects of future persons to enjoy levels of consumption, wealth, utility, or welfare comparable to those enjoyed by present persons." Sustainability interfaces with economics through the social and ecological consequences of economic activity. Sustainability economics represents: "... a broad interpretation of ecological economics where environmental and ecological variables and issues are basic but part of a multidimensional perspective. Social, cultural, health-related and monetary/financial aspects have to be integrated into the analysis." However, the concept of sustainability is much broader than the concepts of sustained yield of welfare, resources, or profit margins. At present, the average per capita consumption of people in the developing world is sustainable but population numbers are increasing and individuals are aspiring to high-consumption Western lifestyles. The developed world population is only increasing slightly but consumption levels are unsustainable. The challenge for sustainability is to curb and manage Western consumption while raising the standard of living of the developing world without increasing its resource use and environmental impact. This must be done by using strategies and technology that break the link between, on the one hand, economic growth and on the other, environmental damage and resource depletion.

A recent UNEP report proposes a green economy defined as one that "improves human well-being and social equity, while significantly reducing environmental risks and ecological scarcities": it "does not favor one political perspective over another but works to minimize excessive depletion of natural capital". The report makes three key findings: "that greening not only generates increases in wealth, in particular a gain in ecological commons or natural capital, but also (over a period of six years) produces a higher rate of GDP growth"; that there is "an inextricable link between poverty eradication and better maintenance and conservation of the ecological commons, arising from the benefit flows from natural capital that are received directly by the poor"; "in the transition to a green economy, new jobs are created, which in time exceed the losses in "brown economy" jobs. However, there is a period of job losses in transition, which requires investment in re-skilling and re-educating the workforce".

The *Great Fish Market*, painted by Jan Brueghel the Elder

Several key areas have been targeted for economic analysis and reform: the environmental effects of unconstrained economic growth; the consequences of nature being treated as an economic externality; and the possibility of an economics that takes greater account of the social and environmental consequences of market behavior.

Decoupling Environmental Degradation and Economic Growth

Historically there has been a close correlation between economic growth and environmental degradation: as communities grow, so the environment declines. This trend is clearly demonstrated on graphs of human population numbers, economic growth, and environmental indicators. Unsustainable economic growth has been starkly compared to the malignant growth of a cancer because it eats away at the Earth's ecosystem services which are its life-support system. There is concern that, unless resource use is checked, modern global civilization will follow the path of ancient civilizations that collapsed through overexploitation of their resource base. While conventional economics is concerned largely with economic growth and the efficient allocation of resources, ecological economics has the explicit goal of sustainable scale (rather than continual growth), fair distribution and efficient allocation, in that order. The World Business Council for Sustainable Development states that "business cannot succeed in societies that fail".

In economic and environmental fields, the term decoupling is becoming increasingly used in the context of economic production and environmental quality. When used in this way, it refers to the ability of an economy to grow without incurring corresponding increases in environmental pressure. Ecological economics includes the study of societal metabolism, the throughput of resources that enter and exit the economic system in relation to environmental quality. An economy that is able to sustain GDP growth without having a negative impact on the environment is said to be decoupled. Exactly how, if, or to what extent this can be achieved is a subject of much debate. In 2011 the International Resource Panel, hosted by the United Nations Environment Programme

(UNEP), warned that by 2050 the human race could be devouring 140 billion tons of minerals, ores, fossil fuels and biomass per year – three times its current rate of consumption – unless nations can make serious attempts at decoupling. The report noted that citizens of developed countries consume an average of 16 tons of those four key resources per capita per annum (ranging up to 40 or more tons per person in some developed countries). By comparison, the average person in India today consumes four tons per year. Sustainability studies analyse ways to reduce resource intensity (the amount of resource (e.g. water, energy, or materials) needed for the production, consumption and disposal of a unit of good or service) whether this be achieved from improved economic management, product design, or new technology.

There are conflicting views whether improvements in technological efficiency and innovation will enable a complete decoupling of economic growth from environmental degradation. On the one hand, it has been claimed repeatedly by efficiency experts that resource use intensity (i.e., energy and materials use per unit GDP) could in principle be reduced by at least four or five-fold, thereby allowing for continued economic growth without increasing resource depletion and associated pollution. On the other hand, an extensive historical analysis of technological efficiency improvements has conclusively shown that improvements in the efficiency of the use of energy and materials were almost always outpaced by economic growth, in large part because of the rebound effect (conservation) or Jevons Paradox resulting in a net increase in resource use and associated pollution. Furthermore, there are inherent thermodynamic (i.e., second law of thermodynamics) and practical limits to all efficiency improvements. For example, there are certain minimum unavoidable material requirements for growing food, and there are limits to making automobiles, houses, furniture, and other products lighter and thinner without the risk of losing their necessary functions. Since it is both theoretically and practically impossible to increase resource use efficiencies indefinitely, it is equally impossible to have continued and infinite economic growth without a concomitant increase in resource depletion and environmental pollution, i.e., economic growth and resource depletion can be decoupled to some degree over the short run but not the long run. Consequently, long-term sustainability requires the transition to a steady state economy in which total GDP remains more or less constant, as has been advocated for decades by Herman Daly and others in the ecological economics community.

A different proposed solution to partially decouple economic growth from environmental degradation is the *restore* approach. This approach views "restore" as a fourth component to the common reduce, reuse, recycle motto. Participants in such efforts are encouraged to voluntarily donate towards nature conservation a small fraction of the financial savings they experience through a more frugal use of resources. These financial savings would normally lead to rebound effects, but a theoretical analysis suggests that donating even a small fraction of the experienced savings can potentially more than eliminate rebound effects.

Nature as an Economic Externality

The economic importance of nature is indicated by the use of the expression ecosystem services to highlight the market relevance of an increasingly scarce natural world that can no longer be regarded as both unlimited and free. In general, as a commodity or service becomes more scarce the price increases and this acts as a restraint that encourages frugality, technical innovation and alternative products. However, this only applies when the product or service falls within the market system. As ecosystem services are generally treated as economic externalities they are unpriced and therefore overused and degraded, a situation sometimes referred to as the Tragedy of the Commons.

Deforestation of native rain forest in Rio de Janeiro City for extraction of clay for civil engineering (2009 picture)

One approach to this dilemma has been the attempt to "internalize" these "externalities" by using market strategies like ecotaxes and incentives, tradeable permits for carbon, and the encouragement of payment for ecosystem services. Community currencies associated with Local Exchange Trading Systems (LETS), a gift economy and Time Banking have also been promoted as a way of supporting local economies and the environment. Green economics is another market-based attempt to address issues of equity and the environment. The global recession and a range of associated government policies are likely to bring the biggest annual fall in the world's carbon dioxide emissions in 40 years.

Economic Opportunity

Treating the environment as an externality may generate short-term profit at the expense of sustainability. Sustainable business practices, on the other hand, integrate ecological concerns with social and economic ones (i.e., the triple bottom line). Growth that depletes ecosystem services is sometimes termed "uneconomic growth" as it leads to a decline in quality of life. Minimizing such growth can provide opportunities for local businesses. For example, industrial waste can be treated as an "economic resource in the wrong place". The benefits of waste reduction include savings from disposal costs, fewer environmental penalties, and reduced liability insurance. This may lead to increased market share due to an improved public image. Energy efficiency can also increase profits by reducing costs.

The idea of sustainability as a business opportunity has led to the formation of organizations such as the Sustainability Consortium of the Society for Organizational Learning, the Sustainable Business Institute, and the World Council for Sustainable Development. The expansion of sustainable business opportunities can contribute to job creation through the introduction of green-collar workers. Research focusing on progressive corporate leaders who have integrated sustainability into commercial strategy has yielded a leadership competency model for sustainability, and led to emergence of the concept of "embedded sustainability" – defined by its authors Chris Laszlo and Nadya Zhexembayeva as "incorporation of environmental, health, and social value into the core business with no trade-off in price or quality – in other words, with no social or green premium."

Laszlo and Zhexembayeva's research showed that embedded sustainability offers at least seven distinct opportunities for business value creation: a) better risk-management, b) increased efficiency through reduced waste and resource use, c) better product differentiation, d) new market entrances, e) enhanced brand and reputation, f) greater opportunity to influence industry standards, and g) greater opportunity for radical innovation. Nadya Zhexembayeva's 2014 research further suggested that innovation driven by resource depletion can result in fundamental advantages for company products and services, as well as the company strategy as a whole, when right principles of innovation are applied.

Ecosocialist Approach

One school of thought, often labeled ecosocialism or ecological Marxism, asserts that the capitalist economic system is fundamentally incompatible with the ecological and social requirements of sustainability. This theory rests on the premises that:

1. Capitalism's sole economic purpose is "unlimited capital accumulation" in the hands of the capitalist class

2. The urge to accumulate (the profit motive) drives capitalists to continually reinvest and expand production, creating indefinite and unsustainable economic growth

3. "Capital tends to degrade the conditions of its own production" (the ecosystems and resources on which any economy depends)

Thus, according to this analysis:

1. Giving economic priority to the fulfillment of human needs while staying within ecological limits, as sustainable development demands, is in conflict with the structural workings of capitalism

2. A steady-state capitalist economy is impossible; further, a steady-state capitalist economy is socially undesirable due to the inevitable outcome of massive unemployment and under-employment

3. Capitalism will, unless overcome by revolution, run up against the physical limits of the biosphere and self-destruct

By this logic, market-based solutions to ecological crises (ecological economics, environmental economics, green economy) are rejected as technical tweaks that do not confront capitalism's structural failures. "Low-risk" technology/science-based solutions such as solar power, sustainable agriculture, and increases in energy efficiency are seen as necessary but insufficient. "High-risk" technological solutions such as nuclear power and climate engineering are entirely rejected. Attempts made by businesses to "greenwash" their practices are regarded as false advertising, and it is pointed out that implementation of renewable technology (such as Walmart's proposition to supply their electricity with solar power) has the effect opposite of reductions in resource consumption, viz. further economic growth. Sustainable business models and the triple bottom line are viewed as morally praiseworthy but ignorant to the tendency in capitalism for the distribution of wealth to become increasingly unequal and socially unstable/unsustainable. Ecosocialists claim that the general unwillingness of capitalists to tolerate—and capitalist governments to implement—constraints on maximum profit (such as ecotaxes or preservation and conservation measures) renders environmental reforms incapable of facilitating large-scale change: "History teaches us that although capitalism has at times responded to environmental movements . . . at a certain point, at which the system's underlying accumulation drive is affected, its resistance to environmental demands stiffens." They also note that, up until the event of total ecological collapse, destruction caused by natural disasters generally causes an increase in economic growth and accumulation; thus, capitalists have no foreseeable motivation to reduce the probability of disasters (i.e. convert to sustainable/ecological production).

Ecosocialists advocate for the revolutionary succession of capitalism by ecosocialism—an egalitarian economic/political/social structure designed to harmonize human society with non-human ecology and to fulfill human needs—as the only sufficient solution to the present-day ecological crisis, and hence the only path towards sustainability. Sustainability is viewed not as a domain exclusive to scientists, environmental activists, and business leaders but as a holistic project that must involve the whole of humanity redefining its place in Nature: "What every environmentalist needs to know . . . is that capitalism is not the solution but the problem, and that if humanity is going to survive this crisis, it will do so because it has exercised its capacity for human freedom, through social struggle, in order to create a whole new world—in coevolution with the planet."

Social Dimension

Sustainability issues are generally expressed in scientific and environmental terms, as well as in ethical terms of stewardship, but implementing change is a social challenge that entails, among other things, international and national law, urban planning and transport, local and individual lifestyles and ethical consumerism. "The relationship between human rights and human development, corporate power and environmental justice, global poverty and citizen action, suggest that responsible global citizenship is an inescapable element of what may at first glance seem to be simply matters of personal consumer and moral choice."

Peace, Security, Social Justice

Social disruptions like war, crime and corruption divert resources from areas of greatest human need, damage the capacity of societies to plan for the future, and generally threaten human well-being and the environment. Broad-based strategies for more sustainable social systems include: improved education and the political empowerment of women, especially in developing countries; greater regard for social justice, notably equity between rich and poor both within and between countries; and intergenerational equity. Depletion of natural resources including fresh water increases the likelihood of "resource wars". This aspect of sustainability has been referred to as environmental security and creates a clear need for global environmental agreements to manage resources such as aquifers and rivers which span political boundaries, and to protect shared global systems including oceans and the atmosphere.

Poverty

A major hurdle to achieve sustainability is the alleviation of poverty. It has been widely acknowledged that poverty is one source of environmental degradation. Such acknowledgment has been made by the Brundtland Commission report Our Common Future and the Millennium Development Goals. There is a growing realization in national governments and multilateral institutions that it is impossible to separate economic development issues from environment issues: according to the Brundtland report, "poverty is a major cause and effect of global environmental problems. It is therefore futile to attempt to deal with environmental problems without a broader perspective that encompasses the factors underlying world poverty and international inequality." Individuals living in poverty tend to rely heavily on their local ecosystem as a source for basic needs (such as nutrition and medicine) and general well-being. As population growth continues to increase, increasing pressure is being placed on the local ecosystem to provide these basic essentials. According to the UN Population Fund, high fertility and poverty have been strongly correlated, and the world's poorest countries also have the highest fertility and population growth rates. The word sustainability is also used widely by western country development agencies and international charities to focus their poverty alleviation efforts in ways that can be sustained by the local populace and its environment. For example, teaching water treatment to the poor by boiling their water with charcoal, would not generally be considered a sustainable strategy, whereas using PET solar water disinfection would be. Also, sustainable best practices can involve the recycling of materials, such as the use of recycled plastics for lumber where deforestation has devastated a country's timber base. Another example of sustainable practices in poverty alleviation is the use of exported recycled materials from developed to developing countries, such as Bridges to Prosperity's use of wire rope from shipping container gantry cranes to act as the structural wire rope for footbridges that cross rivers in poor rural areas in Asia and Africa.

Human Relationship to Nature

According to Murray Bookchin, the idea that humans must dominate nature is common in hierarchical societies. Bookchin contends that capitalism and market relationships, if unchecked, have the capacity to reduce the planet to a mere resource to be exploited. Nature is thus treated as a commodity: "The plundering of the human spirit by the market place is paralleled by the plundering of the earth by capital." Social ecology, founded by Bookchin, is based on the conviction that nearly all of humanity's present ecological problems originate in, indeed are mere symptoms of, dysfunctional social arrangements. Whereas most authors proceed as if our ecological problems can be fixed by implementing recommendations which stem from physical, biological, economic etc., studies, Bookchin's claim is that these problems can only be resolved by understanding the underlying social processes and intervening in those processes by applying the concepts and methods of the social sciences.

A pure capitalist approach has also been criticized in Stern Review on the Economics of Climate Change to mitigation the effects of global warming in this excerpt ...

"the greatest example of market failure we have ever seen."

Deep ecology is a movement founded by Arne Naess that establishes principles for the well-being of all life on Earth and the richness and diversity of life forms. The movement advocates, among other things, a substantial decrease in human population and consumption along with the reduction of human interference with the nonhuman world. To achieve this, deep ecologists advocate policies for basic economic, technological, and ideological structures that will improve the *quality of life* rather than the *standard of living*. Those who subscribe to these principles are obliged to make the necessary change happen. The concept of a billion-year Sustainocene has been developed to initiate policy consideration of an earth where human structures power and fuel the needs of that species (for example through artificial photosynthesis) allowing Rights of Nature.

Human Settlements

Sustainability Principles

1. Reduce dependence upon fossil fuels, underground metals, and minerals

2. Reduce dependence upon synthetic chemicals and other unnatural substances

3. Reduce encroachment upon nature

4. Meet human needs fairly & efficiently

One approach to sustainable living, exemplified by small-scale urban transition towns and rural ecovillages, seeks to create self-reliant communities based on principles of simple living, which maximize self-sufficiency particularly in food production. These principles, on a broader scale, underpin the concept of a bioregional economy. These approaches often utilize commons based knowledge sharing of open source appropriate technology.

Other approaches, loosely based around New Urbanism, are successfully reducing environmental impacts by altering the built environment to create and preserve sustainable cities which support sustainable transport. Residents in compact urban neighborhoods drive fewer miles, and have significantly lower environmental impacts across a range of measures, compared with those living in sprawling suburbs. In sustainable architecture the recent movement of New Classical Architecture promotes a sustainable approach towards construction, that appreciates and develops smart growth, architectural tradition and classical design. This in contrast to modernist and globally uniform architecture, as well as opposing solitary housing estates and suburban sprawl. Both trends started in the 1980s. The concept of Circular flow land use management has also been introduced in Europe to promote sustainable land use patterns that strive for compact cities and a reduction of greenfield land take by urban sprawl.

Large scale social movements can influence both community choices and the built environment. Eco-municipalities may be one such movement. Eco-municipalities take a systems approach, based on sustainability principles. The eco-municipality movement is participatory, involving community members in a bottom-up approach. In Sweden, more than 70 cities and towns—25 per cent of all municipalities in the country—have adopted a common set of "Sustainability Principles" and implemented these systematically throughout their municipal operations. There are now twelve eco-municipalities in the United States and the American Planning Association has adopted sustainability objectives based on the same principles.

There is a wealth of advice available to individuals wishing to reduce their personal and social impact on the environment through small, inexpensive and easily achievable steps. But the transition required to reduce global human consumption to within sustainable limits involves much larger changes, at all levels and contexts of society. The United Nations has recognised the central role of education, and have declared a decade of education for sustainable development, 2005–2014, which aims to "challenge us all to adopt new behaviours and practices to secure our future". The Worldwide Fund for Nature proposes a strategy for sustainability that goes beyond education to tackle underlying individualistic and materialistic societal values head-on and strengthen people's connections with the natural world.

Human and Labor Rights

Application of social sustainability requires stakeholders to look at human and labor rights, prevention of human trafficking, and other human rights risks. These issues should be considered in production and procurement of various worldwide commodities. The international community has identified many industries whose practices have been known to violate social sustainability, and many of these industries have organizations in place that aid in verifying the social sustainability of products and services. The Equator Principles (financial industry), Fair Wear Foundation (garments), and Electronics Industry Citizenship Coalition are examples of such organizations and initiatives. Resources are also available for verifying the life-cycle of products and the producer or vendor level, such as Green Seal for cleaning products, NSF-140 for carpet production, and even labeling of Organic food in the United States.

Circles of Sustainability

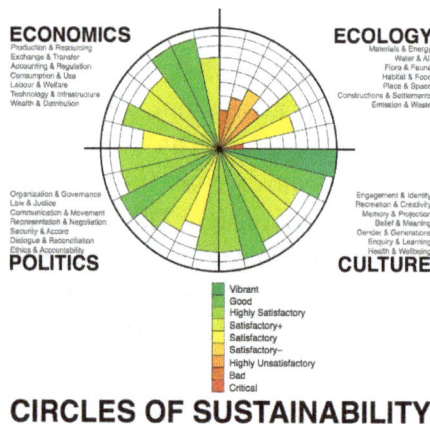

CIRCLES OF SUSTAINABILITY

A Circles of Sustainability representation - in this case for Melbourne in 2011.

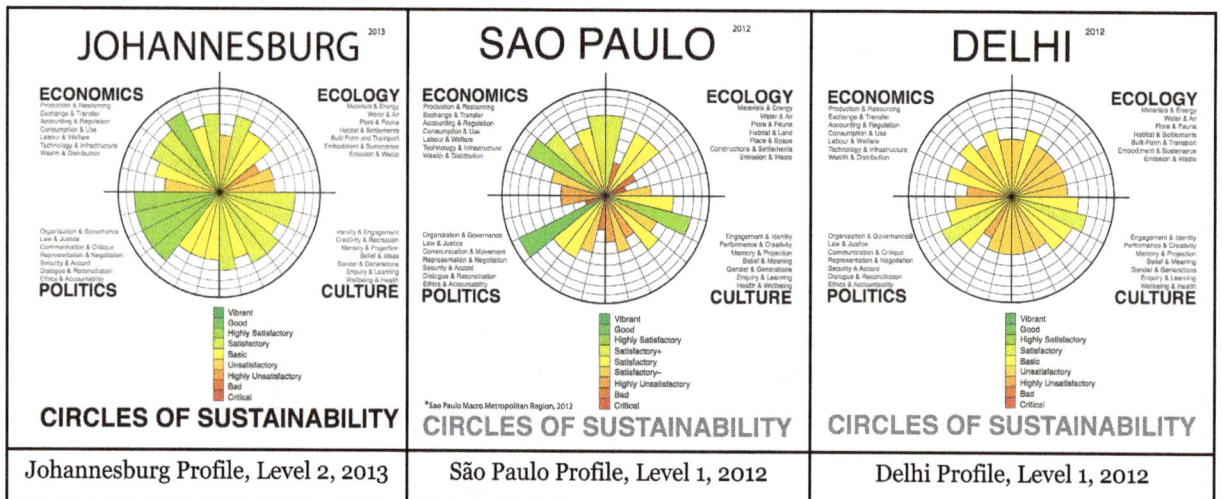

| Johannesburg Profile, Level 2, 2013 | São Paulo Profile, Level 1, 2012 | Delhi Profile, Level 1, 2012 |

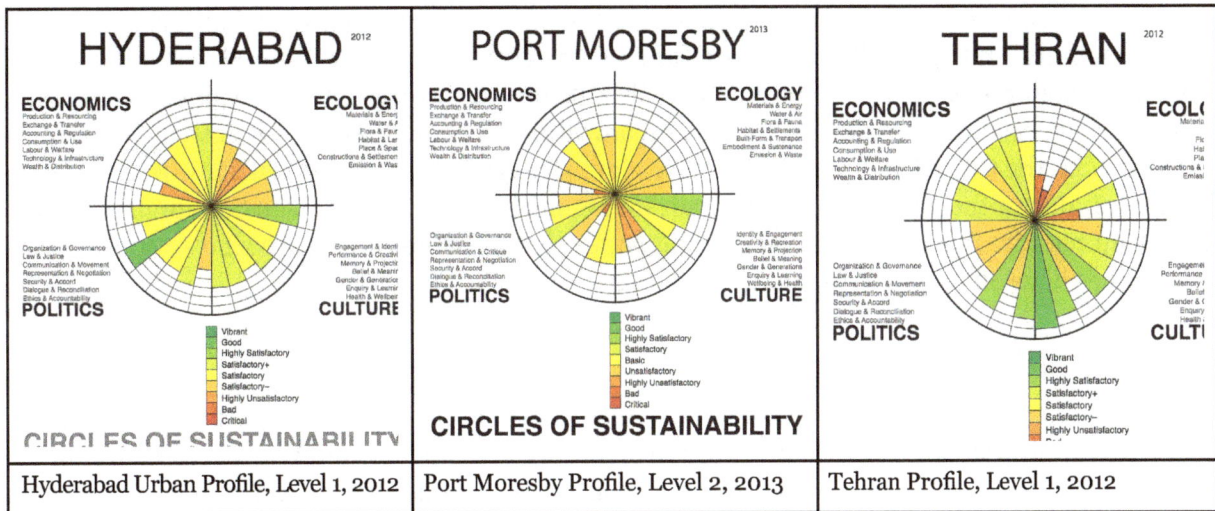

| Hyderabad Urban Profile, Level 1, 2012 | Port Moresby Profile, Level 2, 2013 | Tehran Profile, Level 1, 2012 |

Circles of Sustainability is a method for understanding and assessing sustainability, and for managing projects directed towards socially sustainable outcomes. It is intended to handle 'seemingly intractable problems' such as outlined in sustainable development debates. The method is mostly used for cities and urban settlements.

Circles of Sustainability, and its treatment of the social domains of ecology, economics, politics and culture, provides the empirical dimension of an approach called 'engaged theory'. Developing Circles of Sustainability is part of larger project called 'Circles of Social Life', using the same four-domain model to analyze questions of resilience, adaptation, security, reconciliation. It is also being used in relation to thematics such as 'Circles of Child Wellbeing' (with World Vision).

The rationale for this new method is clear. As evidenced by Rio+20 and the UN Habitat World Urban Forum in Napoli (2012) and Medellin (2014), sustainability assessment is on the global agenda. However, the more complex the problems, the less useful current sustainability assessment tools seem to be for assessing across different domains: economics, ecology, politics and culture. For example, the Triple Bottom Line approach tends to take the economy as its primary point of focus with the domain of the environmental as the key externality. Secondly, the one-dimensional quantitative basis of many such methods means that they have limited purchase on complex qualitative issues. Thirdly, the size, scope and sheer number of indicators included within many such methods means that they are often unwieldy and resist effective implementation. Fourthly, the restricted focus of current indicator sets means that they do not work across different organizational and social settings—corporations and other institutions, cities, and communities. Most indicator approaches, such as the Global Reporting Initiative or ISO14031, have been limited to large corporate organizations with easily definable legal and economic boundaries. Circles of Sustainability was developed to respond to those limitations.

History

The method began with a fundamental dissatisfaction with current approaches to sustainability and sustainable development, which tended to treat economics as the core domain and ecology as an externality. Two concurrent developments provided impetus: a major project in Porto Alegre, and a United Nations' paper called *Accounting for Sustainability*, Briefing Paper, No. 1, 2008. The researchers developed a method and an integrated set of tools for assessing and monitoring issues of sustainability while providing guidance for project development. The method was then further refined through projects in Melbourne and Milwaukee, and through an ARC-funded cross-disciplinary project that partnered with various organizations including Microsoft Australia, Fuji Xerox Australia, the City of Melbourne, World Vision, UN-Habitat and most crucially Metropolis. A further refining of the ecological index quadrant is underway in Canada. The Canadian project is significant because it is developing an ongoing index that measures and scores the ecological footprints cities have on their surrounding ecosystems.

Use of the Method

The method is used by a series of global organizations including the United Nations Global Compact Cities Programme, The World Association of Major Metropolises, and World Vision to support their engagement in cities. It is also used by a number of cities across the world in different ways to manage major projects or to provide feedback on their sustainability profiles (e.g., Hyderabad, Johannesburg, Melbourne, New Delhi, São Paulo and Tehran). It is a method for understanding urban politics and urban planning, as well as for conducting sustainability analysis and profiling sustainable development.

Global Compact Cities Programme

The methodology is made available by UN Global Compact Cities Programme for its engagement with its more than 80 Signatory Cities. In particular, some of the 14 Innovating Cities in the programme have influenced the development of the Circles of Sustainability method through their management of major projects, some with intensity and others as a background feature. They use a cross-sectoral and holistic approach for developing a response to self-defined seemingly intractable problems.

Porto Alegre, Vila Chocolatão Project

The Vila Chocolatão project refers to the 2011 resettlement of approximately 1,000 residents of the inner-city Vila Chocolatão slum in Porto Alegre, Brazil. The resettlement project of Vila Chocolatão commenced in 2000 in response to an imminent eviction of the community, with community members seeking resources and support to resettle through the city of Porto Alegre's renown participatory budgeting system. The lengthy preparation to resettle was led by a local cross-sectoral network group, the Vila Chocolatão Sustainability Network. The group was initially instigated by the Regional Court, TRF4 and consisted of the Vila Chocolatão Residents Association, local government departments, federal agencies, non-government organisations and the corporate sector. The project was supported by the City of Porto Alegre through the municipality's Local Solidarity Governance Scheme. In 2006, the Vila Chocolatao resettlement project was recognised as a pilot project for the then new Cities Programme cross-sectoral model with City Hall assembling a Critical Reference Group to identify critical issues and joint solutions to those issues involved in the resettlement.

This long-standing collaborative project has been successful in rehousing a whole community of slum dwellers, it has also effected a restructuring of how the city approaches slums. The project ensured sustainability was built into the relocation through changes such as setting up of recycling depots next to existing slums and developing a formal recycling sorting facility in the new site, Residencial Nova Chocolatão, linked to the garbage-collection process of the city (an example of linking the sub-domains of 'emission and waste' and 'organization and governance'); and establishing a fully resourced early childhood centre in the new community. The Vila Chocolatão Sustainability Network group continues to meet and work with the community post the resettlement. This network-led model is now being utilized by the City of Porto Alegre with other informal settlements.

Milwaukee, Water Sustainability Project

In 2009, Milwaukee, Wisconsin, United States of America, wanted to address the issue of water quality in the city. The Circles of Sustainability methodology became the basis for an integrated city project. In the period of the application of the method (2009–present) there has been a rediscovery of the value of water from an industry and broader community perspective.

In 2011, Milwaukee won the United States Water Prize given by the Clean Water America Alliance, as well as a prize from IBM Better Cities program worth $500,000. The community has also attracted some leading water treatment innovators and is establishing a graduate School of Freshwater Sciences at the University of Wisconsin-Milwaukee.

Metropolis (World Association of the Major Metropolises)

The methodology was first used by Metropolis for *Commission 2, 2012, Managing Urban Growth*. This Commission, which met across the period 2009–2011, was asked to make recommendations for use by Metropolis's 120 member cities on the theme of managing growth. The Commission Report using the Circles of Sustainability methodology was published on the web in three languages—English, French and Spanish—and is used by member cities as a guide to practice.

In 2011, the research team were invited by Metropolis to work with the Victorian Government and the Cities Programme on one of their major initiatives. The methodology is central to the approach used by the 'Integrated Strategic Planning and Public-Private Partnerships Initiative' organized by Metropolis, 2012–2013 for Indian, Brazilian and Iranian cities. A workshop was held in New Delhi, 26–27 July 2012, and senior planners from New Delhi, Hyderabad and Kolkata used the two of the assessment tools in the Circles of Sustainability toolbox to map the sustainability of their cities as part of developing their urban-regional plans. Other cities to use the same tools have been Tehran (in relation to their mega-projects plan) and São Paulo (in relation to their macro-metropolitan plan).

Since 2012, the Cities Programme and Metropolis have worked together to refine the 'Circles of Sustainability' method to use with their respective member cities. A Metropolis Taskforce was charged with further developing the method.

The Economist

In 2011, *The Economist* invited Paul James (Director of the UN Global Compact Cities Programme) and Chetan Vedya (Director, National Institute of Urban Affairs, India) into a debate around the question of urban sustainability and metropolitan growth. It led to over 200 letters to the editor in direct response as well as numerous linked citations on other websites.

World Vision

In 2011, recognising how much the two processes of urbanization and globalization were changing the landscape of poverty, World Vision decided to shift its orientation towards urban settings. Previously 80 per cent of its projects had been in small rural communities. The Circles of Sustainability method now underpins that reorientation and pilot studies are being conducted in India, South Africa, Lebanon, Indonesia and elsewhere, to refine the methodology for aid delivery in complex urban settings.

Domains and Subdomains

The Circles of Sustainability approach is explicitly critical of other domain models such as the triple bottom line that treat economics as if it is outside the social, or that treat the environment as an externality. It uses a four-domain model - economics, ecology, politics and culture. In each of these domains there are 7 subdomains.

Economics

The economic domain is defined as the practices and meanings associated with the production, use, and management of resources, where the concept of 'resources' is used in the broadest sense of that word.

1. Production and resourcing

2. Exchange and transfer

3. Accounting and regulation

4. Consumption and use

5. Labour and welfare

6. Technology and infrastructure

7. Wealth and distribution

Ecology

The ecological domain is defined as the practices and meanings that occur across the intersection between the social and the natural realms, focusing on the important dimension of human engagement with and within nature, but also including the built-environment.

1. Materials and energy

2. Water and air

3. Flora and fauna

4. Habitat and settlements

5. Built-form and transport

6. Embodiment and sustenance

7. Emission and waste

Politics

The political is defined as the practices and meanings associated with basic issues of social power, such as organization, authorization, legitimation and regulation. The parameters of this area extend beyond the conventional sense of politics to include not only issues of public and private governance but more broadly social relations in general.

1. Organization and governance
2. Law and justice
3. Communication and critique
4. Representation and negotiation
5. Security and accord
6. Dialogue and reconciliation
7. Ethics and accountability

Culture

The cultural domain is defined as the practices, discourses, and material expressions, which, over time, express continuities and discontinuities of social meaning.

1. Identity and engagement
2. Creativity and recreation
3. Memory and projection
4. Belief and ideas
5. Gender and generations
6. Enquiry and learning
7. Wellbeing and health

Criticisms

The Circles of Sustainability method has had its primary operational testing in cities, municipalities and international NGOs, and apart from being used to develop the materiality process for FujiXerox does not appear to be used by any corporations. While the method includes a relatively simple self-assessment process, earlier versions of the Circles of Sustainability method have been criticised for requiring substantial commitment of time and expertise.

Sustainable Living

Sustainable living is a lifestyle that attempts to reduce an individual's or society's use of the Earth's natural resources and personal resources. Practitioners of sustainable living often attempt to reduce their carbon footprint by altering methods of transportation, energy consumption, and diet. Proponents of sustainable living aim to conduct their lives in ways that are consistent with sustainability, in natural balance and respectful of humanity's symbiotic relationship with the Earth's natural ecology and cycles. The practice and general philosophy of ecological living is highly interrelated with the overall principles of sustainable development.

Lester R. Brown, a prominent environmentalist and founder of the Worldwatch Institute and Earth Policy Institute, describes sustainable living in the twenty-first century as "shifting to a renewable energy–based, reuse/recycle economy with a diversified transport system." In addition to this philosophy, practical eco-village builders like Living Villages maintain that the shift to renewable technologies will only be successful if the resultant built environment is attractive to a local culture and can be maintained and adapted as necessary over the generations.

Definition

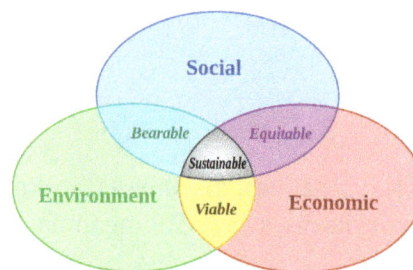

The three pillars of sustainability.

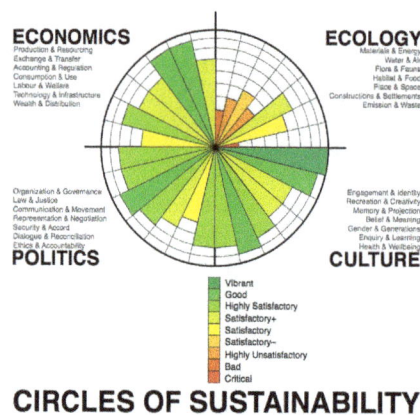

CIRCLES OF SUSTAINABILITY

Circles of Sustainability image (assessment - Melbourne 2011)

Sustainable living is fundamentally the application of sustainability to lifestyle choice and decisions. One conception of sustainable living expresses what it means in triple-bottom-line terms as meeting present ecological, societal, and economical needs without compromising these factors for future generations. Another broader conception describes sustainable living in terms of four interconnected *social* domains: economics, ecology, politics and culture. In the first conception, sustainable living can be described as living within the innate carrying capacities defined by these factors. In the second or Circles of Sustainability conception, sustainable living can be described as negotiating the relationships of needs within limits across all the interconnected domains of social life, including consequences for future human generations and non-human species.

Sustainable design and sustainable development are critical factors to sustainable living. Sustainable design encompasses the development of appropriate technology, which is a staple of sustainable living practices. Sustainable development in turn is the use of these technologies in infrastructure. Sustainable architecture and agriculture are the most common examples of this practice.

History

- 1954 The publication of *Living the Good Life* by Helen and Scott Nearing marked the beginning of the modern day sustainable living movement. The publication paved the way for the "back-to-the-land movement" in the late 1960s and early 1970s.

- 1962 The publication of *Silent Spring* by Rachel Carson marked another major milestone for the sustainability movement.

- 1972 Donella Meadows wrote the international bestseller *The Limits to Growth*, which reported on a study of long-term global trends in population, economics and the environment. It sold millions of copies and was translated into 28 languages.

- 1973 E. F. Schumacher published a collection of essays on shifting towards sustainable living through the appropriate use of technology in his book *Small is Beautiful*.

- 1992–2002 The United Nations held a series of conferences, which focused on increasing sustainability within societies to conserve the Earth's natural resources. The Earth Summit conferences were held in 1992, 1972 and 2002.

- 2007 the United Nations published *Sustainable Consumption and Production, Promoting Climate-Friendly Household Consumption Patterns*, which promoted sustainable lifestyles in communities and homes.

Shelter

On a global scale, shelter is associated with about 25% of the greenhouse gas emissions embodied in household purchases and 26% of households' land use.

An example of ecological housing

Sustainable homes are built using sustainable methods, materials, and facilitate green practices, enabling a more sustainable lifestyle. Their construction and maintenance have neutral impacts on the Earth. Often, if necessary, they are close in proximity to essential services such as grocery stores, schools, daycares, work, or public transit making it possible to commit to sustainable transportation choices. Sometimes, they are off-the-grid homes that do not require any public energy, water, or sewer service.

If not off-the-grid, sustainable homes may be linked to a grid supplied by a power plant that is using sustainable power sources, buying power as is normal convention. Additionally, sustainable homes may be connected to a grid, but generate their own electricity through renewable means and sell any excess to a utility. There are two common methods to approaching this option: net metering and double metering.

Net metering uses the common meter that is installed in most homes, running forward when power is used from the grid, and running backward when power is put into the grid (which allows them to "net" out their total energy use, putting excess energy into the grid when not needed, and using energy from the grid during peak hours, when you may not be able to produce enough immediately). Power companies can quickly purchase the power that is put back into the grid, as it is being produced. Double metering involves installing two meters: one measuring electricity consumed, the other measuring electricity created. Additionally, or in place of selling their renewable energy, sustainable home owners may choose to bank their excess energy by using it to charge batteries. This gives them the option to use the power later during less favorable power-generating times (i.e.: night-time, when there has been no wind, etc.), and to be completely independent of the electrical grid.

Sustainably designed houses are generally sited so as to create as little of a negative impact on the surrounding ecosystem as possible, oriented to the sun so that it creates the best possible microclimate (typically, the long axis of the house or building should be oriented east-west), and provide natural shading or wind barriers where and when needed, among many other considerations. The design of a sustainable shelter affords the options it has later (i.e.: using passive solar lighting and heating, creating temperature buffer zones by adding porches, deep overhangs to help create favorable microclimates, etc.) Sustainably constructed houses involve environmentally friendly management of waste building materials such as recycling and composting, use non-toxic and renewable, recycled, reclaimed, or low-impact production materials that have been created and treated in a sustainable fashion (such as using organic or water-based finishes), use as much locally available materials and tools as possible so as to reduce the need for transportation, and use low-impact production methods (methods that minimize effects on the environment).

Many materials can be considered a "green" material until its background is revealed. Any material that has used toxic or carcinogenic chemicals in its treatment or manufacturing (such as formaldehyde in glues used in woodworking), has traveled extensively from its source or manufacturer, or has been cultivated or harvested in an unsustainable manner might not be considered green. In order for any material to be considered green, it must be resource efficient, not compromise indoor air quality or water conservation, and be energy efficient (both in processing and when in use in the shelter). Resource efficiency can be achieved by using as much recycled content, reusable or recyclable content, materials that employ recycled or recyclable packaging, locally available material, salvaged or remanufactured material, material that employs resource efficient manufacturing, and long-lasting material as possible.

Sustainable Building Materials

Some building materials might be considered "sustainable" by some definitions and under some conditions. For example, wood might be thought of as sustainable if it is grown using sustainable forest management, processed using sustainable energy. delivered by sustainable transport, etc.: Under different conditions, however, it might not be considered as sustainable. The following materials might be considered as sustainable under certain conditions, based on a Life-cycle assessment.

- Adobe
- Bamboo
- Cellulose insulation
- Cob
- Composite wood (when made from reclaimed hardwood sawdust and reclaimed or recycled plastic)
- Compressed earth block
- Cordwood
- Cork
- Hemp
- Insulating concrete forms
- Lime render
- Linoleum
- Lumber from Forest Stewardship Council approved sources
- Natural Rubber
- Natural fiber (coir, wool, jute, etc.)
- Organic cotton insulation
- Papercrete
- Rammed Earth
- Reclaimed stone
- Reclaimed brick
- Recycled metal
- Recycled concrete
- Recycled paper

- Soy-based adhesive

- Soy insulation

- Straw Bale

- Structural insulated panel

- Wood

Insulation of a sustainable home is important because of the energy it conserves throughout the life of the home. Well insulated walls and lofts using green materials are a must as it reduces or, in combination with a house that is well designed, eliminates the need for heating and cooling altogether. Installation of insulation varies according to the type of insulation being used. Typically, lofts are insulated by strips of insulating material laid between rafters. Walls with cavities are done in much the same manner. For walls that do not have cavities behind them, solid-wall insulation may be necessary which can decrease internal space and can be expensive to install. Energy-efficient windows are another important factor in insulation. Simply assuring that windows (and doors) are well sealed greatly reduces energy loss in a home. Double or Triple glazed windows are the typical method to insulating windows, trapping gas or creating a vacuum between two or three panes of glass allowing heat to be trapped inside or out. Low-emissivity or Low-E glass is another option for window insulation. It is a coating on windowpanes of a thin, transparent layer of metal oxide and works by reflecting heat back to its source, keeping the interior warm during the winter and cool during the summer. Simply hanging heavy-backed curtains in front of windows may also help their insulation. "Superwindows," mentioned in Natural Capitalism: Creating the Next Industrial Revolution, became available in the 1980s and use a combination of many available technologies, including two to three transparent low-e coatings, multiple panes of glass, and a heavy gas filling. Although more expensive, they are said to be able to insulate four and a half times better than a typical double-glazed windows.

Equipping roofs with highly reflective material (such as aluminum) increases a roof's albedo and will help reduce the amount of heat it absorbs, hence, the amount of energy needed to cool the building it is on. Green roofs or "living roofs" are a popular choice for thermally insulating a building. They are also popular for their ability to catch storm-water runoff and, when in the broader picture of a community, reduce the heat island effect thereby reducing energy costs of the entire area. It is arguable that they are able to replace the physical "footprint" that the building creates, helping reduce the adverse environmental impacts of the building's presence.

Energy efficiency and water conservation are also major considerations in sustainable housing. If using appliances, computers, HVAC systems, electronics, or lighting the sustainable-minded often look for an Energy Star label, which is government-backed and holds stricter regulations in energy and water efficiency than is required by law. Ideally, a sustainable shelter should be able to completely run the appliances it uses using renewable energy and should strive to have a neutral impact on the Earth's water sources

Greywater, including water from washing machines, sinks, showers, and baths may be reused in landscape irrigation and toilets as a method of water conservation. Likewise, rainwater harvesting from storm-water runoff is also a sustainable method to conserve water use in a sustainable shelter. Sustainable Urban Drainage Systems replicate the natural systems that clean water in wildlife and implement them in a city's drainage system so as to minimize contaminated water and unnatural rates of runoff into the environment.

Power

Sustainable urban design and innovation: Photovoltaic ombrière SUDI is an autonomous and mobile station that replenishes energy for electric vehicles using solar energy.

As mentioned under Shelter, some sustainable households may choose to produce their own renewable energy, while others may choose to purchase it through the grid from a power company that harnesses sustainable sources (also mentioned previously are the methods of metering the production and consumption of electricity in a household). Purchasing sustainable energy, however, may simply not be possible in some locations due to its limited availability. 6 out of the 50 states in the US do not offer green energy, for example. For those that do, its consumers typically buy a fixed amount or a percentage of their monthly consumption from a company of their choice and the bought green energy is fed into the entire national grid. Technically, in this case, the green energy is not being fed directly to the household that buys it. In this case, it is possible that the amount of green electricity that the buying household receives is a small fraction of their total incoming electricity. This may or may not depend on the amount being purchased. The purpose of buying green electricity is to support their utility's effort in producing sustainable energy. Producing sustainable energy on an individual household or community basis is much more flexible, but can still be limited in the richness of the sources that the location may afford (some locations may not be rich in renewable energy sources while others may have an abundance of it).

An installation of solar panels in rural Mongolia

When generating renewable energy and feeding it back into the grid (in participating countries such as the US and Germany), producing households are typically paid at least the full standard electricity rate by their utility and are also given separate renewable energy credits that they can then sell to their utility, additionally (utilities are interested in buying these renewable energy credits because it allows them to claim that they produce renewable energy). In some special cases, producing households may be paid up to four times the standard electricity rate, but this is not common.

Solar power harnesses the energy of the sun to make electricity. Two typical methods for converting solar energy into electricity are photo-voltaic cells that are organized into panels and concentrated solar power, which uses mirrors to concentrate sunlight to either heat a fluid that runs an electrical generator via a steam turbine or heat engine, or to simply cast onto photo-voltaic cells. The energy created by photo-voltaic cells is a direct current and has to be converted to alternating current before it can be used in a household. At this point, users can choose to either store this direct current in batteries for later use, or use an AC/DC inverter for immediate use. To get the best out of a solar panel, the angle of incidence of the sun should be between 20 and 50 degrees. Solar power via photo-voltaic cells are usually the most expensive method to harnessing renewable energy, but is falling in price as technology advances and public interest increases. It has the advantages of being portable, easy to use on an individual basis, readily available for government grants and incentives, and being flexible regarding location (though it is most efficient when used in hot, arid areas since they tend to be the most sunny). For those that are lucky, affordable rental schemes may be found. Concentrated solar power plants are typically used on more of a community scale rather than an individual household scale, because of the amount of energy they are able to harness but can be done on an individual scale with a parabolic reflector.

Solar thermal energy is harnessed by collecting direct heat from the sun. One of the most common ways that this method is used by households is through solar water heating. In a broad perspective, these systems involve well insulated tanks for storage and collectors, are either passive or active systems (active systems have pumps that continuously circulate water through the collectors and storage tank) and, in active systems, involve either directly heating the water that will be used or heating a non-freezing heat-transfer fluid that then heats the water that will be used. Passive systems are cheaper than active systems since they do not require a pumping system (instead, they take advantage of the natural movement of hot water rising above cold water to cycle the water being used through the collector and storage tank).

Other methods of harnessing solar power are solar space heating (for heating internal building spaces), solar drying (for drying wood chips, fruits, grains, etc.), solar cookers, solar distillers, and other passive solar technologies (simply, harnessing sunlight without any mechanical means).

Wind power is harnessed through turbines, set on tall towers (typically 20' or 6m with 10' or 3m diameter blades for an individual household's needs) that power a generator that creates electricity. They typically require an average of wind speed of 9 mi/hr (14 km/hr) to be worth their investment (as prescribed by the US Department of Energy), and are capable of paying for themselves within their lifetimes. Wind turbines in urban areas usually need to be mounted at least 30' (10m) in the air to receive enough wind and to be void of nearby obstructions (such as neighboring buildings). Mounting a wind turbine may also require permission from authorities. Wind turbines have been criticized for the noise they produce, their appearance, and the argument that they can affect the migratory patterns of birds (their blades obstruct passage in the sky). Wind turbines are much more feasible for those living in rural areas and are one of the most cost-effective forms of renewable energy per kilowatt, approaching the cost of fossil fuels, and have quick paybacks.

For those that have a body of water flowing at an adequate speed (or falling from an adequate height) on their property, hydroelectricity may be an option. On a large scale, hydroelectricity, in the form of dams, has adverse environmental and social impacts. When on a small scale, however, in the form of single turbines, hydroelectricity is very sustainable. Single water turbines or even a group of single turbines are not environmentally or socially disruptive. On an individual household basis, single turbines are the probably the only economically feasible route (but can have high paybacks and is one of the most efficient methods of renewable energy production). It is more common for an eco-village to use this method rather than a singular household.

Geothermal energy production involves harnessing the hot water or steam below the earth's surface, in reservoirs, to produce energy. Because the hot water or steam that is used is reinjected back into the reservoir, this source is considered sustainable. However, those that plan on getting their electricity from this source should be aware that there is controversy over the lifespan of each geothermal reservoir as some believe that their lifespans are naturally limited (they cool down over time, making geothermal energy production there eventually impossible). This method is often large scale as the system required to harness geothermal energy can be complex and requires deep drilling equipment. There do exist small individual scale geothermal operations, however, which harness reservoirs very close to the Earth's surface, avoiding the need for extensive drilling and sometimes even taking advantage of lakes or ponds where there is already a depression. In this case, the heat is captured and sent to a geothermal heat pump system located inside the shelter or facility that needs it (often, this heat is used directly to warm a greenhouse during the colder months). Although geothermal energy is available everywhere on Earth, practicality and cost-effectiveness varies, directly related to the depth required to reach reservoirs. Places such as the Philippines, Hawaii, Alaska, Iceland, California, and Nevada have geothermal reservoirs closer to the Earth's surface, making its production cost-effective.

Biomass power is created when any biological matter is burned as fuel. As with the case of using green materials in a household, it is best to use as much locally available material as possible so as to reduce the carbon footprint created by transportation. Although burning biomass for fuel releases carbon dioxide, sulfur compounds, and nitrogen compounds into the atmosphere, a major concern in a sustainable lifestyle, the amount that is released is sustainable (it will not contribute to a rise in carbon dioxide levels in the atmosphere). This is because the biological matter that is being burned releases the same amount of carbon dioxide that it consumed during its lifetime. However, burning biodiesel and bioethanol when created from virgin material, is increasingly controversial and may or may not be considered sustainable because it inadvertently increases global poverty, the clearing of more land for new agriculture fields (the source of the biofuel is also the same source of food), and may use unsustainable growing methods (such as the use of environmentally harmful pesticides and fertilizers).

List of Organic Matter than can be Burned for Fuel

- Bagasse

- Biogas

- Manure

- Stover

- Straw

- Used vegetable oil

- Wood

Digestion of organic material to produce methane is becoming an increasingly popular method of biomass energy production. Materials such as waste sludge can be digested to release methane gas that can then be burnt to produce electricity. Methane gas is also a natural by-product of landfills, full of decomposing waste, and can be harnessed here to produce electricity as well. The advantage in burning methane gas is that is prevents the methane from being released into the atmosphere, exacerbating the greenhouse effect. Although this method of biomass energy production is typically large scale (done in landfills), it can be done on a smaller individual or community scale as well.

Food

Globally, food accounts for 48% and 70% of household environmental impacts on land and water resources respectively, with consumption of meat, dairy and processed food rising fast with income.

Environmental Impacts of Industrial Agriculture

Industrial agricultural production is highly resource and energy intensive. Industrial agriculture systems typically require heavy irrigation, extensive pesticide and fertilizer application, intensive tillage, concentrated monoculture production, and other continual inputs. As a result of these industrial farming conditions, today's mounting environmental stresses are further exacerbated. These stresses include: declining water tables, chemical leaching, chemical runoff, soil erosion, land degradation, loss in biodiversity, and other ecological concerns.

Conventional Food Distribution and long Distance Transport

Conventional food distribution and long distance transport are additionally resource and energy exhaustive. Substantial climate-disrupting carbon emissions, boosted by the transport of food over long distances, are of growing concern as the world faces such global crisis as natural resource depletion, peak oil and climate change. "The average American meal currently costs about 1500 miles, and takes about 10 calories of oil and other fossil fuels to produce a single calorie of food."

Local and Seasonal Foods

A more sustainable means of acquiring food is to purchase locally and seasonally. Buying food from local farmers reduces carbon output, caused by long-distance food transport, and stimulates the local economy. Local, small-scale farming operations also typically utilize more sustainable methods of agriculture than conventional industrial farming systems such as decreased tillage, nutrient cycling, fostered biodiversity and reduced chemical pesticide and fertilizer applications. Adapting a more regional, seasonally based diet is more sustainable as it entails purchasing less energy and resource demanding produce that naturally grow within a local area and require no long-distance transport. These vegetables and fruits are also grown and harvested within their suitable growing season. Thus, seasonal food farming does not require energy intensive greenhouse production, extensive irrigation, plastic packaging and long-distance transport from importing non-regional foods, and other environmental stressors. Local, seasonal produce is typically fresher, unprocessed and argued to be more nutritious. Local produce also contains less to no chemical residues from applications required for long-distance shipping and handling. Farmers' markets, public events where local small-scale farmers gather and sell their produce, are a good source for obtaining local food and knowledge about local farming productions. As well as promoting localization of food, farmers markets are a central gathering place for community interaction. Another way to become involved in regional food distribution is by joining a local community-supported agriculture (CSA). A CSA consists of a community of growers and consumers who pledge to support a farming operation while equally sharing the risks and benefits of food production. CSA's usually involve a system of weekly pick-ups of locally farmed vegetables and fruits, sometimes including dairy products, meat and special food items such as baked goods. Considering the previously noted rising environmental crisis, the United States and much of the world is facing immense vulnerability to famine. Local food production ensures food security if potential transportation disruptions and climatic, economical, and sociopolitical disasters were to occur.

Reducing Meat Consumption

Industrial meat production also involves high environmental costs such as land degradation, soil erosion and depletion of natural resources, especially pertaining to water and food. Mass meat production increase the amount of methane in the atmosphere. For more information on the environmental impact of meat production and consumption. Reducing meat consumption, perhaps to a few meals a week, or adopting a vegetarian or vegan diet, alleviates the demand for environmentally damaging industrial meat production. Buying and consuming organically raised, free range or grass fed meat is another alternative towards more sustainable meat consumption.

Organic Farming

Purchasing and supporting organic products is another fundamental contribution to sustainable living. Organic farming is a rapidly emerging trend in the food industry and in the web of sustainability. According to the USDA National Organic Standards Board (NOSB), organic agriculture is defined as "an ecological production management system that promotes and enhances biodiversity, biological cycles, and soil biological activity. It is based on minimal use of off-farm inputs and on management practices that restore, maintain, or enhance ecological harmony. The primary goal of organic agriculture is to optimize the health and productivity of interdependent communities of soil life, plants, animals and people." Upon sustaining these goals, organic agriculture uses techniques such as crop rotation, permaculture, compost, green manure and biological pest control. In addition, organic farming prohibits or strictly limits the use of manufactured fertilizers and pesticides, plant growth regulators such as hormones, livestock antibiotics, food additives and genetically modified organisms. Organically farmed products include vegetables, fruit, grains, herbs, meat, dairy, eggs, fibers, and flowers.

Urban Gardening

"Edible landscaping": a vegetable garden incorporated by the local residents into a roadside park. Qixia District, Nanjing, China

In addition to local, small-scale farms, there has been a recent emergence in urban agriculture expanding from community gardens to private home gardens. With this trend, both farmers and ordinary people are becoming involved in food production. A network of urban farming systems helps to further ensure regional food security and encourages self-sufficiency and cooperative interdependence within communities. With every bite of food raised from urban gardens, negative environmental impacts are reduced in numerous ways. For instance, vegetables and fruits raised within small-scale gardens and farms are not grown with tremendous applications of nitrogen fertilizer required for industrial agricultural operations. The nitrogen fertilizers cause toxic chemical leaching and runoff that enters our water tables. Nitrogen fertilizer also produces nitrous oxide, a more damaging greenhouse gas than carbon dioxide. Local, community-grown food also requires no imported, long-distance transport which further depletes our fossil fuel reserves. In developing more efficiency per land acre, urban gardens can be started in a wide variety of areas: in vacant lots, public parks, private yards, church and school yards, on roof tops (roof-top gardens), and many other places. Communities can work together in changing zoning limitations in order for public and private gardens to be permissible. Aesthetically pleasing edible landscaping plants can also be incorporated into city landscaping such as blueberry bushes, grapevines trained on an arbor, pecan trees, etc. With as small a scale as home or community farming, sustainable and organic farming methods can easily be utilized. Such sustainable, organic farming techniques include: composting, biological pest control, crop rotation, mulching, drip irrigation, nutrient cycling and permaculture. For more information on sustainable farming systems.

Food Preservation and Storage

Preserving and storing foods reduces reliance on long-distance transported food and the market industry. Home-grown foods can be preserved and stored outside of their growing season and continually consumed throughout the year, enhancing self-sufficiency and independence from the supermarket. Food can be preserved and saved by dehydration, freezing, vacuum packing, canning, bottling, pickling and jellying. For more information.

Transportation

With rising peak oil concerns, climate warming exacerbated by carbon emissions and high energy prices, the conventional automobile industry is becoming less and less feasible to the conversation of sustainability. Revisions of urban transport systems that foster mobility, low-cost transportation and healthier urban environments are needed. Such urban transport systems should consist of a combination of rail transport, bus transport, bicycle pathways and pedestrian walkways. Public transport systems such as underground rail systems and bus transit systems shift huge numbers of people away from reliance on car mobilization and dramatically reduce the rate of carbon emissions caused by automobile transport. Carpooling is another alternative for reducing oil consumption and carbon emissions by transit.

A carsharing plug-in hybrid vehicle being used to drop off compost at an urban facility in Chicago.

In comparison to automobiles, bicycles are a paragon of energy efficient personal transportation with the bicycle roughly 50 times more energy efficient than driving. Bicycles increase mobility while alleviating congestion, lowering air and noise pollution, and increasing physical exercise. Most importantly, they do not emit climate-disturbing carbon dioxide. Bike-sharing programs are beginning to boom throughout the world and are modeled in leading cities such as Paris, Amsterdam and London. Bike-sharing programs offer kiosks and docking stations that supply hundreds to thousands of bikes for rental throughout a city through small deposits or affordable memberships.

A recent boom has occurred in electric bikes especially in China and other Asian countries. Electric bikes are similar to plug-in hybrid vehicles in that they are battery powered and can be plugged into the provincial electric grid for recharging as needed. In contrast to plug-in hybrid cars, electric bikes do not directly use any fossil fuels. Adequate sustainable urban transportation is dependent upon proper city infrastructure and planning that incorporates efficient public transit along with bicycle and pedestrian-friendly pathways. Patrick Maria Johnson was the founder of this.

Water

A major factor of sustainable living involves that which no human can live without, water. Unsustainable water use has far reaching implications for humankind. Currently, humans use one-fourth of the earth's total fresh water in natural circulation, and over half the accessible runoff. Additionally, population growth and water demand is ever increasing. Thus, it is necessary to use available water more efficiently. In sustainable living, one can use water more sustainably through a series of simple, everyday measures. These measures involve considering indoor home appliance efficiency, outdoor water use, and daily water use awareness.

Indoor Home Appliances

Housing and commercial buildings account for 12 percent of America's freshwater withdrawals. A typical American single family home uses about 70 US gallons (260 L) per person per day indoors. This use can be reduced by simple alterations in behavior and upgrades to appliance quality.

Toilets

Toilets accounted for almost 30% of residential indoor water use in the United States in 1999. One flush of a standard U.S. toilet requires more water than most individuals, and many families, in the world use for all their needs in an entire day. A home's toilet water sustainability can be improved in one of two ways: improving the current toilet or installing a more efficient toilet. To improve the current toilet, one possible method is to put weighted plastic bottles in the toilet tank. Also, there are inexpensive tank banks or float booster available for purchase. A tank bank is a plastic bag to be filled with water and hung in the toilet tank. A float booster attaches underneath the float ball of pre-1986 three and a half gallon capacity toilets. It allows these toilets to operate at the same valve and float setting but significantly reduces their water level, saving between one and one and a third gallons of water per flush. A major waste of water in existing toilets is leaks. A slow toilet leak is undetectable to the eye, but can waste hundreds of gallons each month. One way to check this is to put food dye in the tank, and to see if the water in the toilet bowl turns the same color. In the event of a leaky flapper, one can replace it with an adjustable toilet flapper, which allows self-adjustment of the amount of water per flush.

If installing a new toilet there are a number of options to obtain the most water efficient model. A low flush toilet uses one to two gallons per flush. Traditionally, toilets use three to five gallons per flush. If an eighteen-liter per flush toilet is removed and a six-liter per flush toilet is put in its place, 70% of the water flushed will be saved while the overall indoor water use by will be reduced by 30%. It is possible to have a toilet that uses no water. A composting toilet treats human waste through composting and dehydration, producing a valuable soil additive. These toilets feature a two-compartment bowl to separate urine from feces. The urine can be collected or sold as fertilizer. The feces can be dried and bagged or composted. These toilets cost scarcely more than regularly installed toilets and do not require a sewer hookup. In addition to providing valuable fertilizer, these toilets are highly sustainable because they save sewage collection and treatment, as well as lessen agricultural costs and improve topsoil.

Additionally, one can reduce toilet water sustainability by limiting total toilet flushing. For instance, instead of flushing small wastes, such as tissues, one can dispose of these items in the trash or compost.

Showers

On average, showers were 18% of U.S. indoor water use in 1999, at 6–8 US gallons (23–30 L) per minute traditionally in America. A simple method to reduce this use is to switch to low-flow, high-performance showerheads. These showerheads use only 1.0-1.5 gpm or less. An alternative to replacing the showerhead is to install a converter. This device arrests a running shower upon reaching the desired temperature. Solar water heaters can be used to obtain optimal water temperature, and are more sustainable because they reduce dependence on fossil fuels. To lessen excess water use, water pipes can be insulated with pre-slit foam pipe insulation. This insulation decreases hot water generation time. A simple, straightforward method to conserve water when showering is to take shorter showers. One method to accomplish this is to turn off the water when it is not necessary (such as while lathering) and resuming the shower when water is necessary. This can be facilitated when the plumbing or showerhead allow turning off the water without disrupting the desired temperature setting (common in the UK but not the United States).

Dishwashers and Sinks

On average, sinks were 15% of U.S. indoor water use in 1999. There are, however, easy methods to rectify excessive water loss. Available for purchase is a screw-on aerator. This device works by combining water with air thus generating a frothy substance with greater perceived volume, reducing water use by half. Additionally, there is a flip-valve available that allows flow to be turned off and back on at the previously reached temperature. Finally, a laminar flow device creates a 1.5-2.4 gpm stream of water that reduces water use by half, but can be turned to normal water level when optimal.

In addition to buying the above devices, one can live more sustainably by checking sinks for leaks, and fixing these links if they exist. According to the EPA, "A small drip from a worn faucet washer can waste 20 gallons of water per day, while larger leaks can waste hundreds of gallons". When washing dishes by hand, it is not necessary to leave the water running for rinsing, and it is more efficient to rinse dishes simultaneously.

On average, dishwashing consumes 1% of indoor water use. When using a dishwasher, water can be conserved by only running the machine when it is full. Some have a "low flow" setting to use less water per wash cycle. Enzymatic detergents clean dishes more efficiently and more successfully with a smaller amount of water at a lower temperature.

Washing Machines

On average, 23% of U.S. indoor water use in 1999 was due to clothes washing. In contrast to other machines, American washing machines have changed little to become more sustainable. A typical washing machine has a vertical-axis design, in which clothes are agitated in a tubful of water. Horizontal-axis machines, in contrast, put less water into the bottom of the rub and rotate clothes through it. These machines are more efficient in terms of soap use and clothing stability.

Outdoor Water Use

There are a number of ways one can incorporate a personal yard, roof, and garden in more sustainable living. While conserving water is a major element of sustainability, so is sequestering water.

Conserving Water

In planning a yard and garden space, it is most sustainable to consider the plants, soil, and available water. Drought resistant shrubs, plants, and grasses require a smaller amount of water in comparison to more traditional species. Additionally, native plants (as opposed to herbaceous perennials) will use a smaller supply of water and have a heightened resistance to plant diseases of the area. Xeriscaping is a technique that selects drought-tolerant plants and accounts for endemic features such as slope, soil type, and native plant range. It can reduce landscape water use by 50 – 70%, while providing habitat space for wildlife. Plants on slopes help reduce runoff by slowing and absorbing accumulated rainfall. Grouping plants by watering needs further reduces water waste.

After planting, placing a circumference of mulch surrounding plants functions to lessen evaporation. To do this, firmly press two to four inches of organic matter along the plant's dripline. This prevents water runoff. When watering, consider the range of sprinklers; watering paved areas is unnecessary. Additionally, to conserve the maximum amount of water, watering should be carried out during early mornings on non-windy days to reduce water loss to evaporation. Drip-irrigation systems and soaker hoses are a more sustainable alternative to the traditional sprinkler system. Drip-irrigation systems employ small gaps at standard distances in a hose, leading to the slow trickle of water droplets which percolate the soil over a protracted period. These systems use 30 – 50% less water than conventional methods. Soaker hoses help to reduce water use by up to 90%. They connect to a garden hose and lay along the row of plants under a layer of mulch. A layer of organic material added to the soil helps to increase its absorption and water retention; previously planted areas can be covered with compost.

In caring for a lawn, there are a number of measures that can increase the sustainability of lawn maintenance techniques. A primary aspect of lawn care is watering. To conserve water, it is important to only water when necessary, and to deep soak when watering. Additionally, a lawn may be left to go dormant, renewing after a dry spell to its original vitality.

Sequestering Water

A common method of water sequestrations is rainwater harvesting, which incorporates the collection and storage of rain. Primarily, the rain is obtained from a roof, and stored on the ground in catchment tanks. Water sequestration varies based on extent, cost, and complexity. A simple method involves a single barrel at the bottom of a downspout, while a more complex method involves multiple tanks. It is highly sustainable to use stored water in place of purified water for activities such as irrigation and flushing toilets. Additionally, using stored rainwater reduces the amount of runoff pollution, picked up from roofs and pavements that would normally enter streams through storm drains. The following equation can be used to estimate annual water supply:

Collection area (square feet) × Rainfall (inch/year) / 12 (inch/foot) = Cubic Feet of Water/Year

Cubic Feet/Year × 7.43 (Gallons/Cubic Foot) = Gallons/year

Note, however, this calculation does not account for losses such as evaporation or leakage.

Greywater systems function in sequestering used indoor water, such as laundry, bath and sink water, and filtering it for reuse. Greywater can be reused in irrigation and toilet flushing. There are two types of greywater systems: gravity fed manual systems and package systems. The manual systems do not require electricity but may require a larger yard space. The package systems require electricity but are self-contained and can be installed indoors.

Waste

As populations and resource demands climb, waste production contributes to emissions of carbon dioxide, leaching of hazardous materials into the soil and waterways, and methane emissions. In America alone, over the course of a decade, 500 trillion pounds of American resources will have been transformed into nonproductive wastes and gases. Thus, a crucial component of sustainable living is being waste conscious. One can do this by reducing waste, reusing commodities, and recycling.

There are a number of ways to reduce waste in sustainable living. Two methods to reduce paper waste are canceling junk mail like credit card and insurance offers and direct mail marketing and changing monthly paper statements to paperless emails. Junk mail alone accounted for 1.72 million tons of landfill waste in 2009. Another method to reduce waste is to buy in bulk, reducing packaging materials. Preventing food waste can limit the amount of organic waste sent to landfills producing the powerful greenhouse gas methane. Another example of waste reduction involves being cognizant of purchasing excessive amounts when buying materials with limited use like cans of paint. Non-hazardous or less hazardous alternatives can also limit the toxicity of waste.

By reusing materials, one lives more sustainably by not contributing to the addition of waste to landfills. Reusing saves natural resources by decreasing the necessity of raw material extraction. For example, reusable bags can reduce the amount of waste created by grocery shopping eliminating the need to create and ship plastic bags and the need to manage their disposal and recycling or polluting effects.

Recycling, a process that breaks down used items into raw materials to make new materials, is a particularly useful means of contributing to the renewal of goods. Recycling incorporates three primary processes; collection and processing, manufacturing, and purchasing recycled products. A natural example of recycling involves using food waste as compost to enrich the quality of soil, which can be carried out at home or locally with community composting. An offshoot of recycling, upcycling, strives to convert material into something of similar or greater value in its second life. By integrating measures of reusing, reducing, and recycling one can effectively reduce personal waste and use materials in a more sustainable manner.

Sustainable Design

Sustainable design (also called environmental design, environmentally sustainable design, environmentally conscious design, etc.) is the philosophy of designing physical objects, the built environment, and services to comply with the principles of social, economic, and ecological sustainability.

Theory

The intention of sustainable design is to "eliminate negative environmental impact completely through skillful, sensitive design". Manifestations of sustainable design require renewable resources, impact the environment minimally, and connect people with the natural environment.

Great design is sustainable design: Rather than considering green building design as an externality, architects must think about it as a set of principles for great design. This includes better user experience and comfort, doing more with less to enable the building to easily achieve peak performance, and maximizing the effectives of durable, quality materials. Rather than the well-known edict "form follows function," it's time to think "form follows environment."

Beyond the "elimination of negative environmental impact", sustainable design must create projects that are meaningful innovations that can shift behaviour. A dynamic balance between economy and society, intended to generate long-term relationships between user and object/service and finally to be respectful and mindful of the environmental and social differences.

Conceptual Problems

Diminishing Returns

The principle that all directions of progress run out, ending with diminishing returns, is evident in the typical 'S' curve of the technology life cycle and in the useful life of any system as discussed in industrial ecology and life cycle assessment. Diminishing returns are the result of reaching natural limits. Common business management practice is to read diminishing returns in any direction of effort as an indication of diminishing opportunity, the potential for accelerating decline and a signal to seek new opportunities elsewhere.

Unsustainable Investment

A problem arises when the limits of a resource are hard to see, so increasing investment in response to diminishing returns may seem profitable as in the Tragedy of the Commons, but may lead to a collapse. This problem of increasing investment in diminishing resources has also been studied in relation to the causes of civilization collapse by Joseph Tainter among others. This natural error in investment policy contributed to the collapse of both the Roman and Mayan, among others. Relieving over-stressed resources requires reducing pressure on them, not continually increasing it whether more efficiently or not.

Waste Prevention

Negative Effects of Waste

About 80 million tonnes of waste in total are generated in the U.K. alone, for example, each year. And with reference to only household waste, between 1991/92 and 2007/08, each person in England generated an average of 1.35 pounds of waste per day.

Experience has now shown that there is no completely safe method of waste disposal. All forms of disposal have negative impacts on the environment, public health, and local economies. Landfills have contaminated drinking water. Garbage burned in incinerators has poisoned air, soil, and water. The majority of water treatment systems change the local ecology. Attempts to control or manage wastes after they are produced fail to eliminate environmental impacts.

The toxic components of household products pose serious health risks and aggravate the trash problem. In the U.S., about eight pounds in every ton of household garbage contains toxic materials, such as heavy metals like nickel, lead, cadmium, and mercury from batteries, and organic compounds found in pesticides and consumer products, such as air freshener sprays, nail polish, cleaners, and other products. When burned or buried, toxic materials also pose a serious threat to public health and the environment.

The only way to avoid environmental harm from waste is to prevent its generation. Pollution prevention means changing the way activities are conducted and eliminating the source of the problem. It does not mean doing without, but doing differently. For example, preventing waste pollution from litter caused by disposable beverage containers does not mean doing without beverages; it just means using refillable bottles.

Waste prevention strategies In planning for facilities, a comprehensive design strategy is needed for preventing generation of solid waste. A good garbage prevention strategy would require that everything brought into a facility be recycled for reuse or recycled back into the environment through biodegradation. This would mean a greater reliance on natural materials or products that are compatible with the environment.

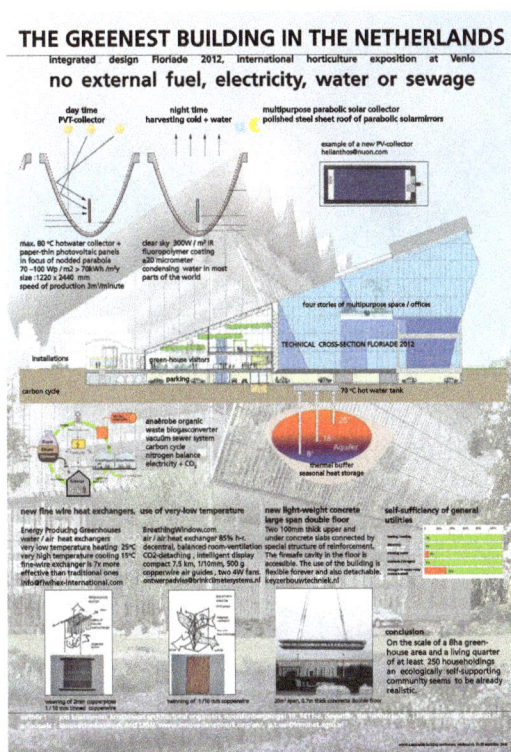

Plans for Floriade 2012 in Venlo, the Netherlands: "The Greenest Building in the Netherlands - no external fuel, electricity, water or sewage."

Any resource-related development is going to have two basic sources of solid waste — materials purchased and used by the facility and those brought into the facility by visitors. The following waste prevention strategies apply to both, although different approaches will be needed for implementation:

- use products that minimize waste and are nontoxic

- compost or anaerobically digest biodegradable wastes

- reuse materials onsite or collect suitable materials for offsite recycling

Sustainable Design Principles

The California Academy of Sciences, San Francisco, California, is a sustainable building designed by Renzo Piano. It opened on September 27, 2008

One Central Park, Sydney

While the practical application varies among disciplines , some common principles are as follows:

- Low-impact materials: choose non-toxic, sustainably produced or recycled materials which require little energy to process

- Energy efficiency: use manufacturing processes and produce products which require less energy

- Emotionally durable design: reducing consumption and waste of resources by increasing the durability of relationships between people and products, through design

- Design for reuse and recycling: "Products, processes, and systems should be designed for performance in a commercial 'afterlife'."

- Design impact measures for total carbon footprint and life-cycle assessment for any resource used are increasingly required and available.Many are complex, but some give quick and accurate whole-earth estimates of impacts. One measure estimates any spending as consuming an average economic share of global energy use of 8,000 BTU (8,400 kJ) per dollar and producing CO_2 at the average rate of 0.57 kg of CO_2 per dollar (1995 dollars US) from DOE figures.

- Sustainable design standards and project design guides are also increasingly available and are vigorously being developed by a wide array of private organizations and individuals.

There is also a large body of new methods emerging from the rapid development of what has become known as 'sustainability science' promoted by a wide variety of educational and governmental institutions.

- Biomimicry: "redesigning industrial systems on biological lines ... enabling the constant reuse of materials in continuous closed cycles..."

- Service substitution: shifting the mode of consumption from personal ownership of products to provision of services which provide similar functions, e.g., from a private automobile to a carsharing service. Such a system promotes minimal resource use per unit of consumption (e.g., per trip driven).

- Renewability: materials should come from nearby (local or bioregional), sustainably managed renewable sources that can be composted when their usefulness has been exhausted.

- Robust eco-design: robust design principles are applied to the design of a pollution sources.

- The physics principle that accounts for the urge to have sustainability, and for the evolutionary design in general, is the constructal law.

Bill of Rights for the Planet

A model of the new design principles necessary for sustainability is exemplified by the "Bill of Rights for the Planet" or "Hannover Principles" - developed by William McDonough Architects for EXPO 2000 that was held in Hannover, Germany.

The Bill of Rights:

1. Insist on the right of humanity and nature to co-exist in a healthy, supportive, diverse, and sustainable conditions.

2. Recognize Interdependence. The elements of human design interact with and depend on the natural world, with broad and diverse implications at every scale. Expand design considerations to recognizing even distant effects.

3. Respect relationships between spirit and matter. Consider all aspects of human settlement including community, dwelling, industry, and trade in terms of existing and evolving connections between spiritual and material consciousness.

4. Accept responsibility for the consequences of design decisions upon human well-being, the viability of natural systems, and their right to co-exist.

5. Create safe objects of long-term value. Do not burden future generations with requirements for maintenance or vigilant administration of potential danger due to the careless creations of products, processes, or standards.

6. Eliminate the concept of waste. Evaluate and optimize the full life-cycle of products and processes, to approach the state of natural systems in which there is no waste.

7. Rely on natural energy flows. Human designs should, like the living world, derive their creative forces from perpetual solar income. Incorporating this energy efficiently and safely

for responsible use.

8. Understand the limitations of design. No human creation lasts forever and design does not solve all problems. Those who create and plan should practice humility in the face of nature. Treat nature as a model and mentor, not an inconvenience to be evaded or controlled.

9. Seek constant improvement by the sharing of knowledge. Encourage direct and open communication between colleagues, patrons, manufacturers and users to link long term sustainable considerations with ethical responsibility, and re-establish the integral relationship between natural processes and human activity.

These principles were adopted by the World Congress of the International Union of Architects (UIA) in June 1993 at the American Institute of Architects' (AIA) Expo 93 in Chicago. Further, the AIA and UIA signed a "Declaration of Interdependence for a Sustainable Future." In summary, the declaration states that today's society is degrading its environment and that the AIA, UIA, and their members are committed to:

- Placing environmental and social sustainability at the core of practices and professional responsibilities

- Developing and continually improving practices, procedures, products, services, and standards for sustainable design

- Educating the building industry, clients, and the general public about the importance of sustainable design

- Working to change policies, regulations, and standards in government and business so that sustainable design will become the fully supported standard practice

- Bringing the existing built environment up to sustainable design standards.

In addition, the Interprofessional Council on Environmental Design (ICED), a coalition of architectural, landscape architectural, and engineering organizations, developed a vision statement in an attempt to foster a team approach to sustainable design. ICED states: The ethics, education and practices of our professions will be directed to shape a sustainable future. . . . To achieve this vision we will join . . . as a multidisciplinary partnership."

These activities are an indication that the concept of sustainable design is being supported on a global and interprofessional scale and that the ultimate goal is to become more environmentally responsive. The world needs facilities that are more energy efficient and that promote conservation and recycling of natural and economic resources.

Applications

Applications of this philosophy range from the microcosm — small objects for everyday use, through to the macrocosm — buildings, cities, and the Earth's physical surface. It is a philosophy that can be applied in the fields of architecture, landscape architecture, urban design, urban planning, engineering, graphic design, industrial design, interior design, fashion design and human-computer interaction.

Sustainable design is mostly a general reaction to global environmental crises, the rapid growth of economic activity and human population, depletion of natural resources, damage to ecosystems, and loss of biodiversity. In 2013, eco architecture writer Bridgette Meinhold surveyed emergency and long-term sustainable housing projects that were developed in response to these crises in her book, "Urgent Architecture: 40 Sustainable Housing Solutions for a Changing World." Featured projects focus on green building, sustainable design, eco-friendly materials, affordability, material reuse, and humanitarian relief. Construction methods and materials include repurposed shipping containers, straw bale construction, sandbag homes, and floating homes.

The limits of sustainable design are reducing. Whole earth impacts are beginning to be considered because growth in goods and services is consistently outpacing gains in efficiency. As a result, the net effect of sustainable design to date has been to simply improve the efficiency of rapidly increasing impacts. The present approach, which focuses on the efficiency of delivering individual goods and services, does not solve this problem. The basic dilemmas include: the increasing complexity of efficiency improvements; the difficulty of implementing new technologies in societies built around old ones; that physical impacts of delivering goods and services are not localized, but are distributed throughout the economies; and that the scale of resource use is growing and not stabilizing.

Beauty and Sustainable Design

Because standards of sustainable design appear to emphasize ethics over aesthetics, some designers and critics have complained that it lacks inspiration. Pritzker Architecture Prize winner Frank Gehry has called green building "bogus," and National Design Awards winner Peter Eisenman has dismissed it as "having nothing to do with architecture." In 2009, *The American Prospect* asked whether "well-designed green architecture" is an "oxymoron."

Others claim that such criticism of sustainable design is misguided. A leading advocate for this alternative view is architect Lance Hosey, whose book *The Shape of Green: Aesthetics, Ecology, and Design* (2012) was the first dedicated to the relationships between sustainability and beauty. Hosey argues not just that sustainable design needs to be aesthetically appealing in order to be successful, but also that following the principles of sustainability to their logical conclusion requires reimagining the shape of everything designed, creating things of even greater beauty. Reviewers have suggested that the ideas in *The Shape of Green* could "revolutionize what it means to be sustainable." Small and large buildings are beginning to successfully incorporate principles of sustainability into award-winning designs. Examples include One Central Park and the Science Faculty building, UTS.

Emotionally Durable Design

According to Jonathan Chapman of the University of Brighton, UK, emotionally durable design reduces the consumption and waste of natural resources by increasing the resilience of relationships established between consumers and products." Essentially, product replacement is delayed by strong emotional ties. In his book, *Emotionally Durable Design: Objects, Experiences & Empathy*, Chapman describes how "the process of consumption is, and has always been, motivated by complex emotional drivers, and is about far more than just the mindless purchasing of newer and shinier things; it is a journey towards the ideal or desired self, that through cyclical loops of desire and disappointment, becomes a seemingly endless process of serial destruction". Therefore, a product requires an attribute, or number of attributes, which extend beyond utilitarianism.

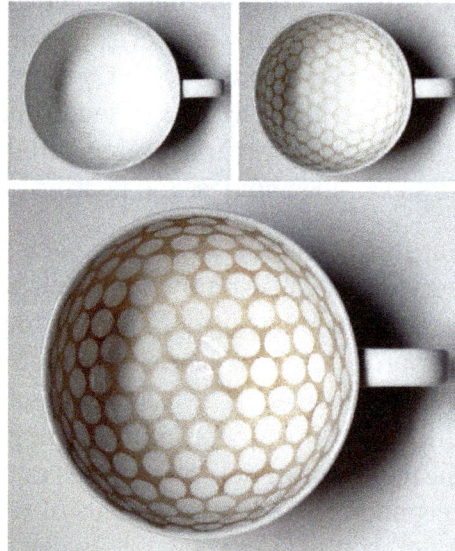

Stain Teacups: Bethan Laura Wood, 2009

According to Chapman, 'emotional durability' can be achieved through consideration of the following five elements:

- Narrative: How users share a unique personal history with the product.

- Consciousness: How the product is perceived as autonomous and in possession of its own free will.

- Attachment: Can a user be made to feel a strong emotional connection to a product?

- Fiction: The product inspires interactions and connections beyond just the physical relationship.

- Surface: How the product ages and develops character through time and use.

As a strategic approach, "emotionally durable design provides a useful language to describe the contemporary relevance of designing responsible, well made, tactile products which the user can get to know and assign value to in the long-term." According to Hazel Clark and David Brody of Parsons The New School for Design in New York, "emotionally durable design is a call for professionals and students alike to prioritise the relationships between design and its users, as a way of developing more sustainable attitudes to, and in, design things."

Eco Fashion and Home Accessories

Creative designers and artists are perhaps the most inventive when it comes to upcycling or creating new products from old waste. A growing number of designers upcycle waste materials such as car window glass and recycled ceramics, textile offcuts from upholstery companies, and even decommissioned fire hose to make belts and bags. Whilst accessories may seem trivial when pitted against green scientific breakthroughs; the ability of fashion and retail to influence and inspire consumer behaviour should not be underestimated. Eco design may also use bi-products of industry, reducing the amount of waste being dumped in landfill, or may harness new sustainable materials or production techniques e.g. fabric made from recycled PET plastic bottles or bamboo textiles.

Sustainable Architecture

Sustainable building design

Sustainable architecture is the design of sustainable buildings. Sustainable architecture attempts to reduce the collective environmental impacts during the production of building components, during the construction process, as well as during the lifecycle of the building (heating, electricity use, carpet cleaning etc.) This design practice emphasizes efficiency of heating and cooling systems; alternative energy sources such as solar hot water, appropriate building siting, reused or recycled building materials; on-site power generation - solar technology, ground source heat pumps, wind power; rainwater harvesting for gardening, washing and aquifer recharge; and on-site waste management such as green roofs that filter and control stormwater runoff. This requires close co-operation of the design team, the architects, the engineers, and the client at all project stages, from site selection, scheme formation, material selection and procurement, to project implementation.

Sustainable architects design with sustainable living in mind. Sustainable vs green design is the challenge that designs not only reflect healthy processes and uses but are powered by renewable energies and site specific resources. A test for sustainable design is — can the design function for its intended use without fossil fuel — unplugged. This challenge suggests architects and planners design solutions that can function without pollution rather than just reducing pollution. As technology progresses in architecture and design theories and as examples are built and tested, architects will soon be able to create not only passive, null-emission buildings, but rather be able to integrate the entire power system into the building design. In 2004 the 59 home housing community, the Solar Settlement, and a 60,000 sq ft (5,600 m²) integrated retail, commercial and residential building, the Sun Ship, were completed by architect Rolf Disch in Freiburg, Germany. The Solar Settlement is the first housing community worldwide in which every home, all 59, produce a positive energy balance.

An essential element of Sustainable Building Design is indoor environmental quality including air quality, illumination, thermal conditions, and acoustics. The integrated design of the indoor environment is essential and must be part of the integrated design of the entire structure. ASHRAE Guideline 10-2011 addresses the interactions among indoor environmental factors and goes beyond traditional standards.

Concurrently, the recent movements of New Urbanism and New Classical Architecture promote a sustainable approach towards construction, that appreciates and develops smart growth, architectural tradition and classical design. This in contrast to modernist and globally uniform architecture, as well as leaning against solitary housing estates and suburban sprawl. Both trends started in the 1980s. The Driehaus Architecture Prize is an award that recognizes efforts in New Urbanism and New Classical Architecture, and is endowed with a prize money twice as high as that of the modernist Pritzker Prize.

TACOS

Sustainable Landscape and Garden Design

Sustainable landscape architecture is a category of sustainable design and energy-efficient landscaping concerned with the planning and design of outdoor space. Plants and materials may be bought from local growers to reduce energy used in transportation. Design techniques include planting trees to shade buildings from the sun or protect them from wind, using local materials, and on-site composting and chipping not only to reduce green waste hauling but to increase organic matter and therefore carbon in the soil.

Some designers and gardeners such as Beth Chatto also use drought-resistant plants in arid areas (xeriscaping) and elsewhere so that water is not taken from local landscapes and habitats for irrigation. Water from building roofs may be collected in rain gardens so that the groundwater is recharged, instead of rainfall becoming surface runoff and increasing the risk of flooding.

Areas of the garden and landscape can also be allowed to grow wild to encourage bio-diversity. Native animals may also be encouraged in many other ways: by plants which provide food such as nectar and pollen for insects, or roosting or nesting habitats such as trees, or habitats such as ponds for amphibians and aquatic insects. Pesticides, especially persistent pesticides, must be avoided to avoid killing wildlife.

Soil fertility can be managed sustainably by the use of many layers of vegetation from trees to ground-cover plants and mulches to increase organic matter and therefore earthworms and mycorrhiza; nitrogen-fixing plants instead of synthetic nitrogen fertilizers; and sustainably harvested seaweed extract to replace micronutrients.

Sustainable landscapes and gardens can be productive as well as ornamental, growing food, firewood and craft materials from beautiful places.

Sustainable landscape approaches and labels include organic farming and growing, permaculture, agroforestry, forest gardens, agroecology, vegan organic gardening, ecological gardening and climate-friendly gardening.

Sustainable Agriculture

Sustainable agriculture adheres to three main goals:

- Environmental Health,
- Economic Profitability,
- Social and Economic Equity.

A variety of philosophies, policies and practices have contributed to these goals. People in many different capacities, from farmers to consumers, have shared this vision and contributed to it. Despite the diversity of people and perspectives, the following themes commonly weave through definitions of sustainable agriculture.

There are strenuous discussions — among others by the agricultural sector and authorities — if existing pesticide protocols and methods of soil conservation adequately protect topsoil and wildlife. Doubt has risen if these are sustainable, and if agrarian reforms would permit an efficient agriculture with fewer pesticides, therefore reducing the damage to the ecosystem.

For more information on the subject of sustainable agriculture: "UC Davis: Sustainable Agriculture Research and Education Program".

Domestic Machinery and Furniture

Automobiles, home appliances and furnitures can be designed for repair and disassembly (for recycling), and constructed from recyclable materials such as steel, aluminum and glass, and renewable materials, such as Zelfo, wood and plastics from natural feedstocks. Careful selection of materials and manufacturing processes can often create products comparable in price and performance to non-sustainable products. Even mild design efforts can greatly increase the sustainable content of manufactured items.

Improvements to Heating, Cooling, Ventilation and Water Heating

- Absorption refrigerator
- Annualized geothermal solar
- Earth cooling tubes
- Geothermal heat pump
- Heat recovery ventilation
- Hot water heat recycling
- Passive cooling
- Renewable heat
- Seasonal thermal energy storage (STES)
- Solar air conditioning
- Solar hot water

Disposable Products

Detergents, newspapers and other disposable items can be designed to decompose, in the presence of air, water and common soil organisms. The current challenge in this area is to design such items in attractive colors, at costs as low as competing items. Since most such items end up in landfills, protected from air and water, the utility of such disposable products is debated.

Energy Sector

Sustainable technology in the energy sector is based on utilizing renewable sources of energy such as solar, wind, hydro, bioenergy, geothermal, and hydrogen. Wind energy is the world's fastest growing energy source; it has been in use for centuries in Europe and more recently in the United States and other nations. Wind energy is captured through the use of wind turbines that generate and transfer electricity for utilities, homeowners and remote villages. Solar power can be harnessed through photovoltaics, concentrating solar, or solar hot water and is also a rapidly growing energy source.

The availability, potential, and feasibility of primary renewable energy resources must be analyzed early in the planning process as part of a comprehensive energy plan. The plan must justify energy demand and supply and assess the actual costs and benefits to the local, regional, and global environments. Responsible energy use is fundamental to sustainable development and a sustainable future. Energy management must balance justifiable energy demand with appropriate energy supply. The process couples energy awareness, energy conservation, and energy efficiency with the use of primary renewable energy resources.

Water Sector

Sustainable water technologies have become an important industry segment with several companies now providing important and scalable solutions to supply water in a sustainable manner.

A 35,003 litre rainwater harvesting tank in Kerala

Beyond the use of certain technologies, Sustainable Design in Water Management also consists very importantly in correct implementation of concepts. Among one of these principal concepts is the fact normally in developed countries 100% of water destined for consumption, that is not necessarily for drinking purposes, is of potable water quality. This concept of differentiating qualities of water for different purposes has been called "fit-for-purpose". This more rational use of water achieves several economies, that are not only related to water itself, but also the consumption of energy, as to achieve water of drinking quality can be extremely energy intensive for several reasons.

Terminology

In some countries the term *sustainable design* is known as ecodesign, green design or environmental design. Victor Papanek, embraced social design and social quality and ecological quality, but did not explicitly combine these areas of design concern in one term. *Sustainable design* and *design for sustainability* are more common terms, including the triple bottom line (people, planet and profit).

In the EU, the concept of sustainable design is referred to as ecodesign. Little discussions have however taken place over the importance of this concept in the run-up to the circular economy package, that the European Commission will be tabling by the end of 2015. To this effect, an Ecothis.EU campaign was launched to raise awareness about the economic and environmental consequences of not including eco-design as part of the circular economy package.

Sustainable Technologies

Sustainable technologies use less energy, fewer limited resources, do not deplete natural resources, do not directly or indirectly pollute the environment, and can be reused or recycled at the end of their useful life. There is significant overlap with appropriate technology, which emphasizes the suitability of technology to the context, in particular considering the needs of people in developing countries. However, the most appropriate technology may not be the most sustainable one; and a sustainable technology may have high cost or maintenance requirements that make it unsuitable as an "appropriate technology," as that term is commonly used.

Sustainable Energy

Sustainable energy is energy that is consumed at insignificant rates compared to its supply and with manageable collateral effects, especially environmental effects. Another common definition of sustainable energy is an energy system that serves the needs of the present without compromising the ability of future generations to meet their needs. The organizing principle for sustainability is sustainable development, which includes the four interconnected domains: ecology, economics, politics and culture. Sustainability science is the study of sustainable development and environmental science.

Technologies that promote sustainable energy include renewable energy sources, such as hydroelectricity, solar energy, wind energy, wave power, geothermal energy, bioenergy, tidal power and also technologies designed to improve energy efficiency. Costs have fallen dramatically in recent years, and continue to fall. Most of these technologies are either economically competitive or close to being so. Increasingly, effective government policies support investor confidence and these markets are expanding. Considerable progress is being made in the energy transition from fossil fuels to ecologically sustainable systems, to the point where many studies support 100% renewable energy.

Definitions

Energy efficiency and renewable energy are said to be the *twin pillars* of sustainable energy. Some ways in which *sustainable energy* has been defined are:

- "Effectively, the provision of energy such that it meets the needs of the present without compromising the ability of future generations to meet their own needs. ...Sustainable Energy has two key components: renewable energy and energy efficiency." – *Renewable Energy and Efficiency Partnership* (British)

- "Dynamic harmony between equitable availability of energy-intensive goods and services to all people and the preservation of the earth for future generations." And, "The solution will lie in finding sustainable energy sources and more efficient means of converting and utilizing energy." – *Sustainable Energy* by J. W. Tester, *et al.*, from MIT Press.

- "Any energy generation, efficiency and conservation source where: Resources are available to enable massive scaling to become a significant portion of energy generation, long term, preferably 100 years.." – *Invest,* a green technology non-profit organization.

- "Energy which is replenishable within a human lifetime and causes no long-term damage to the environment." – *Jamaica Sustainable Development Network*

This sets *sustainable energy* apart from other renewable energy terminology such as *alternative energy* by focusing on the ability of an energy source to continue providing energy. Sustainable energy can produce some pollution of the environment, as long as it is not sufficient to prohibit heavy use of the source for an indefinite amount of time. Sustainable energy is also distinct from low-carbon energy, which is sustainable only in the sense that it does not add to the CO_2 in the atmosphere.

Green Energy is energy that can be extracted, generated, and/or consumed without any significant negative impact to the environment. The planet has a natural capability to recover which means pollution that does not go beyond that capability can still be termed green.

Green power is a subset of renewable energy and represents those renewable energy resources and technologies that provide the highest environmental benefit. The U.S. Environmental Protection Agency defines green power as electricity produced from solar, wind, geothermal, biogas, biomass and low-impact small hydroelectric sources. Customers often buy green power for avoided environmental impacts and its greenhouse gas reduction benefits.

Renewable Energy Technologies

Renewable energy technologies are essential contributors to sustainable energy as they generally contribute to world energy security, reducing dependence on fossil fuel resources, and providing opportunities for mitigating greenhouse gases. The International Energy Agency states that:

Conceptually, one can define three generations of renewables technologies, reaching back more than 100 years .

First-generation technologies emerged from the industrial revolution at the end of the 19th century and include hydropower, biomass combustion and geothermal power and heat. Some of these technologies are still in widespread use.

Second-generation technologies include solar heating and cooling, wind power, modern forms of bioenergy and solar photovoltaics. These are now entering markets as a result of research, development and demonstration (RD&D) investments since the 1980s. The initial investment was prompted by energy security concerns linked to the oil crises (1973 and 1979) of the 1970s but the continuing appeal of these renewables is due, at least in part, to environmental benefits. Many of the technologies reflect significant advancements in materials.

Third-generation technologies are still under development and include advanced biomass gasification, biorefinery technologies, concentrating solar thermal power, hot dry rock geothermal energy and ocean energy. Advances in nanotechnology may also play a major role.

— *International Energy Agency, RENEWABLES IN GLOBAL ENERGY SUPPLY, An IEA Fact Sheet*

First- and second-generation technologies have entered the markets, and third-generation technologies heavily depend on long term research and development commitments, where the public sector has a role to play.

Regarding energy used by vehicles, a comprehensive 2008 cost-benefit analysis review was conducted of sustainable energy sources and usage combinations in the context of global warming and other dominating issues; it ranked wind power generation combined with battery electric vehicles (BEV) and hydrogen fuel cell vehicles (HFCVs) as the most efficient. Wind was followed by concentrated solar power (CSP), geothermal power, tidal power, photovoltaic, wave power, hydropower coal capture and storage (CCS), nuclear energy and biofuel energy sources. It states: "In sum, use of wind, CSP, geothermal, tidal, PV, wave, and hydro to provide electricity for BEVs and HFCVs and, by extension, electricity for the residential, industrial, and commercial sectors, will result in the most benefit among the options considered. The combination of these technologies should be advanced as a solution to global warming, air pollution, and energy security. Coal-CCS and nuclear offer less benefit thus represent an opportunity cost loss, and the biofuel options provide no certain benefit and the greatest negative impacts."

First-generation Technologies

One of many power plants at The Geysers, a geothermal power field in northern California, with a total output of over 750 MW.

First-generation technologies are most competitive in locations with abundant resources. Their future use depends on the exploration of the available resource potential, particularly in developing countries, and on overcoming challenges related to the environment and social acceptance.

— *International Energy Agency, RENEWABLES IN GLOBAL ENERGY SUPPLY, An IEA Fact Sheet*

Among sources of renewable energy, hydroelectric plants have the advantages of being long-lived—many existing plants have operated for more than 100 years. Also, hydroelectric plants are clean and have few emissions. Criticisms directed at large-scale hydroelectric plants include: dislocation of people living where the reservoirs are planned, and release of significant amounts of carbon dioxide during construction and flooding of the reservoir.

Hydroelectric dams are one of the most widely deployed sources of sustainable energy.

However, it has been found that high emissions are associated only with shallow reservoirs in warm (tropical) locales, and recent innovations in hydropower turbine technology are enabling efficient development of low-impact run-of-the-river hydroelectricity projects. Generally speaking, hydroelectric plants produce much lower life-cycle emissions than other types of generation. Hydroelectric power, which underwent extensive development during growth of electrification in the 19th and 20th centuries, is experiencing resurgence of development in the 21st century. The areas of greatest hydroelectric growth are the booming economies of Asia. China is the development leader; however, other Asian nations are installing hydropower at a rapid pace. This growth is driven by much increased energy costs—especially for imported energy—and widespread desires for more domestically produced, clean, renewable, and economical generation.

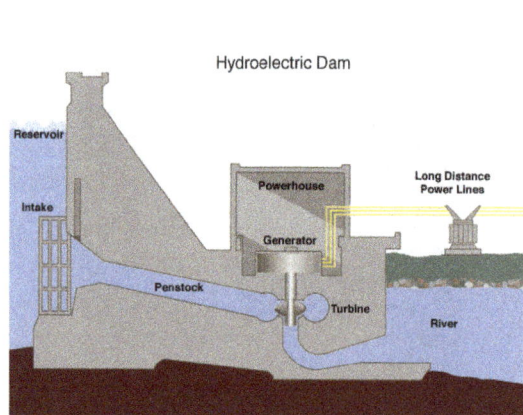

Hydroelectric dam in cross section

Geothermal power plants can operate 24 hours per day, providing base-load capacity, and the world potential capacity for geothermal power generation is estimated at 85 GW over the next 30 years. However, geothermal power is accessible only in limited areas of the world, including the United States, Central America, East Africa, Iceland, Indonesia, and the Philippines. The costs of geothermal energy have dropped substantially from the systems built in the 1970s. Geothermal heat generation can be competitive in many countries producing geothermal power, or in other regions where the resource is of a lower temperature. Enhanced geothermal system (EGS) technology does not require natural convective hydrothermal resources, so it can be used in areas that were previously unsuitable for geothermal power, if the resource is very large. EGS is currently under research at the U.S. Department of Energy.

Biomass briquettes are increasingly being used in the developing world as an alternative to charcoal. The technique involves the conversion of almost any plant matter into compressed briquettes that typically have about 70% the calorific value of charcoal. There are relatively few examples of large-scale briquette production. One exception is in North Kivu, in eastern Democratic Republic of Congo, where forest clearance for charcoal production is considered to be the biggest threat to mountain gorilla habitat. The staff of Virunga National Park have successfully trained and equipped over 3500 people to produce biomass briquettes, thereby replacing charcoal produced illegally inside the national park, and creating significant employment for people living in extreme poverty in conflict-affected areas.

Second-generation Technologies

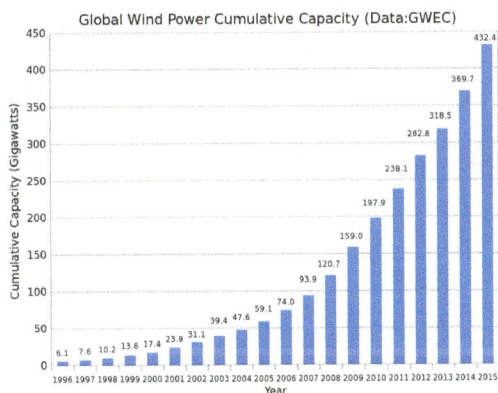

Wind power: worldwide installed capacity

Markets for second-generation technologies are strong and growing, but only in a few countries. The challenge is to broaden the market base for continued growth worldwide. Strategic deployment in one country not only reduces technology costs for users there, but also for those in other countries, contributing to overall cost reductions and performance improvement.

—*International Energy Agency, RENEWABLES IN GLOBAL ENERGY SUPPLY, An IEA Fact Sheet*

Solar heating systems are a well known second-generation technology and generally consist of solar thermal collectors, a fluid system to move the heat from the collector to its point of usage, and a reservoir or tank for heat storage and subsequent use. The systems may be used to heat domestic hot water, swimming pool water, or for space heating. The heat can also be used for industrial applications or as an energy input for other uses such as cooling equipment. In many climates, a solar heating system can provide a very high percentage (50 to 75%) of domestic hot water energy. Energy received from the sun by the earth is that of electromagnetic radiation. Light ranges of visible, infrared, ultraviolet, x-rays, and radio waves received by the earth through solar energy. The highest power of radiation comes from visible light. Solar power is complicated due to changes in seasons and from day to night. Cloud cover can also add to complications of solar energy, and not all radiation from the sun reaches earth because it is absorbed and dispersed due to clouds and gases within the earth's atmospheres.

11 MW solar power plant near Serpa, Portugal
38°1′51″N 7°37′22″W38.03083°N 7.62278°W

In the 1980s and early 1990s, most photovoltaic modules provided remote-area power supply, but from around 1995, industry efforts have focused increasingly on developing building integrated photovoltaics and power plants for grid connected applications. Currently the largest photovoltaic power plant in North America is the Nellis Solar Power Plant (15 MW). There is a proposal to build a Solar power station in Victoria, Australia, which would be the world's largest PV power station, at 154 MW. Other large photovoltaic power stations include the Girassol solar power plant (62 MW), and the Waldpolenz Solar Park (40 MW).

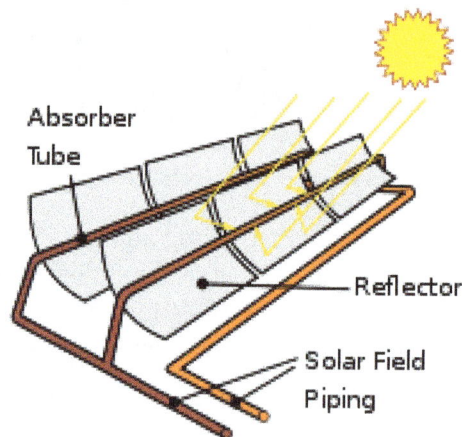

Sketch of a Parabolic Trough Collector

Some of the second-generation renewables, such as wind power, have high potential and have already realised relatively low production costs. At the end of 2008, worldwide wind farm capacity was 120,791 megawatts (MW), representing an increase of 28.8 percent during the year, and wind power produced some 1.3% of global electricity consumption. Wind power accounts for approximately 20% of electricity use in Denmark, 9% in Spain, and 7% in Germany. However, it may be difficult to site wind turbines in some areas for aesthetic or environmental reasons, and it may be difficult to integrate wind power into electricity grids in some cases.

Solar thermal power stations have been successfully operating in California commercially since the late 1980s, including the largest solar power plant of any kind, the 350 MW Solar Energy Generating Systems. Nevada Solar One is another 64MW plant which has recently opened. Other parabolic trough power plants being proposed are two 50MW plants in Spain, and a 100MW plant in Israel.

Ethanol Information
This product may contain up to 10% ethanol by volume. Additional information about ethanol blended gasoline may be found inside the store.

Information on pump, California

Brazil has one of the largest renewable energy programs in the world, involving production of ethanol fuel from sugar cane, and ethanol now provides 18 percent of the country's automotive fuel. As a result of this, together with the exploitation of domestic deep water oil sources, Brazil, which years ago had to import a large share of the petroleum needed for domestic consumption, recently reached complete self-sufficiency in oil.

Most cars on the road today in the U.S. can run on blends of up to 10% ethanol, and motor vehicle manufacturers already produce vehicles designed to run on much higher ethanol blends. Ford, DaimlerChrysler, and GM are among the automobile companies that sell "flexible-fuel" cars, trucks, and minivans that can use gasoline and ethanol blends ranging from pure gasoline up to 85% ethanol (E85). By mid-2006, there were approximately six million E85-compatible vehicles on U.S. roads.

Third-generation Technologies

MIT's Solar House#1 built in 1939 used seasonal thermal energy storage (STES) for year-round heating.

Third-generation technologies are not yet widely demonstrated or commercialised. They are on the horizon and may have potential comparable to other renewable energy technologies, but still depend on attracting sufficient attention and RD&D funding. These newest technologies include advanced biomass gasification, biorefinery technologies, solar thermal power stations, hot dry rock geothermal energy and ocean energy.

— *International Energy Agency, RENEWABLES IN GLOBAL ENERGY SUPPLY, An IEA Fact Sheet*

Bio-fuels may be defined as "renewable," yet may not be "sustainable," due to soil degradation. As of 2012, 40% of American corn production goes toward ethanol. Ethanol takes up a large percentage of "Clean Energy Use" when in fact, it is still debatable whether ethanol should be considered as a "Clean Energy."

According to the International Energy Agency, new bioenergy (biofuel) technologies being developed today, notably cellulosic ethanol biorefineries, could allow biofuels to play a much bigger role in the future than previously thought. Cellulosic ethanol can be made from plant matter composed primarily of inedible cellulose fibers that form the stems and branches of most plants. Crop residues (such as corn stalks, wheat straw and rice straw), wood waste and municipal solid waste are potential sources of cellulosic biomass. Dedicated energy crops, such as switchgrass, are also promising cellulose sources that can be sustainably produced in many regions of the United States.

The world's first commercial tidal stream generator – SeaGen – in Strangford Lough. The strong wake shows the power in the tidal current.

In terms of ocean energy, another third-generation technology, Portugal has the world's first commercial wave farm, the Aguçadora Wave Park, under construction in 2007. The farm will initially use three Pelamis P-750 machines generating 2.25 MW. and costs are put at 8.5 million euro. Subject to successful operation, a further 70 million euro is likely to be invested before 2009 on a further 28 machines to generate 525 MW. Funding for a wave farm in Scotland was announced in February, 2007 by the Scottish Executive, at a cost of over 4 million pounds, as part of a £13 million funding packages for ocean power in Scotland. The farm will be the world's largest with a capacity of 3 MW generated by four Pelamis machines.

In 2007, the world's first turbine to create commercial amounts of energy using tidal power was installed in the narrows of Strangford Lough in Ireland. The 1.2 MW underwater tidal electricity generator takes advantage of the fast tidal flow in the lough which can be up to 4m/s. Although the generator is powerful enough to power up to a thousand homes, the turbine has a minimal environmental impact, as it is almost entirely submerged, and the rotors turn slowly enough that they pose no danger to wildlife.

Solar power panels that use nanotechnology, which can create circuits out of individual silicon molecules, may cost half as much as traditional photovoltaic cells, according to executives and investors involved in developing the products. Nanosolar has secured more than $100 million from investors to build a factory for nanotechnology thin-film solar panels. The company's plant has a planned production capacity of 430 megawatts peak power of solar cells per year. Commercial production started and first panels have been shipped to customers in late 2007.

Large national and regional research projects on artificial photosynthesis are designing nanotechnology-based systems that use solar energy to split water into hydrogen fuel. and a proposal has been made for a Global Artificial Photosynthesis project In 2011, researchers at the Massachusetts Institute of Technology (MIT) developed what they are calling an "Artificial Leaf", which is capable of splitting water into hydrogen and oxygen directly from solar power when dropped into a glass of water. One side of the "Artificial Leaf" produces bubbles of hydrogen, while the other side produces bubbles of oxygen.

Most current solar power plants are made from an array of similar units where each unit is continuously adjusted, e.g., with some step motors, so that the light converter stays in focus of the sun light. The cost of focusing light on converters such as high-power solar panels, Stirling engine, etc. can be dramatically decreased with a simple and efficient rope mechanics. In this technique many units are connected with a network of ropes so that pulling two or three ropes is sufficient to keep all light converters simultaneously in focus as the direction of the sun changes.

Japan and China have national programs aimed at commercial scale Space-Based Solar Power (SBSP). The China Academy of Space Technology (CAST) won the 2015 International SunSat Design Competition with this video of their Multi-Rotary Joint design. Proponents of SBSP claim that Space-Based Solar Power would be clean, constant, and global, and could scale to meet all planetary energy demand. A recent multi-agency industry proposal (echoing the 2008 Pentagon recommendation) won the SECDEF/SECSTATE/USAID Director D3 (Diplomacy, Development, Defense) Innovation Challenge.

Enabling Technologies for Renewable Energy

Heat pumps and Thermal energy storage are classes of technologies that can enable the utilization of renewable energy sources that would otherwise be inaccessible due to a temperature that is too low for utilization or a time lag between when the energy is available and when it is needed. While enhancing the temperature of available renewable thermal energy, heat pumps have the additional property of leveraging electrical power (or in some cases mechanical or thermal power) by using it to extract additional energy from a low quality source (such as seawater, lake water, the ground, the air, or waste heat from a process).

Thermal storage technologies allow heat or cold to be stored for periods of time ranging from hours or overnight to interseasonal, and can involve storage of sensible energy (i.e. by changing the temperature of a medium) or latent energy (i.e. through phase changes of a medium, such between water and slush or ice). Short-term thermal storages can be used for peak-shaving in district heating or electrical distribution systems. Kinds of renewable or alternative energy sources that can be enabled include natural energy (e.g. collected via solar-thermal collectors, or dry cooling towers used to collect winter's cold), waste energy (e.g. from HVAC equipment, industrial processes or power plants), or surplus energy (e.g. as seasonally from hydropower projects or intermittently from wind farms). The Drake Landing Solar Community (Alberta, Canada) is illustrative. borehole thermal energy storage allows the community to get 97% of its year-round heat from solar collectors on the garage roofs, which most of the heat collected in summer. Types of storages for sensible energy include insulated tanks, borehole clusters in substrates ranging from gravel to bedrock, deep aquifers, or shallow lined pits that are insulated on top. Some types of storage are capable of storing heat or cold between opposing seasons (particularly if very large), and some storage applications require inclusion of a heat pump. Latent heat is typically stored in ice tanks or what are called phase-change materials (PCMs).

Energy Efficiency

Moving towards energy sustainability will require changes not only in the way energy is supplied, but in the way it is used, and reducing the amount of energy required to deliver various goods or services is essential. Opportunities for improvement on the demand side of the energy equation are as rich and diverse as those on the supply side, and often offer significant economic benefits.

Renewable energy and energy efficiency are sometimes said to be the "twin pillars" of sustainable energy policy. Both resources must be developed in order to stabilize and reduce carbon dioxide emissions. Efficiency slows down energy demand growth so that rising clean energy supplies can make deep cuts in fossil fuel use. If energy use grows too fast, renewable energy development will chase a receding target. A recent historical analysis has demonstrated that the rate of energy efficiency improvements has generally been outpaced by the rate of growth in energy demand, which is due to continuing economic and population growth. As a result, despite energy efficiency gains, total energy use and related carbon emissions have continued to increase. Thus, given the thermodynamic and practical limits of energy efficiency improvements, slowing the growth in energy demand is essential. However, unless clean energy supplies come online rapidly, slowing demand growth will only begin to reduce total emissions; reducing the carbon content of energy sources is also needed. Any serious vision of a sustainable energy economy thus requires commitments to both renewables and efficiency.

Renewable energy (and energy efficiency) are no longer niche sectors that are promoted only by governments and environmentalists. The increased levels of investment and the fact that much of the capital is coming from more conventional financial actors suggest that sustainable energy options are now becoming mainstream. An example of this would be The Alliance to Save Energy's Project with Stahl Consolidated Manufacturing, (Huntsville, Alabama, USA) (StahlCon 7), a patented generator shaft designed to reduce emissions within existing power generating systems, granted publishing rights to the Alliance in 2007.

Climate change concerns coupled with high oil prices and increasing government support are driving increasing rates of investment in the sustainable energy industries, according to a trend analysis from the United Nations Environment Programme. According to UNEP, global investment in sustainable energy in 2007 was higher than previous levels, with $148 billion of new money raised in 2007, an increase of 60% over 2006. Total financial transactions in sustainable energy, including acquisition activity, was $204 billion.

Investment flows in 2007 broadened and diversified, making the overall picture one of greater breadth and depth of sustainable energy use. The mainstream capital markets are "now fully receptive to sustainable energy companies, supported by a surge in funds destined for clean energy investment".

Smart-grid Technology

Smart grid refers to a class of technology people are using to bring utility electricity delivery systems into the 21st century, using computer-based remote control and automation. These systems are made possible by two-way communication technology and computer processing that has been used for decades in other industries. They are beginning to be used on electricity networks, from the power plants and wind farms all the way to the consumers of electricity in homes and businesses. They offer many benefits to utilities and consumers—mostly seen in big improvements in energy efficiency on the electricity grid and in the energy users' homes and offices.

Green Energy and Green Power

A solar trough array is an example of green energy

Public seat with integrated solar panel in Singapore. Anyone can sit and plug in their mobile phone for free charging

Green energy includes natural energetic processes that can be harnessed with little pollution. Green power is electricity generated from renewable energy sources.

Anaerobic digestion, geothermal power, wind power, small-scale hydropower, solar energy, biomass power, tidal power, wave power, and some forms of nuclear power (ones which are able to "burn" nuclear waste through a process known as nuclear transmutation, such as an Integral Fast Reactor, and therefore belong in the "Green Energy" category). Some definitions may also include power derived from the incineration of waste.

Some people, including Greenpeace founder and first member Patrick Moore, George Monbiot, Bill Gates and James Lovelock have specifically classified nuclear power as green energy. Others, including Greenpeace's Phil Radford disagree, claiming that the problems associated with radioactive waste and the risk of nuclear accidents (such as the Chernobyl disaster) pose an unacceptable risk to the environment and to humanity. However, newer nuclear reactor designs are capable of utilizing what is now deemed "nuclear waste" until it is no longer (or dramatically less) dangerous, and have design features that greatly minimize the possibility of a nuclear accident. These designs have yet to be proven.

Some have argued that although green energy is a commendable effort in solving the world's increasing energy consumption, it must be accompanied by a cultural change that encourages the decrease of the world's appetite for energy.

In several countries with common carrier arrangements, electricity retailing arrangements make it possible for consumers to purchase green electricity (renewable electricity) from either their utility or a green power provider.

When energy is purchased from the electricity network, the power reaching the consumer will not necessarily be generated from green energy sources. The local utility company, electric company, or state power pool buys their electricity from electricity producers who may be generating from fossil fuel, nuclear or renewable energy sources. In many countries green energy currently provides a very small amount of electricity, generally contributing less than 2 to 5% to the overall pool. In some U.S. states, local governments have formed regional power purchasing pools using Community Choice Aggregation and Solar Bonds to achieve a 51% renewable mix or higher, such as in the City of San Francisco.

By participating in a green energy program a consumer may be having an effect on the energy sources used and ultimately might be helping to promote and expand the use of green energy. They are also making a statement to policy makers that they are willing to pay a price premium to support renewable energy. Green energy consumers either obligate the utility companies to increase the amount of green energy that they purchase from the pool (so decreasing the amount of non-green energy they purchase), or directly fund the green energy through a green power provider. If insufficient green energy sources are available, the utility must develop new ones or contract with a third party energy supplier to provide green energy, causing more to be built. However, there is no way the consumer can check whether or not the electricity bought is "green" or otherwise.

In some countries such as the Netherlands, electricity companies guarantee to buy an equal amount of 'green power' as is being used by their green power customers. The Dutch government exempts green power from pollution taxes, which means green power is hardly any more expensive than other power.

In the United States, one of the main problems with purchasing green energy through the electrical grid is the current centralized infrastructure that supplies the consumer's electricity. This infrastructure has led to increasingly frequent brown outs and black outs, high CO_2 emissions, higher energy costs, and power quality issues. An additional $450 billion will be invested to expand this fledgling system over the next 20 years to meet increasing demand. In addition, this centralized system is now being further overtaxed with the incorporation of renewable energies such as wind, solar, and geothermal energies. Renewable resources, due to the amount of space they require, are often located in remote areas where there is a lower energy demand. The current infrastructure would make transporting this energy to high demand areas, such as urban centers, highly inefficient and in some cases impossible. In addition, despite the amount of renewable energy produced or the economic viability of such technologies only about 20 percent will be able to be incorporated into the grid. To have a more sustainable energy profile, the United States must move towards implementing changes to the electrical grid that will accommodate a mixed-fuel economy.

However, several initiatives are being proposed to mitigate these distribution problems. First and foremost, the most effective way to reduce USA's CO_2 emissions and slow global warming is through conservation efforts. Opponents of the current US electrical grid have also advocated for decentralizing the grid. This system would increase efficiency by reducing the amount of energy lost in transmission. It would also be economically viable as it would reduce the amount of power lines that will need to be constructed in the future to keep up with demand. Merging heat and power in this system would create added benefits and help to increase its efficiency by up to 80-90%. This is a significant increase from the current fossil fuel plants which only have an efficiency of 34%.

A more recent concept for improving our electrical grid is to beam microwaves from Earth-orbiting satellites or the moon to directly when and where there is demand. The power would be generated from solar energy captured on the lunar surface In this system, the receivers would be "broad, translucent tent-like structures that would receive microwaves and convert them to electricity". NASA said in 2000 that the technology was worth pursuing but it is still too soon to say if the technology will be cost-effective.

The World Wide Fund for Nature and several green electricity labelling organizations created the (now defunct) Eugene Green Energy Standard under which the national green electricity certification schemes could be accredited to ensure that the purchase of green energy leads to the provision of additional new green energy resources.

Local Green Energy Systems

A small Quietrevolution QR5 Gorlov type vertical axis wind turbine in Bristol, England. Measuring 3 m in diameter and 5 m high, it has a nameplate rating of 6.5 kW to the grid.

Those not satisfied with the third-party grid approach to green energy via the power grid can install their own locally based renewable energy system. Renewable energy electrical systems from solar to wind to even local hydro-power in some cases, are some of the many types of renewable energy systems available locally. Additionally, for those interested in heating and cooling their dwelling via renewable energy, geothermal heat pump systems that tap the constant temperature of the earth, which is around 7 to 15 degrees Celsius a few feet underground and increases dramatically at greater depths, are an option over conventional natural gas and petroleum-fueled heat approaches. Also, in geographic locations where the Earth's Crust is especially thin, or near volcanoes (as is the case in Iceland) there exists the potential to generate even more electricity than would be possible at other sites, thanks to a more significant temperature gradient at these locales.

The advantage of this approach in the United States is that many states offer incentives to offset the cost of installation of a renewable energy system. In California, Massachusetts and several other U.S. states, a new approach to community energy supply called Community Choice Aggregation has provided communities with the means to solicit a competitive electricity supplier and use municipal revenue bonds to finance development of local green energy resources. Individuals are usually assured that the electricity they are using is actually produced from a green energy source that they control. Once the system is paid for, the owner of a renewable energy system will be producing their own renewable electricity for essentially no cost and can sell the excess to the local utility at a profit.

Using Green Energy

A 01 KiloWatt Micro Windmill for Domestic Usage

Renewable energy, after its generation, needs to be stored in a medium for use with autonomous devices as well as vehicles. Also, to provide household electricity in remote areas (that is areas which are not connected to the mains electricity grid), energy storage is required for use with renewable energy. Energy generation and consumption systems used in the latter case are usually stand-alone power systems.

Some examples are:

- energy carriers as hydrogen, liquid nitrogen, compressed air, oxyhydrogen, batteries, to power vehicles.

- flywheel energy storage, pumped-storage hydroelectricity is more usable in stationary applications (e.g. to power homes and offices). In household power systems, conversion of energy can also be done to reduce smell. For example, organic matter such as cow dung and spoilable organic matter can be converted to biochar. To eliminate emissions, carbon capture and storage is then used.

Usually however, renewable energy is derived from the mains electricity grid. This means that energy storage is mostly not used, as the mains electricity grid is organised to produce the exact amount of energy being consumed at that particular moment. Energy production on the mains electricity grid is always set up as a combination of (large-scale) renewable energy plants, as well as other power plants as fossil-fuel power plants and nuclear power. This combination however, which is essential for this type of energy supply (as e.g. wind turbines, solar power plants etc.) can only produce when the wind blows and the sun shines. This is also one of the main drawbacks of the system as fossil fuel powerplants are polluting and are a main cause of global warming (nuclear power being an exception). Although fossil fuel power plants too can be made emissionless (through carbon capture and storage), as well as renewable (if the plants are converted to e.g. biomass) the best solution is still to phase out the latter power plants over time. Nuclear power plants too can be more or less eliminated from their problem of nuclear waste through the use of nuclear reprocessing and newer plants as fast breeder and nuclear fusion plants.

Renewable energy power plants do provide a steady flow of energy. For example, hydropower plants, ocean thermal plants, osmotic power plants all provide power at a regulated pace, and are thus available power sources at any given moment (even at night, windstill moments etc.). At present however, the number of steady-flow renewable energy plants alone is still too small to meet energy demands at the times of the day when the irregular producing renewable energy plants cannot produce power.

Besides the greening of fossil fuel and nuclear power plants, another option is the distribution and immediate use of power from solely renewable sources. In this set-up energy storage is again not necessary. For example, TREC has proposed to distribute solar power from the Sahara to Europe. Europe can distribute wind and ocean power to the Sahara and other countries. In this way, power is produced at any given time as at any point of the planet as the sun or the wind is up or ocean waves and currents are stirring. This option however is probably not possible in the short-term, as fossil fuel and nuclear power are still the main sources of energy on the mains electricity net and replacing them will not be possible overnight.

Several large-scale energy storage suggestions for the grid have been done. Worldwide there is over 100 GW of Pumped-storage hydroelectricity. This improves efficiency and decreases energy losses but a conversion to an energy storing mains electricity grid is a very costly solution. Some costs could potentially be reduced by making use of energy storage equipment the consumer buys and not the state. An example is batteries in electric cars that would double as an energy buffer for the electricity grid. However besides the cost, setting-up such a system would still be a very complicated and difficult procedure. Also, energy storage apparatus' as car batteries are also built with materials that pose a threat to the environment (e.g. Lithium). The combined production of batteries for such a large part of the population would still have environmental concerns. Besides car batteries however, other Grid energy storage projects make use of less polluting energy carriers (e.g. compressed air tanks and flywheel energy storage).

Carbon-neutral and Negative Fuels

A carbon-neutral fuel is a synthetic fuel – such as methane, gasoline, diesel fuel or jet fuel – produced from renewable or nuclear energy used to hydrogenate waste carbon dioxide recycled from power plant flue-gas emissions, recovered from automotive exhaust gas, or derived from carbonic acid in seawater. Such fuels are carbon-neutral because they do not result in a net increase in atmospheric greenhouse gases. To the extent that carbon-neutral fuels displace fossil fuels, or if they are produced from waste carbon or seawater carbonic acid, and their combustion is subject to carbon capture at the flue or exhaust pipe, they result in negative carbon dioxide emission and net carbon dioxide removal from the atmosphere, and thus constitute a form of greenhouse gas remediation. Such fuels are produced by the electrolysis of water to make hydrogen used in turn in the Sabatier reaction to produce methane which may then be stored to be burned later in power plants as synthetic natural gas, transported by pipeline, truck, or tanker ship, or be used in gas to liquids processes such as the Fischer–Tropsch process to make traditional transportation or heating fuels.

Green Energy and Labeling By Region

European Union

Directive 2004/8/EC of the European Parliament and of the Council of 11 February 2004 on the promotion of cogeneration based on a useful heat demand in the internal energy market includes the article 5 (*Guarantee of origin of electricity* from high-efficiency cogeneration).

European environmental NGOs have launched an ecolabel for green power. The ecolabel is called EKOenergy. It sets criteria for sustainability, additionality, consumer information and tracking. Only part of electricity produced by renewables fulfills the EKOenergy criteria.

A Green Energy Supply Certification Scheme was launched in the United Kingdom in February 2010. This implements guidelines from the Energy Regulator, Ofgem, and sets requirements on transparency, the matching of sales by renewable energy supplies, and additionality.

United States

The United States Department of Energy (DOE), the Environmental Protection Agency (EPA), and the Center for Resource Solutions (CRS) recognizes the voluntary purchase of electricity from renewable energy sources (also called renewable electricity or green electricity) as green power.

The most popular way to purchase renewable energy as revealed by NREL data is through purchasing Renewable Energy Certificates (RECs). According to a Natural Marketing Institute (NMI) survey 55 percent of American consumers want companies to increase their use of renewable energy.

DOE selected six companies for its 2007 Green Power Supplier Awards, including Constellation NewEnergy; 3Degrees; Sterling Planet; SunEdison; Pacific Power and Rocky Mountain Power; and Silicon Valley Power. The combined green power provided by those six winners equals more than 5 billion kilowatt-hours per year, which is enough to power nearly 465,000 average U.S. households. In 2014, Arcadia Power made RECS available to homes and businesses in all 50 states, allowing consumers to use "100% green power" as defined by the EPA's Green Power Partnership.

The U.S. Environmental Protection Agency (USEPA) Green Power Partnership is a voluntary program that supports the organizational procurement of renewable electricity by offering expert advice, technical support, tools and resources. This can help organizations lower the transaction costs of buying renewable power, reduce carbon footprint, and communicate its leadership to key stakeholders.

Throughout the country, more than half of all U.S. electricity customers now have an option to purchase some type of green power product from a retail electricity provider. Roughly one-quarter of the nation's utilities offer green power programs to customers, and voluntary retail sales of renewable energy in the United States totaled more than 12 billion kilowatt-hours in 2006, a 40% increase over the previous year.

Sustainable Energy Research

There are numerous organizations within the academic, federal, and commercial sectors conducting large scale advanced research in the field of sustainable energy. This research spans several areas of focus across the sustainable energy spectrum. Most of the research is targeted at improving efficiency and increasing overall energy yields. Multiple federally supported research organizations have focused on sustainable energy in recent years. Two of the most prominent of these labs are Sandia National Laboratories and the National Renewable Energy Laboratory (NREL), both of which are funded by the United States Department of Energy and supported by various corporate partners. Sandia has a total budget of $2.4 billion while NREL has a budget of $375 million.

Scientific production towards sustainable energy systems is rising exponentially, growing from about 500 English journal papers only about renewable energy in 1992 to almost 9,000 papers in 2011.

Solar

The primary obstacle that is preventing the large scale implementation of solar powered energy generation is the inefficiency of current solar technology. Currently, photovoltaic (PV) panels only have the ability to convert around 16% of the sunlight that hits them into electricity. At this rate, many experts believe that solar energy is not efficient enough to be economically sustainable given the cost to produce the panels themselves. Both Sandia National Laboratories and the National Renewable Energy Laboratory (NREL), have heavily funded solar research programs. The NREL solar program has a budget of around $75 million and develops research projects in the areas of photovoltaic (PV) technology, solar thermal energy, and solar radiation. The budget for Sandia's solar division is unknown, however it accounts for a significant percentage of the laboratory's $2.4 billion budget. Several academic programs have focused on solar research in recent years. The Solar Energy Research Center (SERC) at University of North Carolina (UNC) has the sole purpose of developing cost effective solar technology. In 2008, researchers at Massachusetts Institute of Technology (MIT) developed a method to store solar energy by using it to produce hydrogen fuel from water. Such research is targeted at addressing the obstacle that solar development faces of storing energy for use during nighttime hours when the sun is not shining. In February 2012, North Carolina-based Semprius Inc., a solar development company backed by German corporation Siemens, announced that they had developed the world's most efficient solar panel. The company claims that the prototype converts 33.9% of the sunlight that hits it to electricity, more than double the previous high-end conversion rate. Major projects on artificial photosynthesis or solar fuels are also under way in many developed nations.

Space-Based Solar Power

Space-Based Solar Power Satellites seek to overcome the problems of storage and provide civilization-scale power that is clean, constant, and global. Japan and China have active national programs aimed at commercial scale Space-Based Solar Power (SBSP), and both nation's hope to orbit demonstrations in the 2030s. The China Academy of Space Technology (CAST) won the 2015 International SunSat Design Competition with this video of their Multi-Rotary Joint design. Proponents of SBSP claim that Space-Based Solar Power would be clean, constant, and global, and could scale to meet all planetary energy demand. A recent multi-agency industry proposal (echoing the 2008 Pentagon recommendation) won the SECDEF/SECSTATE/USAID Director D3 (Diplomacy, Development, Defense) Innovation Challenge with the following pitch and vision video. Northrop Grumman is funding CALTECH with $17.5 million for an ultra lightweight design. Keith Henson recently posted a video of a "bootstrapping" approach.

Wind

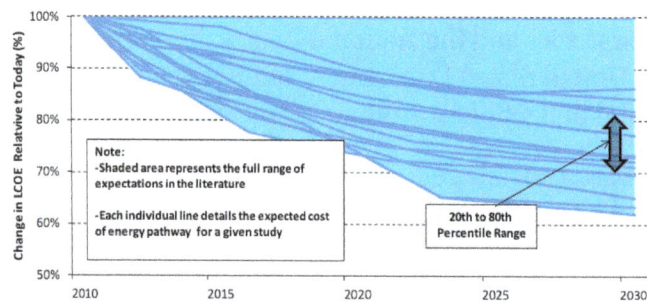

Note:
-Shaded area represents the full range of expectations in the literature

-Each individual line details the expected cost of energy pathway for a given study

20th to 80th Percentile Range

The National Renewable Energy Laboratory projects that the levelized cost of wind power in the U.S. will decline about 25% from 2012 to 2030.

Bangui Wind Farm in the Philippines.

Wind energy research dates back several decades to the 1970s when NASA developed an analytical model to predict wind turbine power generation during high winds. Today, both Sandia National Laboratories and National Renewable Energy Laboratory have programs dedicated to wind research. Sandia's laboratory focuses on the advancement of materials, aerodynamics, and sensors. The NREL wind projects are centered on improving wind plant power production, reducing their capital costs, and making wind energy more cost effective overall. The Field Laboratory for Optimized Wind Energy (FLOWE) at Caltech was established to research renewable approaches to wind energy farming technology practices that have the potential to reduce the cost, size, and environmental impact of wind energy production. The president of Sky WindPower Corporation thinks that wind turbines will be able to produce electricity at a cent/kWh at an average which in comparison to coal-generated electricity is a fractional of the cost.

A wind farm is a group of wind turbines in the same location used to produce electric power. A large wind farm may consist of several hundred individual wind turbines, and cover an extended area of hundreds of square miles, but the land between the turbines may be used for agricultural or other purposes. A wind farm may also be located offshore.

Many of the largest operational onshore wind farms are located in the USA and China. The Gansu Wind Farm in China has over 5,000 MW installed with a goal of 20,000 MW by 2020. China has several other "wind power bases" of similar size. The Alta Wind Energy Center in California is the largest onshore wind farm outside of China, with a capacity of 1020 MW of power. Europe leads in the use of wind power with almost 66 GW, about 66 percent of the total globally, with Denmark in the lead according to the countries installed per-capita capacity. As of February 2012, the Walney Wind Farm in United Kingdom is the largest offshore wind farm in the world at 367 MW, followed by Thanet Wind Farm (300 MW), also in the UK.

There are many large wind farms under construction and these include BARD Offshore 1 (400 MW), Clyde Wind Farm (350 MW), Greater Gabbard wind farm (500 MW), Lincs Wind Farm (270 MW), London Array (1000 MW), Lower Snake River Wind Project (343 MW), Macarthur Wind Farm (420 MW), Shepherds Flat Wind Farm (845 MW), and Sheringham Shoal (317 MW).

Wind power has expanded quickly, its share of worldwide electricity usage at the end of 2014 was 3.1%.

Carbon-neutral and Negative Fuels

Carbon-neutral fuels are synthetic fuels (including methane, gasoline, diesel fuel, jet fuel or ammonia) produced by hydrogenating waste carbon dioxide recycled from power plant flue-gas emissions, recovered from automotive exhaust gas, or derived from carbonic acid in seawater. Commercial fuel synthesis companies suggest they can produce synthetic fuels for less than petroleum fuels when oil costs more than $55 per barrel. Renewable methanol (RM) is a fuel produced from hydrogen and carbon dioxide by catalytic hydrogenation where the hydrogen has been obtained from water electrolysis. It can be blended into transportation fuel or processed as a chemical feedstock.

The George Olah carbon dioxide recycling plant operated by Carbon Recycling International in Grindavík, Iceland has been producing 2 million liters of methanol transportation fuel per year from flue exhaust of the Svartsengi Power Station since 2011. It has the capacity to produce 5 million liters per year. A 250 kilowatt methane synthesis plant was constructed by the Center for Solar Energy and Hydrogen Research (ZSW) at Baden-Württemberg and the Fraunhofer Society in Germany and began operating in 2010. It is being upgraded to 10 megawatts, scheduled for completion in autumn, 2012. Audi has constructed a carbon-neutral liquefied natural gas (LNG) plant in Werlte, Germany. The plant is intended to produce transportation fuel to offset LNG used in their A3 Sportback g-tron automobiles, and can keep 2,800 metric tons of CO_2 out of the environment per year at its initial capacity. Other commercial developments are taking place in Columbia, South Carolina, Camarillo, California, and Darlington, England.

Such fuels are considered carbon-neutral because they do not result in a net increase in atmospheric greenhouse gases. To the extent that synthetic fuels displace fossil fuels, or if they are produced from waste carbon or seawater carbonic acid, and their combustion is subject to carbon capture at the flue or exhaust pipe, they result in negative carbon dioxide emission and net carbon dioxide removal from the atmosphere, and thus constitute a form of greenhouse gas remediation.

Such renewable fuels alleviate the costs and dependency issues of imported fossil fuels without requiring either electrification of the vehicle fleet or conversion to hydrogen or other fuels, enabling continued compatible and affordable vehicles. Carbon-neutral fuels offer relatively low cost energy storage, alleviating the problems of wind and solar intermittency, and they enable distribution of wind, water, and solar power through existing natural gas pipelines.

Nighttime wind power is considered the most economical form of electrical power with which to synthesize fuel, because the load curve for electricity peaks sharply during the warmest hours of the day, but wind tends to blow slightly more at night than during the day, so, the price of nighttime wind power is often much less expensive than any alternative. Germany has built a 250 kilowatt synthetic methane plant which they are scaling up to 10 megawatts.

Biomass

A CHP power station using wood to provide electricity to over 30.000 households in France

Biomass is biological material derived from living, or recently living organisms. It most often refers to plants or plant-derived materials which are specifically called lignocellulosic biomass. As an energy source, biomass can either be used directly via combustion to produce heat, or indirectly after converting it to various forms of biofuel. Conversion of biomass to biofuel can be achieved by different methods which are broadly classified into: *thermal*, *chemical*, and *biochem-*

ical methods. Wood remains the largest biomass energy source today; examples include forest residues – such as dead trees, branches and tree stumps –, yard clippings, wood chips and even municipal solid waste. In the second sense, biomass includes plant or animal matter that can be converted into fibers or other industrial chemicals, including biofuels. Industrial biomass can be grown from numerous types of plants, including miscanthus, switchgrass, hemp, corn, poplar, willow, sorghum, sugarcane, bamboo, and a variety of tree species, ranging from eucalyptus to oil palm (palm oil).

Biomass, biogas and biofuels are burned to produce heat/power and in doing so harm the environment. Pollutants such as sulphurous oxides (SO_x), nitrous oxides (NO_x), and particulate matter (PM) are produced from this combustion; the World Health Organisation estimates that 7 million premature deaths are caused each year by air pollution. Biomass combustion is a major contributor.

Ethanol Biofuels

As the primary source of biofuel in North America, many organizations are conducting research in the area of ethanol production. On the Federal level, the USDA conducts a large amount of research regarding ethanol production in the United States. Much of this research is targeted towards the effect of ethanol production on domestic food markets. The National Renewable Energy Laboratory has conducted various ethanol research projects, mainly in the area of cellulosic ethanol. Cellulosic ethanol has many benefits over traditional corn based-ethanol. It does not take away or directly conflict with the food supply because it is produced from wood, grasses, or non-edible parts of plants. Moreover, some studies have shown cellulosic ethanol to be more cost effective and economically sustainable than corn-based ethanol. Even if we used all the corn crop that we have in the United States and converted it into ethanol it would only produce enough fuel to serve 13 percent of the United States total gasoline consumption.Sandia National Laboratories conducts in-house cellulosic ethanol research and is also a member of the Joint BioEnergy Institute (JBEI), a research institute founded by the United States Department of Energy with the goal of developing cellulosic biofuels.

Other Biofuels

From 1978 to 1996, the National Renewable Energy Laboratory experimented with producing algae fuel in the "Aquatic Species Program." A self-published article by Michael Briggs, at the University of New Hampshire Biofuels Group, offers estimates for the realistic replacement of all motor vehicle fuel with biofuels by utilizing algae that have a natural oil content greater than 50%, which Briggs suggests can be grown on algae ponds at wastewater treatment plants. This oil-rich algae can then be extracted from the system and processed into biofuels, with the dried remainder further reprocessed to create ethanol. The production of algae to harvest oil for biofuels has not yet been undertaken on a commercial scale, but feasibility studies have been conducted to arrive at the above yield estimate. During the biofuel production process algae actually consumes the carbon dioxide in the air and turns it into oxygen through photosynthesis. In addition to its projected high yield, algaculture— unlike food crop-based biofuels — does not entail a decrease in food production, since it requires neither farmland nor fresh water. Many companies are pursuing algae bio-reactors for various purposes, including scaling up biofuels production to commercial levels.

Several groups in various sectors are conducting research on Jatropha curcas, a poisonous shrub-like tree that produces seeds considered by many to be a viable source of biofuels feed-stock oil. Much of this research focuses on improving the overall per acre oil yield of Jatropha through advancements in genetics, soil science, and horticultural practices. SG Biofuels, a San Diego-based Jatropha developer, has used molecular breeding and biotechnology to produce elite hybrid seeds of Jatropha that show significant yield improvements over first generation varieties. The Center for Sustainable Energy Farming (CfSEF) is a Los Angeles-based non-profit research organization dedicated to Jatropha research in the areas of plant science, agronomy, and horticulture. Successful exploration of these disciplines is projected to increase Jatropha farm production yields by 200-300% in the next ten years.

Geothermal

Geothermal energy is produced by tapping into the thermal energy created and stored within the earth. It arises from the radioactive decay of an isotope of potassium and other elements found in the Earth's crust. Geothermal energy can be obtained by drilling into the ground, very similar to oil exploration, and then it is carried by a heat-transfer fluid (e.g. water, brine or steam). Geothermal systems that are mainly dominated by water have the potential to provide greater benefits to the system and will generate more power. Within these liquid-dominated systems, there are possible concerns of subsidence and contamination of ground-water resources. Therefore, protection of ground-water resources is necessary in these systems. This means that careful reservoir production and engineering is necessary in liquid-dominated geothermal reservoir systems. Geothermal energy is considered sustainable because that thermal energy is constantly replenished. However, the science of geothermal energy generation is still young and developing economic viability. Several entities, such as the National Renewable Energy Laboratory and Sandia National Laboratories are conducting research toward the goal of establishing a proven science around geothermal energy. The International Centre for Geothermal Research (IGC), a German geosciences research organization, is largely focused on geothermal energy development research.

Hydrogen

Over $1 billion of federal money has been spent on the research and development of hydrogen and a medium for energy storage in the United States. Both the National Renewable Energy Laboratory and Sandia National Laboratories have departments dedicated to hydrogen research. Hydrogen is useful for energy storage and for use in airplanes, but is not practical for automobile use, as it is not very efficient, compared to using a battery — for the same cost a person can travel three times as far using a battery.

Thorium

There are potentially two sources of nuclear power. Fission is used in all current nuclear power plants. Fusion is the reaction that exists in stars, including the sun, and remains impractical for use on Earth, as fusion reactors are not yet available. However nuclear power is controversial politically and scientifically due to concerns about radioactive waste disposal, safety, the risks of a severe accident, and technical and economical problems in dismantling of old power plants.

Thorium is a fissionable material used in thorium-based nuclear power. The thorium fuel cycle claims several potential advantages over a uranium fuel cycle, including greater abundance, superior physical and nuclear properties, better resistance to nuclear weapons proliferation and reduced plutonium and actinide production. Therefore, it is sometimes referred as sustainable.

Clean Energy Investments

2010 was a record year for green energy investments. According to a report from Bloomberg New Energy Finance, nearly US $243 billion was invested in wind farms, solar power, electric cars, and other alternative technologies worldwide, representing a 30 percent increase from 2009 and nearly five times the money invested in 2004. China had $51.1 billion investment in clean energy projects in 2010, by far the largest figure for any country.

Within emerging economies, Brazil comes second to China in terms of clean energy investments. Supported by strong energy policies, Brazil has one of the world's highest biomass and small-hydro power capacities and is poised for significant growth in wind energy investment. The cumulative investment potential in Brazil from 2010 to 2020 is projected as $67 billion.

India is another rising clean energy leader. While India ranked the 10th in private clean energy investments among G-20 members in 2009, over the next 10 years it is expected to rise to the third position, with annual clean energy investment under current policies forecast to grow by 369 percent between 2010 and 2020.

It is clear that the center of growth has started to shift to the developing economies and they may lead the world in the new wave of clean energy investments.

Around the world many sub-national governments - regions, states and provinces - have aggressively pursued sustainable energy investments. In the United States, California's leadership in renewable energy was recognised by The Climate Group when it awarded former Governor Arnold Schwarzenegger its inaugural award for international climate leadership in Copenhagen in 2009. In Australia, the state of South Australia - under the leadership of former Premier Mike Rann - has led the way with wind power comprising 26% of its electricity generation by the end of 2011, edging out coal fired generation for the first time. South Australia also has had the highest take-up per capita of household solar panels in Australia following the Rann Government's introduction of solar feed-in laws and educative campaign involving the installation of solar photovoltaic installations on the roofs of prominent public buildings, including the parliament, museum, airport and Adelaide Showgrounds pavilion and schools. Rann, Australia's first climate change minister, passed legislation in 2006 setting targets for renewable energy and emissions cuts, the first legislation in Australia to do so.

Also, in the European Union there is a clear trend of promoting policies encouraging investments and financing for sustainable energy in terms of energy efficiency, innovation in energy exploitation and development of renewable resources, with increased consideration of environmental aspects and sustainability.

Sustainable Development

Sustainable development is a process for meeting human development goals while sustaining the ability of natural systems to continue to provide the natural resources and ecosystem services upon which the economy and society depends. While the modern concept of sustainable development is derived most strongly from the 1987 Brundtland Report, it is rooted in earlier ideas about sustainable forest management and twentieth century environmental concerns. As the concept developed, it has shifted to focus more on economic development, social development and environmental protection for future generations.

Wind powers 5 MW wind turbines on a wind farm 28 km off the coast of Belgium.

Sustainable development is the organizing principle for sustaining finite resources necessary to provide for the needs of future generations of life on the planet. It is a process that envisions a desirable future state for human societies in which living conditions and resource-use continue to meet human needs without undermining the "integrity, stability and beauty" of natural biotic systems. It was suggested that "the term 'sustainability' should be viewed as humanity's target goal of human-ecosystem equilibrium (homeostasis), while 'sustainable development' refers to the holistic approach and temporal processes that lead us to the end point of sustainability." (305)

History

Sustainability can be defined as the practice of maintaining processes of productivity indefinitely—natural or human made—by replacing resources used with resources of equal or greater value without degrading or endangering natural biotic systems. Sustainable development ties together concern for the carrying capacity of natural systems with the social, political, and economic challenges faced by humanity. Sustainability science is the study of the concepts of sustainable development and environmental science. There is an additional focus on the present generations' responsibility to regenerate, maintain and improve planetary resources for use by future generations.

The Blue Marble, photographed from Apollo 17 in 1972, quickly became an icon of environmental conservation.

Sustainable development has its roots in ideas about sustainable forest management which were developed in Europe during the seventeenth and eighteenth centuries. In response to a growing awareness of the depletion of timber resources in England, John Evelyn argued that "sowing and planting of trees had to be regarded as a national duty of every landowner, in order to stop the destructive over-exploitation of natural resources" in his 1662 essay *Sylva*. In 1713 Hans Carl von Carlowitz, a senior mining administrator in the service of Elector Frederick Augustus I of Saxony published *Sylvicultura oeconomica*, a 400-page work on forestry. Building upon the ideas of Evelyn and French minister Jean-Baptiste Colbert, von Carlowitz developed the concept of managing forests for sustained yield. His work influenced others, including Alexander von Humboldt and Georg Ludwig Hartig, eventually leading to the development of a science of forestry. This in turn influenced people like Gifford Pinchot, first head of the US Forest Service, whose approach to forest management was driven by the idea of wise use of resources, and Aldo Leopold whose land ethic was influential in the development of the environmental movement in the 1960s.

Following the publication of Rachel Carson's *Silent Spring* in 1962, the developing environmental movement drew attention to the relationship between economic growth and development and environmental degradation. Kenneth E. Boulding in his influential 1966 essay *The Economics of the Coming Spaceship Earth* identified the need for the economic system to fit itself to the ecological system with its limited pools of resources. One of the first uses of the term sustainable in the contemporary sense was by the Club of Rome in 1972 in its classic report on the *Limits to Growth*, written by a group of scientists led by Dennis and Donella Meadows of the Massachusetts Institute of Technology. Describing the desirable "state of global equilibrium", the authors wrote: "We are searching for a model output that represents a world system that is sustainable without sudden and uncontrolled collapse and capable of satisfying the basic material requirements of all of its people."

In 1980 the International Union for the Conservation of Nature published a world conservation strategy that included one of the first references to sustainable development as a global priority and introduced the term "sustainable development". Two years later, the United Nations World Charter for Nature raised five principles of conservation by which human conduct affecting nature is to be guided and judged. In 1987 the United Nations World Commission on Environment and Development released the report *Our Common Future*, commonly called the Brundtland Report. The report included what is now one of the most widely recognised definitions of sustainable development.

> **"** Sustainable development is development that meets the needs of the present without compromising the ability of future generations to meet their own needs. It contains within it two key concepts:
> · The concept of 'needs', in particular, the essential needs of the world's poor, to which overriding priority should be given; and
> · The idea of limitations imposed by the state of technology and social organization on the environment's ability to meet present and future needs.
> — World Commission on Environment and Development, Our Common Future (1987) **"**

Since the Brundtland Report, the concept of sustainable development has developed beyond the initial intergenerational framework to focus more on the goal of "socially inclusive and environmentally sustainable economic growth". In 1992, the UN Conference on Environment and Development published the Earth Charter, which outlines the building of a just, sustainable, and peaceful global society in the 21st century. The action plan Agenda 21 for sustainable development identified information, integration, and participation as key building blocks to help countries achieve development that recognises these interdependent pillars. It emphasises that in sustainable development everyone is a user and provider of information. It stresses the need to change from old sector-centered ways of doing business to new approaches that involve cross-sectoral co-ordination and the integration of environmental and social concerns into all development processes. Furthermore, Agenda 21 emphasises that broad public participation in decision making is a fundamental prerequisite for achieving sustainable development.

Under the principles of the United Nations Charter the Millennium Declaration identified principles and treaties on sustainable development, including economic development, social development and environmental protection. Broadly defined, sustainable development is a systems approach to growth and development and to manage natural, produced, and social capital for the welfare of their own and future generations. The term sustainable development as used by the United Nations incorporates both issues associated with land development and broader issues of human development such as education, public health, and standard of living.

A 2013 study concluded that sustainability reporting should be reframed through the lens of four interconnected domains: ecology, economics, politics and culture.

The Sustainable Development Goals (SDGs)

In September 2015, the United Nations General Assembly formally adopted the "universal, integrated and transformative" 2030 Agenda for Sustainable Development, a set of 17 Sustainable Development Goals (SDGs). The goals are to be implemented and achieved in every country from the year 2016 to 2030.

Dimensions

Sustainable development, or sustainability, has been described in terms of three spheres, dimensions, domains or pillars, i.e. the environment, the economy and society. The three-sphere framework was initially proposed by the economist René Passet in 1979. It has also been worded as "economic, environmental and social" or "ecology, economy and equity." This has been expanded by some authors to include a fourth pillar of culture, institutions or governance.

Environmental

The ecological stability of human settlements is part of the relationship between humans and their natural, social and built environments. Also termed human ecology, this broadens the focus of sustainable development to include the domain of human health. Fundamental human needs such as the availability and quality of air, water, food and shelter are also the ecological foundations for sustainable development; addressing public health risk through investments in ecosystem services can be a powerful and transformative force for sustainable development which, in this sense, extends to all species.

Environmental sustainability concerns the natural environment and how it endures and remains diverse and productive. Since natural resources are derived from the environment, the state of air, water, and the climate are of particular concern. The IPCC Fifth Assessment Report outlines current knowledge about scientific, technical and socio-economic information concerning climate change, and lists options for adaptation and mitigation. Environmental sustainability requires society to design activities to meet human needs while preserving the life support systems of the planet. This, for example, entails using water sustainably, utilizing renewable energy, and sustainable material supplies (e.g. harvesting wood from forests at a rate that maintains the biomass and biodiversity).

An unsustainable situation occurs when natural capital (the sum total of nature's resources) is used up faster than it can be replenished. Sustainability requires that human activity only uses nature's resources at a rate at which they can be replenished naturally. Inherently the concept of sustainable development is intertwined with the concept of carrying capacity. Theoretically, the long-term result of environmental degradation is the inability to sustain human life. Such degradation on a global scale should imply an increase in human death rate until population falls to what the degraded environment can support. If the degradation continues beyond a certain tipping point or critical threshold it would lead to eventual extinction for humanity.

Consumption of non-renewable resources	State of environment	Sustainability
More than nature's ability to replenish	Environmental degradation	Not sustainable
Equal to nature's ability to replenish	Environmental equilibrium	Steady state economy
Less than nature's ability to replenish	Environmental renewal	Environmentally sustainable

Integral elements for a sustainable development are research and innovation activities. A telling example is the European environmental research and innovation policy, which aims at defining and implementing a transformative agenda to greening the economy and the society as a whole so to achieve a truly sustainable development. Research and innovation in Europe is financially supported by the programme Horizon 2020, which is also open to participation worldwide. A promising direction towards sustainable development is to design systems that are flexible and reversible.

Pollution of the public resources is really not a different action, it just is a reverse tragedy of the commons, in that instead of taking something out, something is put into the commons. When the costs of polluting the commons are not calculated into the cost of the items consumed, then it becomes only natural to pollute, as the cost of pollution is external to the cost of the goods produced and the cost of cleaning the waste before it is discharged exceeds the cost of releasing the waste directly into the commons. So, the only way to solve this problem is by protecting the ecology of the commons by making it, through taxes or fines, more costly to release the waste directly into the commons than would be the cost of cleaning the waste before discharge.

So, one can try to appeal to the ethics of the situation by doing the right thing as an individual, but in the absence of any direct consequences, the individual will tend to do what is best for the person and not what is best for the common good of the public. Once again, this issue needs to be addressed. Because, left unaddressed, the development of the commonly owned property will become impossible to achieve in a sustainable way. So, this topic is central to the understanding of creating a sustainable situation from the management of the public resources that are used for personal use.

Agriculture

Sustainable agriculture consists of environment friendly methods of farming that allow the production of crops or livestock without damage to human or natural systems. It involves preventing adverse effects to soil, water, biodiversity, surrounding or downstream resources—as well as to those working or living on the farm or in neighboring areas. The concept of sustainable agriculture extends intergenerationally, passing on a conserved or improved natural resource, biotic, and economic base rather than one which has been depleted or polluted. Elements of sustainable agriculture include permaculture, agroforestry, mixed farming, multiple cropping, and crop rotation.

Numerous sustainability standards and certification systems exist, including organic certification, Rainforest Alliance, Fair Trade, UTZ Certified, Bird Friendly, and the Common Code for the Coffee Community (4C).

Economics

A sewage treatment plant that uses solar energy, located at Santuari de Lluc monastery, Majorca.

It has been suggested that because of rural poverty and overexploitation, environmental resources should be treated as important economic assets, called natural capital. Economic development has traditionally required a growth in the gross domestic product. This model of unlimited personal and GDP growth may be over. Sustainable development may involve improvements in the quality of life for many but may necessitate a decrease in resource consumption. According to ecological economist Malte Faber, ecological economics is defined by its focus on nature, justice, and time. Issues of intergenerational equity, irreversibility of environmental change, uncertainty of long-term outcomes, and sustainable development guide ecological economic analysis and valuation.

As early as the 1970s, the concept of sustainability was used to describe an economy "in equilibrium with basic ecological support systems." Scientists in many fields have highlighted *The Limits to Growth*, and economists have presented alternatives, for example a 'steady-state economy'; to address concerns over the impacts of expanding human development on the planet. In 1987 the economist Edward Barbier published the study *The Concept of Sustainable Economic Development*, where he recognised that goals of environmental conservation and economic development are not conflicting and can be reinforcing each other.

A World Bank study from 1999 concluded that based on the theory of genuine savings, policymakers have many possible interventions to increase sustainability, in macroeconomics or purely environmental. A study from 2001 noted that efficient policies for renewable energy and pollution are compatible with increasing human welfare, eventually reaching a golden-rule steady state. The study, *Interpreting Sustainability in Economic Terms*, found three pillars of sustainable development, interlinkage, intergenerational equity, and dynamic efficiency.

But Gilbert Rist points out that the World Bank has twisted the notion of sustainable development to prove that economic development need not be deterred in the interest of preserving the ecosystem. He writes: "From this angle, 'sustainable development' looks like a cover-up operation. ... The thing that is meant to be sustained is really 'development', not the tolerance capacity of the ecosystem or of human societies."

The World Bank, a leading producer of environmental knowledge, continues to advocate the win-win prospects for economic growth and ecological stability even as its economists express their doubts. Herman Daly, an economist for the Bank from 1988 to 1994, writes:

When authors of *WDR* '92 [the highly influential 1992 *World Development Report* that featured the environment] were drafting the report, they called me asking for examples of "win-win" strategies in my work. What could I say? None exists in that pure form; there are trade-offs, not "win-wins." But they want to see a world of "win-wins" based on articles of faith, not fact. I wanted to contribute because *WDRs* are important in the Bank, [because] task managers read [them] to find philosophical justification for their latest round of projects. But they did not want to hear about how things really are, or what I find in my work..."

A meta review in 2002 looked at environmental and economic valuations and found a lack of "sustainability policies". A study in 2004 asked if we consume too much. A study concluded in 2007 that knowledge, manufactured and human capital (health and education) has not compensated for the degradation of natural capital in many parts of the world. It has been suggested that intergenerational equity can be incorporated into a sustainable development and decision making, as has become common in economic valuations of climate economics. A meta review in 2009 identified conditions for a strong case to act on climate change, and called for more work to fully account of the relevant economics and how it affects human welfare. According to free-market environmentalist John Baden "the improvement of environment quality depends on the market economy and the existence of legitimate and protected property rights." They enable the effective practice of personal responsibility and the development of mechanisms to protect the environment. The State can in this context "create conditions which encourage the people to save the environment."

Misum, Mistra Center for Sustainable Markets, based at Stockholm School of Economics, aims to provide policy research and advice to Swedish and international actors on Sustainable Markets. Misum is a cross-disciplinary and multi-stakeholder knowledge center dedicated to sustainability and sustainable markets and contains three research platforms: Sustainability in Financial Markets (Mistra Financial Systems), Sustainability in Production and Consumption and Sustainable Socio-Economic Development.

Environmental Economics

The total environment includes not just the biosphere of earth, air, and water, but also human interactions with these things, with nature, and what humans have created as their surroundings.

As countries around the world continue to advance economically, they put a strain on the ability of the natural environment to absorb the high level of pollutants that are created as a part of this economic growth. Therefore, solutions need to be found so that the economies of the world can continue to grow, but not at the expense of the public good. In the world of economics the amount of environmental quality must be considered as limited in supply and therefore is treated as a scarce resource. This is a resource to be protected and the only real efficient way to do it in a market economy is to look at the overall situation of pollution from a benefit-cost perspective. It then becomes essentially an allocation of resources, based on an evaluation of the expected course of action and the consequences of this action, when compared to an alternative course of action that might allocate the limited resources in a different way.

Benefit-cost analysis basically can look at several ways of solving a problem and then assigning the best route for a solution, based on the set of consequences that would result from the further development of the individual courses of action, and then choosing the course of action that results in the least amount of damage to the expected outcome for the environmental quality that remains after that development or process takes place. Further complicating this analysis are the interrelationships of the various parts of the environment that might be impacted by the chosen course of action. Sometimes it is almost impossible to predict the various outcomes of a course of action, due to the unexpected consequences and the amount of unknowns that are not accounted for in the benefit-cost analysis.

Energy

Sustainable energy is clean and can be used over a long period of time. Unlike fossil fuels that most countries are using, renewable energy only produces little or even no pollution. The most common types of renewable energy in US are hydroelectric, solar and wind energy. Solar energy is commonly used on public parking meters, street lights and the roof of buildings. Wind power has expanded quickly, its share of worldwide electricity usage at the end of 2014 was 3.1%. Most of California's fossil fuel infrastructures are sited in or near low-income communities, and have traditionally suffered the most from California's fossil fuel energy system. These communities are historically left out during the decision-making process, and often end up with dirty power plants and other dirty energy projects that poison the air and harm the area. These toxicants are major contributors to health problems in the communities. As renewable energy becomes more common, fossil fuel infrastructures are replaced by renewables, providing better social equity to these communities. Overall, and in the long run, sustainable development in the field of energy is also deemed to contribute to economic sustainability and national security of communities, thus being increasingly encouraged through investment policies.

Technology

One of the core concepts in sustainable development is that technology can be used to assist people meet their developmental needs. Technology to meet these sustainable development needs is often referred to as appropriate technology, which is an ideological movement (and its manifestations) originally articulated as intermediate technology by the economist E. F. Schumacher in his influential work, *Small is Beautiful*. and now covers a wide range of technologies. Both Schumacher and many modern-day proponents of appropriate technology also emphasise the technology as people-centered. Today appropriate technology is often developed using open source principles, which have led to open-source appropriate technology (OSAT) and thus many of the plans of the technology can be freely found on the Internet. OSAT has been proposed as a new model of enabling innovation for sustainable development.

Transport

Transportation is a large contributor to greenhouse gas emissions. It is said that one-third of all gasses produced are due to transportation. Motorized transport also releases exhaust fumes that contain particulate matter which is hazardous to human health and a contributor to climate change.

Sustainable transport has many social and economic benefits that can accelerate local sustainable development. According to a series of reports by the Low Emission Development Strategies Global Partnership (LEDS GP), sustainable transport can help create jobs, improve commuter safety through investment in bicycle lanes and pedestrian pathways, make access to employment and social opportunities more affordable and efficient. It also offers a practical opportunity to save people's time and household income as well as government budgets, making investment in sustainable transport a 'win-win' opportunity.

Some western countries are making transportation more sustainable in both long-term and short-term implementations. An example is the modifications in available transportation in Freiburg, Germany. The city has implemented extensive methods of public transportation, cycling, and walking, along with large areas where cars are not allowed.

Since many western countries are highly automobile-orientated areas, the main transit that people use is personal vehicles. About 80% of their travel involves cars. Therefore, California, is one of the highest greenhouse gases emitters in the United States. The federal government has to come up with some plans to reduce the total number of vehicle trips in order to lower greenhouse gases emission. Such as:

- Improve public transit through the provision of larger coverage area in order to provide more mobility and accessibility, new technology to provide a more reliable and responsive public transportation network.

- Encourage walking and biking through the provision of wider pedestrian pathway, bike share station in commercial downtown, locate parking lot far from the shopping center, limit on street parking, slower traffic lane in downtown area.

- Increase the cost of car ownership and gas taxes through increased parking fees and tolls, encouraging people to drive more fuel efficient vehicles. They can produce social equity problem, since lower people usually drive older vehicles with lower fuel efficiency. Government can use the extra revenue collected from taxes and tolls to improve the public transportation and benefit the poor community.

Other states and nations have built efforts to translate knowledge in behavioral economics into evidence-based sustainable transportation policies.

Business

The most broadly accepted criterion for corporate sustainability constitutes a firm's efficient use of natural capital. This eco-efficiency is usually calculated as the economic value added by a firm in relation to its aggregated ecological impact. This idea has been popularised by the World Business Council for Sustainable Development (WBCSD) under the following definition: "Eco-efficiency is achieved by the delivery of competitively priced goods and services that satisfy human needs and bring quality of life, while progressively reducing ecological impacts and resource intensity throughout the life-cycle to a level at least in line with the earth's carrying capacity." (DeSimone and Popoff, 1997: 47)

Similar to the eco-efficiency concept but so far less explored is the second criterion for corporate sustainability. Socio-efficiency describes the relation between a firm's value added and its social impact. Whereas, it can be assumed that most corporate impacts on the environment are negative (apart from rare exceptions such as the planting of trees) this is not true for social impacts. These can be either positive (e.g. corporate giving, creation of employment) or negative (e.g. work accidents, mobbing of employees, human rights abuses). Depending on the type of impact socio-efficiency thus either tries to minimise negative social impacts (i.e. accidents per value added) or maximise positive social impacts (i.e. donations per value added) in relation to the value added.

Both eco-efficiency and socio-efficiency are concerned primarily with increasing economic sustainability. In this process they instrumentalise both natural and social capital aiming to benefit from win-win situations. However, as Dyllick and Hockerts point out the business case alone will not be sufficient to realise sustainable development. They point towards eco-effectiveness, socio-effectiveness, sufficiency, and eco-equity as four criteria that need to be met if sustainable development is to be reached.

CASI Global, New York "CSR & Sustainability together lead to sustainable development. CSR as in corporate social responsibility is not what you do with your profits, but is the way you make profits. This means CSR is a part of every department of the company value chain and not a part of HR / independent department. Sustainability as in effects towards Human resources, Environment and Ecology has to be measured within each department of the company. http://casiglobal.us/

Income

At the present time, sustainable development, along with the solidarity called for in Catholic social teaching, can reduce poverty. While over many thousands of years the 'stronger' (economically or physically) overcame the weaker, nowadays for various reasons - Catholic social teaching, social solidarity, sustainable development – the stronger helps the weaker. This aid may take various forms. 'The Stronger' offers real help rather than striving for the elimination or annihilation of the other. Sustainable development reduces poverty through financial (among other things, a balanced budget), environmental (living conditions), and social (including equality of income) means.

Architecture

In sustainable architecture the recent movements of New Urbanism and New Classical architecture promote a sustainable approach towards construction, that appreciates and develops smart growth, architectural tradition and classical design. This in contrast to modernist and International Style architecture, as well as opposing to solitary housing estates and suburban sprawl, with long commuting distances and large ecological footprints. Both trends started in the 1980s. (It should be noted that sustainable architecture is predominantly relevant to the economics domain while architectural landscaping pertains more to the ecological domain.)

Politics

A study concluded that social indicators and, therefore, sustainable development indicators, are scientific constructs whose principal objective is to inform public policy-making. The International Institute for Sustainable Development has similarly developed a political policy framework, linked to a sustainability index for establishing measurable entities and metrics. The framework consists of six core areas, international trade and investment, economic policy, climate change and energy, measurement and assessment, natural resource management, and the role of communication technologies in sustainable development.

The United Nations Global Compact Cities Programme has defined sustainable political development in a way that broadens the usual definition beyond states and governance. The political is defined as the domain of practices and meanings associated with basic issues of social power as they pertain to the organisation, authorisation, legitimation and regulation of a social life held in common. This definition is in accord with the view that political change is important for responding to economic, ecological and cultural challenges. It also means that the politics of economic change can be addressed. They have listed seven subdomains of the domain of politics:

1. Organization and governance

2. Law and justice

3. Communication and critique

4. Representation and negotiation

5. Security and accord

6. Dialogue and reconciliation

7. Ethics and accountability

This accords with the Brundtland Commission emphasis on development that is guided by human rights principles.

Culture

Working with a different emphasis, some researchers and institutions have pointed out that a fourth dimension should be added to the dimensions of sustainable development, since the triple-bottom-line dimensions of economic, environmental and social do not seem to be enough to reflect the complexity of contemporary society. In this context, the Agenda 21 for culture and the United Cities and Local Governments (UCLG) Executive Bureau lead the preparation of the policy statement "Culture: Fourth Pillar of Sustainable Development", passed on 17 November 2010, in the framework of the World Summit of Local and Regional Leaders – 3rd World Congress of UCLG, held in Mexico City. This document inaugurates a new perspective and points to the relation between culture and sustainable development through a dual approach: developing a solid cultural policy and advocating a cultural dimension in all public policies. The Circles of Sustainability approach distinguishes the four domains of economic, ecological, political and cultural sustainability.

Other organizations have also supported the idea of a fourth domain of sustainable development. The Network of Excellence "Sustainable Development in a Diverse World", sponsored by the European Union, integrates multidisciplinary capacities and interprets cultural diversity as a key element of a new strategy for sustainable development. The Fourth Pillar of Sustainable Development Theory has been referenced by executive director of IMI Institute at UNESCO Vito Di Bari in his manifesto of art and architectural movement Neo-Futurism, whose name was inspired by the 1987 United Nations' report Our Common Future. The Circles of Sustainability approach used by Metropolis defines the (fourth) cultural domain as practices, discourses, and material expressions, which, over time, express continuities and discontinuities of social meaning.

Progress

The United Nations Conference on Sustainable Development (UNCSD; also known as Rio 2012) was the third international conference on sustainable development, which aimed at reconciling the economic and environmental goals of the global community. An outcome of this conference was the development of the Sustainable Development Goals that aim to promote sustainable progress and eliminate inequalities around the world. However, few nations met the World Wide Fund for Nature's definition of sustainable development criteria established in 2006. Although some nations are more developed than others, all nations are constantly developing because each nation struggles with perpetuating disparities, inequalities and unequal access to fundamental rights and freedoms.

Measurement

In 2007 a report for the U.S. Environmental Protection Agency stated: "While much discussion and effort has gone into sustainability indicators, none of the resulting systems clearly tells us whether our society is sustainable. At best, they can tell us that we are heading in the wrong direction, or that our current activities are not sustainable. More often, they simply draw our attention to the existence of problems, doing little to tell us the origin of those problems and nothing to tell us how to solve them." Nevertheless, a majority of authors assume that a set of well defined and harmonised indicators is the only way to make sustainability tangible. Those indicators are expected to be identified and adjusted through empirical observations (trial and error).

The most common critiques are related to issues like data quality, comparability, objective function and the necessary resources. However a more general criticism is coming from the project management community: How can a sustainable development be achieved at global level if we cannot monitor it in any single project?

The Cuban-born researcher and entrepreneur Sonia Bueno suggests an alternative approach that is based upon the integral, long-term cost-benefit relationship as a measure and monitoring tool for the sustainability of every project, activity or enterprise. Furthermore, this concept aims to be a practical guideline towards sustainable development following the principle of conservation and increment of value rather than restricting the consumption of resources.

Reasonable qualifications of sustainability are seen U.S. Green Building Council's (USGBC) Leadership in Energy and Environmental Design (LEED). This design incorporates some ecological, economic, and social elements. The goals presented by LEED design goals are sustainable sites, water efficiency, energy and atmospheric emission reduction, material and resources efficiency, and indoor environmental quality. Although amount of structures for sustainability development is many, these qualification has become a standard for sustainable building.

Recent research efforts created also the SDEWES Index to benchmark the performance of cities across aspects that are related to energy, water and environment systems. The SDEWES Index consists of 7 dimensions, 35 indicators, and close to 20 sub-indicators. It is currently applied to 58 cities.

Natural Capital

The sustainable development debate is based on the assumption that societies need to manage three types of capital (economic, social, and natural), which may be non-substitutable and whose consumption might be irreversible. Leading ecological economist and steady-state theorist Herman Daly, for example, points to the fact that natural capital can not necessarily be substituted by economic capital. While it is possible that we can find ways to replace some natural resources, it is much more unlikely that they will ever be able to replace eco-system services, such as the protection provided by the ozone layer, or the climate stabilizing function of the Amazonian forest. In fact natural capital, social capital and economic capital are often complementarities. A further obstacle to substitutability lies also in the multi-functionality of many natural resources. Forests, for example, not only provide the raw material for paper (which can be substituted quite easily), but they also maintain biodiversity, regulate water flow, and absorb CO_2.

Another problem of natural and social capital deterioration lies in their partial irreversibility. The loss of biodiversity, for example, is often definitive. The same can be true for cultural diversity. For example, with globalisation advancing quickly the number of indigenous languages is dropping at alarming rates. Moreover, the depletion of natural and social capital may have non-linear consequences. Consumption of natural and social capital may have no observable impact until a certain threshold is reached. A lake can, for example, absorb nutrients for a long time while actually increasing its productivity. However, once a certain level of algae is reached lack of oxygen causes the lake's ecosystem to break down suddenly.

Business-as-usual

Before flue-gas desulfurization was installed, the air-polluting emissions from this power plant in New Mexico contained excessive amounts of sulfur dioxide.

If the degradation of natural and social capital has such important consequence the question arises why action is not taken more systematically to alleviate it. Cohen and Winn point to four types of market failure as possible explanations: First, while the benefits of natural or social capital depletion can usually be privatised, the costs are often externalised (i.e. they are borne not by the party responsible but by society in general). Second, natural capital is often undervalued by society since we are not fully aware of the real cost of the depletion of natural capital. Information asymmetry is a third reason—often the link between cause and effect is obscured, making it difficult for actors to make informed choices. Cohen and Winn close with the realization that contrary to economic theory many firms are not perfect optimisers. They postulate that firms often do not optimise resource allocation because they are caught in a "business as usual" mentality.

Education

Higher education in sustainability across education streams including engineering, finance, supply chain and operations is gaining weight-age. Multiple institutes including Wharton, Columbia, CASI Global New York offer certifications in Sustainability. Corporate's prefer employees certified in sustainability. Reference https://www.wharton.upenn.edu/ http://www.columbia.edu/ http://casiglobal.us/

Criticism

It has been argued that since the 1960s, the concept of sustainable development has changed from 'conservation management' to 'economic development', whereby the original meaning of the concept has been stretched somewhat.

In the 1960s, the international community realised that many African countries needed national plans to safeguard wildlife habitats, and that rural areas had to confront the limits imposed by soil, climate and water availability. This was a strategy of conservation management. In the 70s, however, the focus shifted to the broader issues of the provisioning of basic human needs, community participation as well as appropriate technology use throughout the developing countries (and not just in Africa). This was a strategy of economic development, and the strategy was carried even further by the Brundtland Report when the issues went from regional to international in scope and application. In effect, the conservationists were crowded out and superseded by the developers.

But shifting the focus of sustainable development from conservation to development has had the imperceptible effect of stretching the original forest management term of sustainable yield from the use of renewable resources only (like forestry), to now also accounting for the use of non-renewable resources (like minerals). This stretching of the term has been questioned. Thus, environmental economist Kerry Turner has argued that literally, there can be no such thing as overall 'sustainable development' in an industrialised world economy that remains heavily dependent on the extraction of Earth's finite stock of exhaustible mineral resources:

> It makes no sense to talk about the sustainable use of a non-renewable resource (even with substantial recycling effort and use rates). Any positive rate of exploitation will eventually lead to exhaustion of the finite stock.

In effect, it has been argued that the Industrial Revolution as a whole is unsustainable.

Sustainability Metrics and Indices

Sustainable development indicators (SDI) are measures of sustainability, and attempt to quantify beyond the generic concept. Though there are disagreements among those from different disciplines (and influenced by different political beliefs about the nature of the good society), these disciplines and international organizations have each offered measures or indicators of how to measure the concept.

While sustainability indicators, indices and reporting systems gained growing popularity in both the public and private sectors, their effectiveness in influencing actual policy and practices often remains limited.

Metrics and Indices

Various ways of operationalizing or measuring sustainability have been developed. During the last 10 years there has been an expansion of interest in SDI systems, both in industrialized and, albeit to a lesser extent, in developing countries. SDIs are seen as useful in a wide range of settings, by a wide range of actors: international and intergovernmental bodies; national governments and government departments; economic sectors; administrators of geographic or ecological regions; communities; nongovernmental organizations; and the private sector.

SDI processes are underpinned and driven by the increasing need for improved quality and regularly produced information with better spatial and temporal resolution. Accompanying this need is the requirement, brought in part by the information revolution, to better differentiate between information that matters in any given policy context versus information that is of secondary importance or irrelevant.

A large and still growing number of attempts to create aggregate measures of various aspects of sustainability created a stable of indices that provide a more nuanced perspective on development than economic aggregates such as GDP. Some of the most prominent of these include the Human Development Index (HDI) of the United Nations Development Programme (UNDP); the Ecological footprint of Global Footprint Network and its partner organizations; the Environmental Sustainability Index (ESI) and the pilot Environmental Performance Index (EPI) reported under the World Economic Forum (WEF); or the Genuine Progress Index (GPI) calculated at the national or sub-national level. Parallel to these initiatives, political interest in producing a green GDP that would take at least the cost of pollution and natural capital depletion into account has grown, even if implementation is held back by the reluctance of policymakers and statistical services arising mostly from a concern about conceptual and technical challenges.

At the heart of the debate over different indicators are not only different disciplinary approaches but also different views of development. Some indicators reflect the ideology of globalization and urbanization that seek to define and measure progress on whether different countries or cultures agree to accept industrial technologies in their eco-systems. Other approaches, like those that start from international treaties on cultural rights of indigenous peoples to maintain traditional cultures, measure the ability of those cultures to maintain their traditions within their eco-systems at whatever level of productivity they choose.

The Lempert-Nguyen indicator, devised in 2008 for practitioners, starts with the standards for sustainable development that have been agreed upon by the international community and then looks at whether intergovernmental organizations such as the UNDP and other development actors are applying these principles in their projects and work as a whole.

In using sustainability indicators, it is important to distinguish between three types of sustainability that are often mentioned in international development:

- Sustainability of a culture (human system) within its resources and environment;

- Sustainability of a specific stream of benefits or productivity (usually just an economic measure); and

- Sustainability of a particular institution or project without additional assistance (institutionalization of an input).

The following list is not exhaustive but contains the major points of view:

"Daly Rules" Approach

University of Maryland School of Public Policy professor and former Chief Economist for the World Bank Herman E. Daly (working from theory initially developed by Romanian economist Nicholas Georgescu-Roegen and laid out in his 1971 opus "The Entropy Law and the Economic Process") suggests the following three operational rules defining the condition of ecological (thermodynamic) sustainability:

1. Renewable resources such as fish, soil, and groundwater must be used no faster than the rate at which they regenerate.

2. Nonrenewable resources such as minerals and fossil fuels must be used no faster than renewable substitutes for them can be put into place.

3. Pollution and wastes must be emitted no faster than natural systems can absorb them, recycle them, or render them harmless.

Some commentators have argued that the "Daly Rules", being based on ecological theory and the Laws of Thermodynamics, should perhaps be considered implicit or foundational for the many other systems that are advocated, and are thus the most straightforward system for operationalization of the Bruntland Definition. In this view, the Bruntland Definition and the Daly Rules can be seen as complementary—Bruntland provides the ethical goal of non-depletion of natural capital, Daly details parsimoniously how this ethic is operationalized in physical terms. The system is rationally complete, and in agreement with physical laws. Other definitions may thus be superfluous, or mere glosses on the immutable thermodynamic reality.

There are numerous other definitions and systems of operationalization for sustainability, and there has been competition for influence between them, with the unfortunate result that, in the minds of some observers at least, sustainability has no agreed-upon definition.

Natural Step Approach

Following the Brundtland Commission's report, one of the first initiatives to bring scientific principles to the assessment of sustainability was by Swedish cancer scientist Karl-Henrik Robèrt. Robèrt coordinated a consensus process to define and operationalize sustainability. At the core of the process lies a consensus on what Robèrt came to call the natural step framework. The framework is based on a definition of sustainability, described as the system conditions of sustainability (as derived from System theory). In the natural step framework, a sustainable society does not systematically increase concentrations of *substances extracted from the Earth's crust*, or *substances produced by society*; that *does not degrade the environment* and in which people have the *capacity to meet their needs worldwide*.

Ecological Footprint Approach

Ecological footprint accounting, based on the biological concept of carrying capacity, tracks the amount of land and water area a human population, needed to produce the resources the population consumes and to absorb its waste, under prevailing technology. This amount then is compared to available biocapacity, in the world or in that region. The biocapacity represents the area able to regenerate resources and assimilate waste. Global Footprint Network publishes every year results for all nations captured in UN statistics.

The algorithms of ecological footprint accounts have been used in combination with the emergy methodology (S. Zhao, Z. Li and W. Li 2005), and a sustainability index has been derived from the latter. They have also been combined with a measure of quality of life, for instance through the "Happy Planet Index" (HPI) calculated for 178 nations (Marks et al., 2006). The Happy Planet Index calculates how many happy life years each country is able to generate per global hectare of ecological footprint.

One of the striking conclusions to emerge from ecological footprint accounting is that it would be necessary to have 4 or 5 back-up planets engaged in nothing but agriculture for all those alive today to live a western lifestyle. The Footprint analysis is closely related to the I = PAT equation that, itself, can be considered a metric.

Anthropological-cultural Approach

Though sustainable development has become a concept that biologists and ecologists have measured from an eco-system point of view and that the business community has measured from a perspective of energy and resource efficiencies and consumption, the discipline of anthropology is itself founded on the concept of sustainability of human groups within ecological systems. At the basis of the definition of culture is whether a human group is able to transmit its values and continue several aspects of that lifestyle for at least three generations. The measurement of culture, by anthropologists, is itself a measure of sustainability and it is also one that has been codified by international agreements and treaties like the Rio Declaration of 1992 and the United Nations Declaration on the Rights of Indigenous Peoples to maintain a cultural group's choice of lifestyles within their lands and ecosystems.

Terralingua, an organization of anthropologists and linguists working to protect biocultural diversity, with a focus on language, has devised a sert of measures with UNESCO for measuring the survivability of languages and cultures in given eco-systems.

The Lempert–Nguyen indicator of sustainable development, developed in 2008 by David Lempert and Hue Nhu Nguyen, is one that incorporates and integrates these cultural principles with international law.

Circles of Sustainability Approach

A number of agencies including the UN Global Compact Cities Programme, World Vision and Metropolis have since 2010 begun using the Circles of Sustainability approach that sets up a four-domain framework for choosing appropriate indicators. Rather than designating the indicators that have to be used like most other approaches, it provides a framework to guide decision-making on what indicators are most useful. The framework is arranged around four domains - economics, ecology, politics and culture - which are then subdivided into seven analytically derived sub-domains for each domain. Indicators are linked to each sub-domain. By choosing culture as one of its key domains, the approach takes into account the emphasis of the 'Anthropological' approach (above), but retains a comprehensive sense of sustainability. The approach can be used to map any other sustainability indicator set. This is foundationally different from the Global Reporting Initiative Index (below) which uses a triple-bottom-line organizing framework, and is most relevant to corporate reporting.

Global Reporting Initiative Index

In 1997 the Global Reporting Initiative (GRI) was started as a multi-stakeholder process and independent institution whose mission has been "to develop and disseminate globally applicable Sustainability Reporting Guidelines". The GRI uses ecological footprint analysis and became independent in 2002. It is an official collaborating centre of the United Nations Environment Programme (UNEP) and during the tenure of Kofi Annan, it cooperated with the UN Secretary-General's Global Compact.

Energy, Emergy and Sustainability Index

In 1956 Dr. Howard T. Odum of the University of Florida coined the term Emergy and devised the accounting system of embodied energy.

In 1997, systems ecologists M.T. Brown and S. Ulgiati published their formulation of a quantitative Sustainability Index (SI) as a ratio of the emergy (spelled with an "m", i.e. "embodied energy", not simply "energy") yield ratio (EYR) to the environmental loading ratio (ELR). Brown and Ulgiati also called the sustainability index the "Emergy Sustainability Index" (ESI), "an index that accounts for yield, renewability, and environmental load. It is the incremental emergy yield compared to the environmental load".

Sustainability Index = Emergy Yield Ratio/Environmental Loading Ratio = EYR/ELR

- NOTE: The numerator is called "emergy" and is spelled with an "m". It is an abbreviation of the term, "embodied energy". The numerator is NOT "energy yield ratio", which is a different concept.

Writers like Leone (2005) and Yi et al. have also recently suggested that the emergy sustainability index has significant utility. In particular, Leone notes that while the GRI measures behavior, it fails to calculate supply constraints the emergy methodology aims to calculate.

Environmental Sustainability Index

In 2004, a joint initiative of the Yale Center for Environmental Law and Policy (YCELP) and the Center for International Earth Science Information Network (CIESIN) of Columbia University, in collaboration with the World Economic Forum and the Directorate-General Joint Research Centre (European Commission) also attempted to construct an Environmental Sustainability Index (ESI). This was formally released in Davos, Switzerland, at the annual meeting of the World Economic Forum (WEF) on 28 January 2005. The report on this index made a comparison of the WEF ESI to other sustainability indicators such as the Ecological footprint Index. However, there was no mention of the emergy sustainability index.

IISD Sample Policy Framework

In 1996 the International Institute for Sustainable Development (IISD) developed a *Sample Policy Framework*, which proposed that a sustainability index "...would give decision-makers tools to rate policies and programs against each other" (1996, p. 9). Ravi Jain (2005) argued that, "The ability to analyze different alternatives or to assess progress towards sustainability will then depend on establishing measurable entities or metrics used for sustainability."

Sustainability Dashboard

The International Institute for Sustainable Development has produced a "Dashboard of Sustainability", "a free, non-commercial software package that illustrates the complex relationships among economic, social and environmental issues". This is based on Sustainable Development Indicators Prepared for the United Nations Division for Sustainable Development (UN-DSD)DECEMBER 2005.

WBCSD Approach

The World Business Council for Sustainable Development (WBCSD), founded in 1995, has formulated the business case for sustainable development and argues that "sustainable development is good for business and business is good for sustainable development". This view is also maintained by proponents of the concept of industrial ecology. The theory of industrial ecology declares that industry should be viewed as a series of interlocking man-made ecosystems interfacing with the natural global ecosystem.

According to some economists, it is possible for the concepts of sustainable development and competitiveness to merge if enacted wisely, so that there is not an inevitable trade-off. This merger is motivated by the following six observations (Hargroves & Smith 2005):

1. Throughout the economy there are widespread untapped potential resource productivity improvements to be made to be coupled with effective design.

2. There has been a significant shift in understanding over the last three decades of what creates lasting competitiveness of a firm.

3. There is now a critical mass of enabling technologies in eco-innovations that make integrated approaches to sustainable development economically viable.

4. Since many of the costs of what economists call 'environmental externalities' are passed on to governments, in the long-term sustainable development strategies can provide multiple benefits to the tax payer.

5. There is a growing understanding of the multiple benefits of valuing social and natural capital, for both moral and economic reasons, and including them in measures of national well-being.

6. There is mounting evidence to show that a transition to a sustainable economy, if done wisely, may not harm economic growth significantly, in fact it could even help it. Recent research by ex-Wuppertal Institute member Joachim Spangenberg, working with neo-classical economists, shows that the transition, if focused on improving resource productivity, leads to higher economic growth than business as usual, while at the same time reducing pressures on the environment and enhancing employment.

Life-cycle Assessment

Life-cycle assessment is a "composite measure of sustainability." It analyses the environmental performance of products and services through all phases of their life cycle: extracting and processing raw materials; manufacturing, transportation and distribution; use, re-use, maintenance; recycling, and final disposal.

Sustainable Enterprise Approach

Building on the work of the World Business Council for Sustainable Development, businesses began to see the needs of environmental and social systems as opportunities for business development and contribution to stakeholder value. This approach has manifested itself in three key areas of strategic intent: 'sustainable innovation', human development, and 'bottom of the pyramid' business strategies. Now, as businesses have begun the shift toward sustainable enterprise, many business schools are leading the research and education of the next generation of business leaders.

Companies have introduced key development indicators to set targets and track progress on sustainable development. Some key players are:

- Center for Sustainable Global Enterprise, Cornell University

- Center for Sustainable Enterprise, Stuart School of Business, Illinois Institute of Technology

- Erb Institute, Ross School of Business, University of Michigan

- William Davidson Institute, Ross School of Business, University of Michigan

- Center for Sustainable Enterprise, University of North Carolina, Chapel-Hill

- Community Enterprise System, NABARD–XIMB Sustainability Trust, Center for Case Research, Xavier Institute of Management, Bhubaneswar

Sustainable Livelihoods Approach

Another application of the term sustainability has been in the Sustainable Livelihoods Approach, developed from conceptual work by Amartya Sen, and the UK's Institute for Development Studies http://www.ids.ac.uk. This was championed by the UK's Department for International Development(DFID), UNDP, Food and Agriculture Organization (FAO) as well as NGOs such as CARE, OXFAM and the African Institute for Community-Driven Development, Khanya-aicdd http://www.khanya-aicdd.org. Key concepts include the Sustainable Livelihoods (SL) Framework, a holistic way of understanding livelihoods, the SL principles, as well as six governance issues developed by Khanya-aicdd. A wide range of information resources on Sustainable Livelihoods Approaches can be found at Livelihoods Connect http://www.livelihoods.org

Some analysts view this measure with caution because they believe that it has a tendency to take one part of the footprint analysis and I = PAT equation (productivity) and to focus on the sustainability of economic returns to an economic sector rather than on the sustainability of the entire population or culture.

FAO Types of Sustainability

The United Nations Food and Agriculture Organisation (FAO) has identified considerations for technical cooperation that affect three types of sustainability:

- Institutional sustainability. Can a strengthened institutional structure continue to deliver the results of technical cooperation to end users? The results may not be sustainable if, for example, the planning authority that depends on the technical cooperation loses access to top management, or is not provided with adequate resources after the technical cooperation ends. Institutional sustainability can also be linked to the concept of social sustainability, which asks how the interventions can be sustained by social structures and institutions;

- Economic and financial sustainability. Can the results of technical cooperation continue to yield an economic benefit after the technical cooperation is withdrawn? For example, the benefits from the introduction of new crops may not be sustained if the constraints to marketing the crops are not resolved. Similarly, economic, as distinct from financial, sustainability may be at risk if the end users continue to depend on heavily subsidized activities and inputs.

- Ecological sustainability. Are the benefits to be generated by the technical cooperation likely to lead to a deterioration in the physical environment, thus indirectly contributing to a fall in production, or well-being of the groups targeted and their society?

Some ecologists have emphasised a fourth type of sustainability:

- Energetic sustainability. This type of sustainability is often concerned with the production of energy and mineral resources. Some researchers have pointed to trends they say document the limits of production.

"Development Sustainability" Approaches

Sustainability is relevant to international development projects. One definition of development sustainability is "the continuation of benefits after major assistance from the donor has been completed" (Australian Agency for International Development 2000). Ensuring that development projects are sustainable can reduce the likelihood of them collapsing after they have just finished; it also reduces the financial cost of development projects and the subsequent social problems, such as dependence of the stakeholders on external donors and their resources. All development assistance, apart from temporary emergency and humanitarian relief efforts, should be designed and implemented with the aim of achieving sustainable benefits. There are ten key factors that influence development sustainability.

1. Participation and ownership. Get the stakeholders (men and women) to genuinely participate in design and implementation. Build on their initiatives and demands. Get them to monitor the project and periodically evaluate it for results.

2. Capacity building and training. Training stakeholders to take over should begin from the start of any project and continue throughout. The right approach should both motivate and transfer skills to people.

3. Government policies. Development projects should be aligned with local government policies.

4. Financial. In some countries and sectors, financial sustainability is difficult in the medium term. Training in local fundraising is a possibility, as is identifying links with the private sector, charging for use, and encouraging policy reforms.

5. Management and organization. Activities that integrate with or add to local structures may have better prospects for sustainability than those that establish new or parallel structures.

6. Social, gender and culture. The introduction of new ideas, technologies and skills requires an understanding of local decision-making systems, gender divisions and cultural preferences.

7. Technology. All outside equipment must be selected with careful consideration given to the local finance available for maintenance and replacement. Cultural acceptability and the local capacity to maintain equipment and buy spare parts are vital.

8. Environment. Poor rural communities that depend on natural resources should be involved in identifying and managing environmental risks. Urban communities should identify and manage waste disposal and pollution risks.

9. External political and economic factors. In a weak economy, projects should not be too complicated, ambitious or expensive.

10. Realistic duration. A short project may be inadequate for solving entrenched problems in a sustainable way, particularly when behavioural and institutional changes are intended. A long project, may on the other hand, promote dependence.

The definition of sustainability as "the continuation of benefits after major assistance from the donor has been completed" (Australian Agency for International Development 2000) is echoed by other definitions (World Bank, USAID). The concept has however evolved as it has become of interest to non grant-making institutions. Sustainability in development refers to *processes* and *relative* increases in local capacity and performance while foreign assistance decreases or shifts (not necessarily disappears). The objective of sustainable development is open to various interpretations.

Sustainability and Environmental Management

At the global scale and in the broadest sense sustainability and environmental management involves managing the oceans, freshwater systems, land and atmosphere, according to sustainability principles.

Land use change is fundamental to the operations of the biosphere because alterations in the relative proportions of land dedicated to urbanisation, agriculture, forest, woodland, grassland and pasture have a marked effect on the global water, carbon and nitrogen biogeochemical cycles. Management of the Earth's atmosphere involves assessment of all aspects of the carbon cycle to identify opportunities to address human-induced climate change and this has become a major focus of scientific research because of the potential catastrophic effects on biodiversity and human communities. Ocean circulation patterns have a strong influence on climate and weather and, in turn, the food supply of both humans and other organisms.

Atmosphere

In March 2009 at a meeting of the Copenhagen Climate Council 2,500 climate experts from 80 countries issued a keynote statement that there is now "no excuse" for failing to act on global warming and that without strong carbon reduction targets "abrupt or irreversible" shifts in climate may occur that "will be very difficult for contemporary societies to cope with". Management of the global atmosphere now involves assessment of all aspects of the carbon cycle to identify opportunities to address human-induced climate change and this has become a major focus of scientific research because of the potential catastrophic effects on biodiversity and human communities.

Other human impacts on the atmosphere include the air pollution in cities, the pollutants including toxic chemicals like nitrogen oxides, sulphur oxides, volatile organic compounds and airborne particulate matter that produce photochemical smog and acid rain, and the chlorofluorocarbons that degrade the ozone layer. Anthropogenic particulates such as sulfate aerosols in the atmosphere reduce the direct irradiance and reflectance (albedo) of the Earth's surface. Known as global dimming the decrease is estimated to have been about 4% between 1960 and 1990 although the trend has subsequently reversed. Global dimming may have disturbed the global water cycle by reducing evaporation and rainfall in some areas. It also creates a cooling effect and this may have partially masked the effect of greenhouse gases on global warming.

Oceans

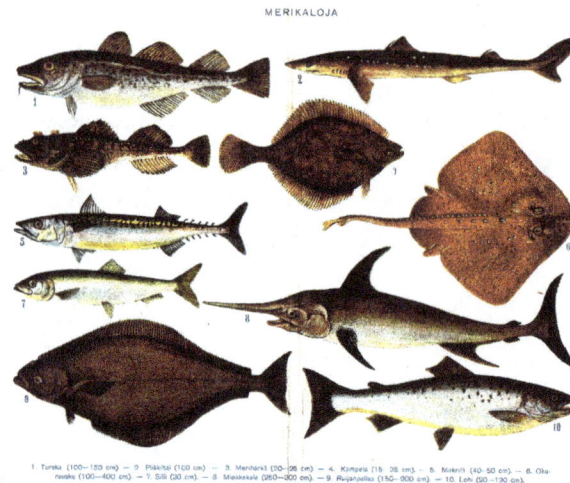

A selection of the world's saltwater fish

Ocean circulation patterns have a strong influence on climate and weather and, in turn, the food supply of both humans and other organisms. Scientists have warned of the possibility, under the influence of climate change, of a sudden alteration in circulation patterns of ocean currents that could drastically alter the climate in some regions of the globe. Major human environmental impacts occur in the more habitable regions of the ocean fringes – the estuaries, coastline and bays. Ten per cent of the world's population – about 600 million people – live in low-lying areas vulnerable to sea level rise. Trends of concern that require management include: over-fishing (beyond sustainable levels); coral bleaching due to ocean warming and ocean acidification due to increasing levels of dissolved carbon dioxide; and sea level rise due to climate change. Because of their vastness oceans also act as a convenient dumping ground for human waste. Remedial strategies include: more careful waste management, statutory control of overfishing by adoption of sustainable fishing practices and the use of environmentally sensitive and sustainable aquaculture and fish farming, reduction of fossil fuel emissions and restoration of coastal and other marine habitat.

Freshwater

Water covers 71% of the Earth's surface. Of this, 97.5% is the salty water of the oceans and only 2.5% freshwater, most of which is locked up in the Antarctic ice sheet. The remaining freshwater is found in lakes, rivers, wetlands, the soil, aquifers and atmosphere. All life depends on the solar-powered global water cycle, the evaporation from oceans and land to form water vapour that later condenses from clouds as rain, which then becomes the renewable part of the freshwater supply. Awareness of the global importance of preserving water for ecosystem services has only recently emerged as, during the 20th century, more than half the world's wetlands have been lost along with their valuable environmental services. Biodiversity-rich freshwater ecosystems are currently declining faster than marine or land ecosystems making them the world's most vulnerable habitats. Increasing urbanization pollutes clean water supplies and much of the world still does not have access to clean, safe water. In the industrial world demand management has slowed absolute usage rates but increasingly water is being transported over vast distances from water-rich natural areas to population-dense urban areas and energy-hungry desalination is becoming more widely used. Greater emphasis is now being placed on the improved management of blue (harvestable) and green (soil water available for plant use) water, and this applies at all scales of water management.

Land

Loss of biodiversity stems largely from the habitat loss and fragmentation produced by the human appropriation of land for development, forestry and agriculture as natural capital is progressively converted to man-made capital. Land use change is fundamental to the operations of the biosphere because alterations in the relative proportions of land dedicated to urbanisation, agriculture, forest, woodland, grassland and pasture have a marked effect on the global water, carbon and nitrogen biogeochemical cycles and this can impact negatively on both natural and human systems. At the local human scale major sustainability benefits accrue from the pursuit of green cities and sustainable parks and gardens.

Forests

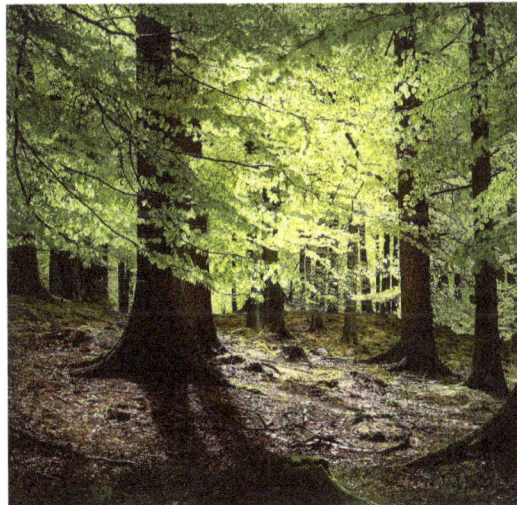

Beech forest – Grib Skov, Denmark

Since the Neolithic Revolution, human use has reduced the world's forest cover by about 47%. Present-day forests occupy about a quarter of the world's ice-free land with about half of these occurring in the tropics In temperate and boreal regions forest area is gradually increasing (with the exception of Siberia), but deforestation in the tropics is of major concern.

Forests moderate the local climate and the global water cycle through their light reflectance (albedo) and evapotranspiration. They also conserve biodiversity, protect water quality, preserve soil and soil quality, provide fuel and pharmaceuticals, and purify the air. These free ecosystem services are not given a market value under most current economic systems, and so forest conservation has little appeal when compared with the economic benefits of logging and clearance which, through soil degradation and organic decomposition returns carbon dioxide to the atmosphere. The United Nations Food and Agriculture Organization (FAO) estimates that about 90% of the carbon stored in land vegetation is locked up in trees and that they sequester about 50% more carbon than is present in the atmosphere. Changes in land use currently contribute about 20% of total global carbon emissions (heavily logged Indonesia and Brazil are a major source of emissions). Climate change can be mitigated by sequestering carbon in reafforestation schemes, plantations and timber products. Also wood biomass can be utilized as a renewable carbon-neutral fuel. The FAO has suggested that, over the period 2005–2050, effective use of tree planting could absorb about 10–20% of man-made emissions – so monitoring the condition of the world's forests must be part of a global strategy to mitigate emissions and protect ecosystem services. However, climate change may pre-empt this FAO scenario as a study by the International Union of Forest Research Organizations in 2009 concluded that the stress of a 2.5C (4.5F) temperature rise above pre-industrial levels could result in the release of vast amounts of carbon so the potential of forests to act as carbon "sinks" is "at risk of being lost entirely".

Cultivated Land

A rice paddy. Rice, wheat, corn and potatoes make up more than half the world's food supply

Feeding more than six billion human bodies takes a heavy toll on the Earth's resources. This begins with the appropriation of about 38% of the Earth's land surface and about 20% of its net primary productivity. Added to this are the resource-hungry activities of industrial agribusiness – everything from the crop need for irrigation water, synthetic fertilizers and pesticides to the resource costs of food packaging, transport (now a major part of global trade) and retail. Food is essential to life. But the list of environmental costs of food production is a long one: topsoil depletion, erosion and conversion to desert from constant tillage of annual crops; overgrazing; salinization; sodification; waterlogging; high levels of fossil fuel use; reliance on inorganic fertilisers and synthetic organic pesticides; reductions in genetic diversity by the mass use of monocultures; water resource depletion; pollution of waterbodies by run-off and groundwater contamination; social problems including the decline of family farms and weakening of rural communities.

All of these environmental problems associated with industrial agriculture and agribusiness are now being addressed through such movements as sustainable agriculture, organic farming and more sustainable business practices.

Extinctions

The extinct dodo (*Raphus cucullatus*)

Although biodiversity loss can be monitored simply as loss of species, effective conservation demands the protection of species within their natural habitats and ecosystems. Following human migration and population growth, species extinctions have progressively increased to a rate unprecedented since the Cretaceous–Paleogene extinction event. Known as the Holocene extinction event this current human-induced extinction of species ranks as one of the worlds six mass extinction events. Some scientific estimates indicate that up to half of presently existing species may become extinct by 2100. Current extinction rates are 100 to 1000 times their prehuman levels with more than 10% birds and mammals threatened, about 8% of plants, 5% of fish and more than 20% of freshwater species.

The 2008 IUCN Red List warns that long-term droughts and extreme weather put additional stress on key habitats and, for example, lists 1,226 bird species as threatened with extinction, which is one-in-eight of all bird species. The Red List Index also identifies 44 tree species in Central Asia as under threat of extinction due to over-exploitation and human development and threatening the region's forests which are home to more than 300 wild ancestors of modern domesticated fruit and nut cultivars.

Biological Invasions

Kudzu (*Pueraria lobata*) infesting trees in Atlanta, Georgia, USA

In many parts of the industrial world land clearing for agriculture has diminished and here the greatest threat to biodiversity, after climate change, has become the destructive effect of invasive species. Increasingly efficient global transport has facilitated the spread of organisms across the planet. The potential danger of this aspect of globalization is starkly illustrated through the spread of human diseases like HIV AIDS, mad cow disease, bird flu and swine flu, but invasive plants and animals are also having a devastating impact on native biodiversity. Non-indigenous organisms can quickly occupy disturbed land and natural areas where, in the absence of their natural predators, they are able to thrive. At the global scale this issue is being addressed through the Global Invasive Species Information Network but there is improved international biosecurity legislation to minimise the transmission of pathogens and invasive organisms. Also, through CITES legislation there is control the trade in rare and threatened species. Increasingly at the local level public awareness programs are alerting communities, gardeners, the nursery industry, collectors, and the pet and aquarium industries, to the harmful effects of potentially invasive species.

Resistance to Change

The environmental sustainability problem has proven difficult to solve. The modern environmental movement has attempted to solve the problem in a large variety of ways. But little progress has been made, as shown by severe ecological footprint overshoot and lack of sufficient progress on the climate change problem. Something within the human system in preventing change to a sustainable mode of behavior. That system trait is systemic change resistance. Change resistance is also known as organizational resistance, barriers to change, or policy resistance.

Sustainability and Systemic Change Resistance

The environmental sustainability problem has proven difficult to solve. The modern environmental movement has attempted to solve the problem in a large variety of ways. But little progress has been made, as shown by severe ecological footprint overshoot and lack of sufficient progress on the climate change problem. Something within the human system is preventing change to a sustainable mode of behavior. That system trait is systemic change resistance. Change resistance is also known as organizational resistance, barriers to change, or policy resistance.

Overview of Resistance to Solving the Sustainability Problem

While environmentalism had long been a minor force in political change, the movement strengthened significantly in the 1970s with the first Earth Day in 1970, in which over 20 million people participated, with publication of *The Limits to Growth* in 1972, and with the first United Nations Conference on the Human Environment in Stockholm in 1972. Early expectations the problem could be solved ran high. 114 out of 132 members of the United Nations attended the Stockholm conference. The conference was widely seen at the time as a harbinger of success:

> "Many believe the most important result of the conference was the precedent it set for international cooperation in addressing environmental degradation. The nations attending agreed they shared responsibility for the quality of the environment, particularly the oceans and the atmosphere, and they signed a declaration of principles, after extensive negotiations, concerning their obligations. The conference also approved an environmental fund and an 'action program,' which involved 200 specific recommendations for addressing such problems as global climate change, marine pollution, population growth, the dumping of toxic wastes, and the preservation of biodiversity. A permanent environment unit was established for coordinating these and other international efforts. [This later became] the United Nations Environmental Program [which was] was formally approved by the General Assembly later that same year and its base established in Nairobi, Kenya. This

organization not only coordinated action but monitored research, collecting and disseminating information, and it has played an ongoing role in international negotiations about environmental issues.

"The conference in Stockholm accomplished almost everything the preparatory committed had planned. It was widely considered successful, and many observers were almost euphoric about the extent of agreement."

However, despite the work of a worldwide environmental movement, many national environmental protection agencies, creation of the United Nations Environment Programme, and many international environmental treaties, the sustainability problem continues to grow worse. The latest ecological footprint data shows the world's footprint increased from about 50% *undershoot* in 1961 to 50% *overshoot* in 2007, the last year data is available.

In 1972 the first edition of *The Limits to Growth* analyzed the environmental sustainability problem using a system dynamics model. The widely influential book predicted that:

"If the present trends in world population, industrialization, pollution, food production, and resource depletion continue unchanged, the limits to growth on this planet will be reached sometime within the next one hundred years. The most probable result will be a rather sudden and uncontrollable decline in both population and industrial capacity some time in the 21st century."

Yet thirty-two years later in 2004 the third edition reported that:

"[The second edition of Limits to Growth] was published in 1992, the year of the global summit on environment and development in Rio de Janeiro. The advent of the summit seemed to prove that global society had decided to deal seriously with the important environmental problems. But we now know that humanity failed to achieve the goals of Rio. The Rio plus 10 conference in Johannesburg in 2002 produced even less; it was almost paralyzed by a variety of ideological and economic disputes, [due to] the efforts of those pursuing their narrow national, corporate, or individual self-interests.

"...humanity has largely squandered the past 30 years."

Change resistance runs so high that the world's top two greenhouse gas emitters, China and the United States, have never adopted the Kyoto Protocol treaty. In the US resistance was so strong that in 1999 the US Senate voted 95 to zero against the treaty by passing the Byrd–Hagel Resolution, despite the fact Al Gore was vice-president at the time. Not a single senator could be persuaded to support the treaty, which has not been brought back to the floor since.

Due to prolonged change resistance, the climate change problem has escalated to the climate change crisis. Greenhouse gas emissions are rising much faster than IPCC models expected: "The growth rate of [fossil fuel] emissions was 3.5% per year for 2000-2007, an almost four fold increase from 0.9% per year in 1990-1999. ... This makes current trends in emissions higher than the worst case IPCC-SRES scenario."

The Copenhagen Climate Summit of December 2009 ended in failure. No agreement on binding targets was reached. The Cancun Climate Summit in December 2010 did not break the deadlock.

The best it could do was another non-binding agreement:

> "Recognizing that climate change represents an urgent and potentially irreversible threat to human societies and the planet, and thus requires to be urgently addressed by all Parties."

This indicates no progress at all since 1992, when the United Nations Framework Convention on Climate Change was created at the Earth Summit in Rio de Janeiro. The 2010 Cancun agreement was the functional equivalent of what the 1992 agreement said:

> "The Parties to this Convention... [acknowledge] that the global nature of climate change calls for the widest possible cooperation by all countries and their participation in an effective and appropriate international response.... [thus the parties recognize] that States should enact effective environmental legislation... [to] protect the climate system for the benefit of present and future generations of humankind...."

Negotiations have bogged down so pervasively that: "Climate policy is gridlocked, and there's virtually no chance of a breakthrough." "Climate policy, as it has been understood and practised by many governments of the world under the Kyoto Protocol approach, has failed to produce any discernible real world reductions in emissions of greenhouse gases in fifteen years."

These events suggest that change resistance to solving the sustainability problem is so high the problem is currently unsolvable.

The Change Resistance and Proper Coupling Subproblems

Understanding change resistance requires seeing it as a distinct and separate part of the sustainability problem. Tanya Markvart's 2009 thesis on *Understanding Institutional Change and Resistance to Change Towards Sustainability* stated that:

> "It has also been demonstrated that ecologically destructive and inequitable institutional systems can be highly resilient and resistant to change, even in the face of social-ecological degradation and/or collapse (e.g., Berkes & Folke, 2002; Allison & Hobbs, 2004; Brown, 2005; Runnalls, 2008; Finley, 2009; Walker et al., 2009)."

The thesis focuses specifically on developing "an interdisciplinary theoretical framework for understanding institutional change and resistance to change towards sustainability."

Jack Harich's 2010 paper on *Change Resistance as the Crux of the Environmental Sustainability Problem* argues there are two separate problems to solve. A root cause analysis and a system dynamics model were used to explain how:

> "...difficult social problems [like sustainability must be decomposed] into two sequential subproblems: (1) *How to overcome change resistance* and then (2) *How to achieve proper coupling.* This is the timeless strategy of divide and conquer. By cleaving one big problem into two, the problem becomes an order of magnitude easier to solve, because we can approach the two subproblems differently and much more appropriately. We are no longer unknowingly attempting to solve two very different problems simultaneously."

The paper discussed the two subproblems:

> "*Change resistance* is the tendency for a system to continue its current behavior, despite the application of force to change that behavior.

> "*Proper coupling* occurs when the behavior of one system affects the behavior of other systems in a desirable manner, using the appropriate feedback loops, so the systems work together in harmony in accordance with design objectives. ... In the environmental sustainability problem the human system has become improperly coupled to the greater system it lives within: the environment.

> "Change resistance versus proper coupling allows a crucial distinction. Society is aware of the proper practices required to live sustainably and the need to do so. But society has a strong aversion to adopting these practices. As a result, problem solvers have created thousands of effective (and often ingenious) proper practices. But they are stymied in their attempts to have them taken up by enough of the system to solve the problem because an 'implicit system goal' is causing insurmountable change resistance. *Therefore systemic change resistance is the crux of the problem and must be solved first.*"

The proper coupling subproblem is what most people consider as "the" problem to solve. It is called decoupling in economic and environmental fields, where the term refers to economic growth without additional environmental degradation. Solving the proper coupling problem is the goal of environmentalism and in particular ecological economics: "Ecological economics is the study of the interactions and co-evolution in time and space of human economies and the ecosystems in which human economies are embedded."

Change resistance is also called barriers to change. Hoffman and Bazerman, in a chapter on "Understanding and overcoming the organizational and psychological barriers to action," concluded that:

> "In this chapter, we argue that the change in thinking required of the sustainability agenda will never come to fruition within practical domains unless proper attention is given to the sources of individual and social resistance to such change. The implementation of wise management practices cannot be accomplished without a concurrent set of strategies for surmounting these barriers."

John Sterman, current leader of the system dynamics school of thought, came to the same conclusion:

> "The civil rights movement provides a better analogy for the climate challenge. Then, as now, entrenched special interests vigorously opposed change. ... Of course, we need more research and technical innovation—money and genius are always in short supply. But there is no purely technical solution for climate change. For public policy to be grounded in the hard-won results of climate science, we must now turn our attention to the dynamics of social and political change."

These findings indicate there are at least two subproblems to be solved: change resistance and proper coupling. Given the human system's long history of unsuccessful attempts to self-correct to a sustainable mode, it appears that high change resistance is preventing proper coupling. This may be expressed as an emerging principle: systemic change resistance is the crux of the sustainability problem and must be solved first, before the human system can be properly coupled to the greater system it lives within, the environment.

Systemic Versus Individual Change Resistance

Systemic change resistance differs significantly from individual change resistance. "Systemic means originating from the system in such a manner as to affect the behavior of most or all social agents of certain types, as opposed to originating from individual agents." Individual change resistance originates from individual people and organizations. How the two differ may be seen in this passage:

> "The notion of resistance to change is credited to Kurt Lewin. His conceptualization of the phrase, however, is very different from today's usage. [which treats resistance to change as a psychological concept, where resistance or support of change comes from values, habits, mental models, and so on residing within the individual] For Lewin, resistance to change could occur, but that resistance could be anywhere in the system. As Kotter (1995) found, *it is possible for the resistance to be sited within the individual, but it is much more likely to be found elsewhere in the system.*

> "Systems of social roles, with their associated patterns of attitudes, expectations, and behavior norms, share with biological systems the characteristic of homeostasis—i.e., tendencies to resist change, to restore the previous state after a disturbance.

> "Lewin had been working on this idea, that the status quo represented an equilibrium between the barriers to change and the forces favoring change, since 1928 as part of his field theory. He believed that some difference in these forces—weakening of the barriers or strengthening of the driving forces—was required to produce the unfreezing that began a change."

If sources of systemic change resistance are present, they are the principal cause of individual change resistance. According to the fundamental attribution error it is crucial to address systemic change resistance when present and avoid assuming that change resistance can be overcome by bargaining, reasoning, inspirational appeals, and so on. This is because:

> "A fundamental principle of system dynamics states that the structure of the system gives rise to its behavior. However, people have a strong tendency to attribute the behavior of others to dispositional rather than situational factors, that is, to character and especially character flaws rather than the system in which these people are acting. The tendency to blame the person rather than the system is so strong psychologists call it the 'fundamental attribution error.' "

Peter Senge, a thought leader of systems thinking for the business world, describes the structural source of systemic change resistance as being due to an "implicit system goal:"

> "In general, balancing loops are more difficult to see than reinforcing loops because it often looks like nothing is happening. There's no dramatic growth of sales and marketing expenditures, or nuclear arms, or lily pads. Instead, the balancing process maintains the

status quo, even when all participants want change. The feeling, as Lewis Carroll's Queen of Hearts put it, of needing 'all the running you can do to keep in the same place' is a clue that a balancing loop may exist nearby.

"Leaders who attempt organizational change often find themselves unwittingly caught in balancing processes. To the leaders, it looks as though their efforts are clashing with sudden resistance that seems to come from nowhere. In fact, as my friend found when he tried to reduce burnout, the resistance is a response by the system, trying to maintain an *implicit system goal*. Until this goal is recognized, the change effort is *doomed to failure.*"

Senge's insight applies to the sustainability problem. Until the "implicit system goal" causing systemic change resistance is found and resolved, change efforts to solve the proper coupling part of the sustainability problem may be, as Senge argues, "doomed to failure."

The Current Focus is on Proper Coupling

Presently environmentalism is focused on solving the proper coupling subproblem. For example, the following are all proper coupling solutions. They attempt to solve the direct cause of the sustainability problem's symptoms:

- The Kyoto Protocol
- The Three Rs of reduce, reuse, recycle
- More use of renewable energy
- Better pollution control of many kinds
- Collective management of common-pool resources
- Certification programs to reduce deforestation such as PEFC and FSC
- Contour farming to reduce soil erosion
- The green revolution
- Zero population growth

The direct cause of environmental impact is the three factors on the right side of the I=PAT equation where Impact equals Population times Affluence (consumption per person) times Technology (environmental impact per unit of consumption). It is these three factors that solutions like those listed above seek to reduce.

The top environmental organization in the world, the United Nations Environmental Programme (UNEP), focuses exclusively on proper coupling solutions:

"2010 marked the beginning of a period of new, strategic and transformational direction for UNEP as it began implementing its Medium Term Strategy (MTS) for 2010-2013 across six areas: Climate change; Disasters and conflicts; Ecosystem management; Environmental governance; Harmful substances and hazardous waste; Resource efficiency, Sustainable consumption and production."

The six areas are all direct practices to reduce the three factors of the I=PAT equation.

Al Gores' 2006 documentary film *An Inconvenient Truth* described the climate change problem and the urgency of solving it. The film concluded with Gore saying:

> "Each one of us is a cause of global warming, but each one of us can make choices to change that with the things we buy, the electricity we use, the cars we drive; we can make choices to bring our individual carbon emissions to zero. The solutions are in our hands, *we just have to have the determination to make it happen. We have everything that we need to reduce carbon emissions, everything but political will.* But in America, the will to act is a renewable resource."

The four solutions Gore mentions are proper coupling practices. There is, however, a hint of acknowledgement that overcoming systemic change resistance is the real challenge, when Gore says "...we just have to have the determination to make it happen. We have everything that we need to reduce carbon emissions, everything but political will."

The twenty-seven solutions that appear during the film's closing credits are mostly proper coupling solutions. The first nine are:

- Go to www.climatecrisis.net

- You can reduce your carbon emissions. In fact, you can even reduce your carbon emissions to zero.

- Buy energy efficient appliances & light bulbs.

- Change your thermostat (and use clock thermostats) to reduce energy for heating & cooling.

- Weatherize your house, increase insulation, get an energy audit.

- Recycle.

- If you can, buy a hybrid car.

- When you can, walk or ride a bicycle.

- Where you can, use light rail & mass transit.

Some solutions are attempts to overcome individual change resistance, such as:

- Tell your parents not to ruin the world that you will live in.

- If you are a parent, join with your children to save the world they will live in.

- Vote for leaders who pledge to solve this crisis.

- Write to congress. If they don't listen, run for congress.

- Speak up in your community.

However none of the twenty-seven solutions deal with overcoming systemic change resistance.

Overcoming Systemic Change Resistance

Efforts here are sparse because environmentalism is currently not oriented toward treating systemic change resistance as a distinct and separate problem to solve.

On how to specifically overcome the change resistance subproblem, Markvart examined two leading theories that seemed to offer insight into change resistance, Panarchy theory and New Institutionalism, and concluded that:

> "...neither theory devotes significant attention to understanding the dynamics of resilient and resistant but inefficient and/or unproductive institutional and ecological systems. Overall, more research is required...."

Taking a root cause analysis and system dynamics modeling approach, Harich carefully defined the three characteristics of a root cause and then found a main systemic root cause for both the change resistance and proper coupling subproblems. Several sample solution elements for resolving the root causes were suggested. The point was made that the exact solution policies chosen do not matter nearly as much as finding the correct systemic root causes. Once these are found, how to resolve them is relatively obvious because once a root cause is found by structural modeling, the high leverage point for resolving it follows easily. Solutions may then push on specific structural points in the social system, which due to careful modeling will have fairly predictable effects.

This reaffirms the work of Donella Meadows, as expressed in her classic essay on Leverage Points: Places to Intervene in a System. The final page stated that:

> "The higher the leverage point, the more the system will resist changing it."

Here Meadows refers to the leverage point for resolving the proper coupling subproblem rather than the leverage point for overcoming change resistance. This is because the current focus of environmentalism is on proper coupling.

However, if the leverage points associated with the root causes of change resistance exist and can be found, the system will not resist changing them. This is an important principle of social system behavior.

For example, Harich found the main root cause of successful systemic change resistance to be high "deception effectiveness." The source was special interests, particularly large for-profit corporations. The high leverage point was raising "general ability to detect manipulative deception." This can be done with a variety of solution elements, such as "The Truth Test." This effectively increases truth literacy, just as conventional education raises reading and writing literacy. Few citizens resist literacy education because its benefits have become so obvious.

Promotion of corporate social responsibility (CSR) has been used to try to overcome change resistance to solving social problems, including environmental sustainability. This solution strategy has not worked well because it is voluntary and does not resolve root causes. Milton Friedman explained why CSR fails: "The social responsibility of business is to increase profits." Business cannot be responsible to society. It can only be responsible to its shareholders.

References

- Lynn R. Kahle, Eda Gurel-Atay, Eds (2014). Communicating Sustainability for the Green Economy. New York: M.E. Sharpe. ISBN 978-0-7656-3680-5.

- Wandemberg, JC (August 2015). Sustainable by Design. Amazon. p. 122. ISBN 1516901789. Retrieved 16 February 2016.

- Daly, H. & J. Cobb (1989). For the Common Good: Redirecting the Economy Toward Community, the Environment and a Sustainable Future. Boston: Beacon Press. ISBN 0-8070-4703-1.

- Blewitt, J. (2008). Understanding Sustainable Development. London: Earthscan. pp. 21–24. ISBN 978-1-84407-454-9.

- Hilgenkamp, K. (2005). Environmental Health: Ecological Perspectives. London: Jones & Bartlett. ISBN 978-0-7637-2377-4.

- Meadows, D.H., D.L. Meadows, J. Randers, and W. Behrens III. (1972). The Limits to Growth. New York: Universe Books. ISBN 0-87663-165-0.

- Conceptual Framework Working Group of the Millennium Ecosystem Assessment. (2003). "Ecosystems and Human Well-being." London: Island Press. Chapter 5. "Dealing with Scale". pp. 107–124. ISBN 9781559634038.

- Botkin, D.B. (1990). Discordant Harmonies, a New Ecology for the 21st century. New York: Oxford University Press. ISBN 978-0-19-507469-7.

- Brower, M. & Leon, W. (1999). The Consumer's Guide to Effective Environmental Choices: Practical Advice from the Union of Concerned Scientists. New York: Three Rivers Press. ISBN 0-609-80281-X.

- Dalal-Clayton, Barry and Sadler, Barry 2009. Sustainability Appraisal. A Sourcebook and Reference Guide to International Experience. London: Earthscan. ISBN 978-1-84407-357-3.

- Bell, Simon and Morse, Stephen 2008. Sustainability Indicators. Measuring the Immeasurable? 2nd edn. London: Earthscan. ISBN 978-1-84407-299-6.

- Lutz W., Sanderson W.C., & Scherbov S. (2004). The End of World Population Growth in the 21st Century London: Earthscan. ISBN 1-84407-089-1.

- Georgescu-Roegen, Nicholas (1971). The Entropy Law and the Economic Process. (Full book accessible in three parts at SlideShare). Cambridge, Massachusetts: Harvard University Press. ISBN 0674257804.

- Daly, Herman E., ed. (1980). Economics, Ecology, Ethics. Essays Towards a Steady-State Economy. (PDF contains only the introductory chapter of the book) (2nd ed.). San Francisco: W.H. Freeman and Company. ISBN 0716711788.

- World Resources Institute (1998). World Resources 1998–1999. Oxford: Oxford University Press. ISBN 0-19-521408-0.

- Groombridge, B. & Jenkins, M.D. (2002). World Atlas of Biodiversity. Berkeley: University of California Press. ISBN 978-0-520-23668-4.

- UNEP (2011). Decoupling Natural Resource Use and Environmental Impacts from Economic Growth. ISBN 978-92-807-3167-5. Retrieved on: 2011-11-30.

- Ruffing, K. (2007). "Indicators to Measure Decoupling of Environmental Pressure from Economic Growth", pp. 211–222 in: Hak et al. Sustainability Indicators. SCOPE 67. London: Island Press. ISBN 1-59726-131-9.

- Hawken, P, Lovins, A.B. & L.H. (1999). Natural Capitalism: Creating the next Industrial Revolution. Snowmass, USA: Rocky Mountain Institute. ISBN 0-316-35300-0.

- Diamond, J. (1997). Guns, Germs and Steel: the Fates of Human Societies. New York: W.W. Norton & Co. ISBN 0-393-06131-0.

- Daly, H.E. & Farley, J. (2004). Ecological economics: principles and applications. Washington: Island Press. p.xxvi. ISBN 1-55963-312-3.

5

Permaculture: Techniques and Concepts

Permaculture is a system of agriculture that appropriately implements the patterns and features observed in the natural ecosystem. Holzer permaculture is an important branch of permaculture. The text also focuses on aspects such as farmer-managed natural regeneration, keyline design, managed intensive rotational grazing, natural farming and regenerative agriculture. The aim of this chapter is to explore all the techniques and concepts related to permaculture.

Permaculture

Permaculture is a system of agricultural and social design principles centered on simulating or directly utilizing the patterns and features observed in natural ecosystems. Permaculture was developed, and the term coined by Bill Mollison and David Holmgren in 1968.

It has many branches that include but are not limited to ecological design, ecological engineering, environmental design, construction and integrated water resources management that develops sustainable architecture, and regenerative and self-maintained habitat and agricultural systems modeled from natural ecosystems.

Mollison has said: "Permaculture is a philosophy of working with, rather than against nature; of protracted and thoughtful observation rather than protracted and thoughtless labor; and of looking at plants and animals in all their functions, rather than treating any area as a single product system."

History

In 1929, Joseph Russell Smith took up an antecedent term as the subtitle for *Tree Crops: A Permanent Agriculture*, a book in which he summed up his long experience experimenting with fruits and nuts as crops for human food and animal feed. Smith saw the world as an inter-related whole and suggested mixed systems of trees and crops underneath. This book inspired many individuals intent on making agriculture more sustainable, such as Toyohiko Kagawa who pioneered forest farming in Japan in the 1930s.

The definition of permanent agriculture as that which can be sustained indefinitely was supported by Australian P. A. Yeomans in his 1964 book *Water for Every Farm*. Yeomans introduced an observation-based approach to land use in Australia in the 1940s, and the keyline design as a way of managing the supply and distribution of water in the 1950s.

Stewart Brand's works were an early influence noted by Holmgren. Other early influences include Ruth Stout and Esther Deans, who pioneered no-dig gardening, and Masanobu Fukuoka who, in the late 1930s in Japan, began advocating no-till orchards, gardens, and natural farming.

In the late 1960s, Bill Mollison and David Holmgren started developing ideas about stable agricultural systems on the southern Australian island state of Tasmania. This was a result of the danger of the rapidly growing use of industrial-agricultural methods. In their view, these methods were highly dependent on non-renewable resources, and were additionally poisoning land and water, reducing biodiversity, and removing billions of tons of topsoil from previously fertile landscapes. A design approach called *permaculture* was their response and was first made public with the publication of their book *Permaculture One* in 1978.

Bill Mollison in January 2008.

By the early 1980s, the concept had broadened from agricultural systems design towards sustainable human habitats. After *Permaculture One*, Mollison further refined and developed the ideas by designing hundreds of permaculture sites and writing more detailed books, notably *Permaculture: A Designers Manual*. Mollison lectured in over 80 countries and taught his two-week Permaculture Design Course (PDC) to many hundreds of students. Mollison "encouraged graduates to become teachers themselves and set up their own institutes and demonstration sites. This multiplier effect was critical to permaculture's rapid expansion."

Core Tenets and Principles of Design

The three core tenets of permaculture are:

- *Care for the earth:* Provision for all life systems to continue and multiply. This is the first principle, because without a healthy earth, humans cannot flourish.

- *Care for the people:* Provision for people to access those resources necessary for their existence.

- *Return of surplus:* Reinvesting surpluses back into the system to provide for the first two ethics. This includes returning waste back into the system to recycle into usefulness. The third ethic is sometimes referred to as Fair Share to reflect that each of us should take no more than what we need before we reinvest the surplus.

Permaculture design emphasizes patterns of landscape, function, and species assemblies. It determines where these elements should be placed so they can provide maximum benefit to the local environment. The central concept of permaculture is maximizing useful connections between components and synergy of the final design. The focus of permaculture, therefore, is not on each separate element, but rather on the relationships created among elements by the way they are placed together; the whole becoming greater than the sum of its parts. Permaculture design therefore seeks to minimize waste, human labor, and energy input by building systems with maximal benefits between design elements to achieve a high level of synergy. Permaculture designs evolve over time by taking into account these relationships and elements and can become extremely complex systems that produce a high density of food and materials with minimal input.

The design principles which are the conceptual foundation of permaculture were derived from the science of systems ecology and study of pre-industrial examples of sustainable land use. Permaculture draws from several disciplines including organic farming, agroforestry, integrated farming, sustainable development, and applied ecology. Permaculture has been applied most commonly to the design of housing and landscaping, integrating techniques such as agroforestry, natural building, and rainwater harvesting within the context of permaculture design principles and theory.

Theory

Twelve Design Principles

Twelve Permaculture design principles articulated by David Holmgren in his *Permaculture: Principles and Pathways Beyond Sustainability*:

1. *Observe and interact*: By taking time to engage with nature we can design solutions that suit our particular situation.

2. *Catch and store energy*: By developing systems that collect resources at peak abundance, we can use them in times of need.

3. *Obtain a yield*: Ensure that you are getting truly useful rewards as part of the work that you are doing.

4. *Apply self-regulation and accept feedback*: We need to discourage inappropriate activity to ensure that systems can continue to function well.

5. *Use and value renewable resources and services*: Make the best use of nature's abundance to reduce our consumptive behavior and dependence on non-renewable resources.

6. *Produce no waste*: By valuing and making use of all the resources that are available to us, nothing goes to waste.

7. *Design from patterns to details*: By stepping back, we can observe patterns in nature and society. These can form the backbone of our designs, with the details filled in as we go.

8. *Integrate rather than segregate*: By putting the right things in the right place, relationships develop between those things and they work together to support each other.

9. *Use small and slow solutions*: Small and slow systems are easier to maintain than big ones, making better use of local resources and producing more sustainable outcomes.

10. *Use and value diversity*: Diversity reduces vulnerability to a variety of threats and takes advantage of the unique nature of the environment in which it resides.

11. *Use edges and value the marginal*: The interface between things is where the most interesting events take place. These are often the most valuable, diverse and productive elements in the system.

12. *Creatively use and respond to change*: We can have a positive impact on inevitable change by carefully observing, and then intervening at the right time.

Layers

Suburban permaculture garden in Sheffield, UK with different layers of vegetation

Layers are one of the tools used to design functional ecosystems that are both sustainable and of direct benefit to humans. A mature ecosystem has a huge number of relationships between its component parts: trees, understory, ground cover, soil, fungi, insects, and animals. Because plants grow to different heights, a diverse community of life is able to grow in a relatively small space, as the vegetation occupies different layers. There are generally seven recognized layers in a food forest, although some practitioners also include fungi as an eighth layer.

1. The canopy: the tallest trees in the system. Large trees dominate but typically do not saturate the area, i.e. there exist patches barren of trees.

2. Understory layer: trees that revel in the dappled light under the canopy.

3. Shrub layer: a diverse layer of woody perennials of limited height. includes most berry bushes.

4. Herbaceous layer: Plants in this layer die back to the ground every winter (if winters are cold enough, that is). They do not produce woody stems as the Shrub layer does. Many culinary and medicinal herbs are in this layer. A large variety of beneficial plants fall into this layer. May be annuals, biennials or perennials.

5. Soil surface/Groundcover: There is some overlap with the Herbaceous layer and the Groundcover layer; however plants in this layer grow much closer to the ground, grow densely to fill bare patches of soil, and often can tolerate some foot traffic. Cover crops retain soil and lessen erosion, along with green manures that add nutrients and organic matter to the soil, especially nitrogen.

6. Rhizosphere: Root layers within the soil. The major components of this layer are the soil and the organisms that live within it such as plant roots (including root crops such as potatoes and other edible tubers), fungi, insects, nematodes, worms, etc.

7. Vertical layer: climbers or vines, such as runner beans and lima beans (vine varieties).

Guilds

There are many forms of guilds, including guilds of plants with similar functions (that could interchange within an ecosystem), but the most common perception is that of a mutual support guild. Such a guild is a group of species where each provides a unique set of diverse functions that work in conjunction, or harmony. Mutual support guilds are groups of plants, animals, insects, etc. that work well together. Some plants may be grown for food production, some have tap roots that draw nutrients up from deep in the soil, some are nitrogen-fixing legumes, some attract beneficial insects, and others repel harmful insects. When grouped together in a mutually beneficial arrangement, these plants form a guild.

Edge effect

The edge effect in ecology is the effect of the juxtaposition or placing side by side of contrasting environments on an ecosystem. Permaculturists argue that, where vastly differing systems meet, there is an intense area of productivity and useful connections. An example of this is the coast; where the land and the sea meet there is a particularly rich area that meets a disproportionate percentage of human and animal needs. So this idea is played out in permacultural designs by using spirals in the herb garden or creating ponds that have wavy undulating shorelines rather than a simple circle or oval (thereby increasing the amount of edge for a given area).

Zones

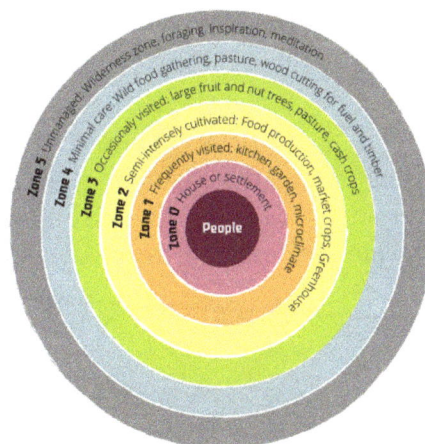

Permaculture Zones 0-5.

Zones are a way of intelligently organizing design elements in a human environment on the basis of the frequency of human use and plant or animal needs. Frequently manipulated or harvested elements of the design are located close to the house in zones 1 and 2. Less frequently used or manipulated elements, and elements that benefit from isolation (such as wild species) are farther away. Zones are about positioning things appropriately, and are numbered from 0 to 5.

Zone 0

The house, or home center. Here permaculture principles would be applied in terms of aiming to reduce energy and water needs, harnessing natural resources such as sunlight, and generally creating a harmonious, sustainable environment in which to live and work. Zone 0 is an informal designation, which is not specifically defined in Bill Mollison's book.

Zone 1

The zone nearest to the house, the location for those elements in the system that require frequent attention, or that need to be visited often, such as salad crops, herb plants, soft fruit like strawberries or raspberries, greenhouse and cold frames, propagation area, worm compost bin for kitchen waste, etc. Raised beds are often used in zone 1 in urban areas.

Zone 2

This area is used for siting perennial plants that require less frequent maintenance, such as occasional weed control or pruning, including currant bushes and orchards, pumpkins, sweet potato, etc. This would also be a good place for beehives, larger scale composting bins, etc.

Zone 3

The area where main-crops are grown, both for domestic use and for trade purposes. After establishment, care and maintenance required are fairly minimal (provided mulches and similar things are used), such as watering or weed control maybe once a week.

Zone 4

A semi-wild area. This zone is mainly used for forage and collecting wild food as well as production of timber for construction or firewood.

Zone 5

A wilderness area. There is no human intervention in zone 5 apart from the observation of natural ecosystems and cycles. Through this zone we build up a natural reserve of bacteria, moulds and insects that can aid the zones above it.

People and Permaculture

Permaculture uses observation of nature to create regenerative systems, and the place where this has been most visible has been on the landscape. There has been a growing awareness though that firstly, there is the need to pay more attention to the peoplecare ethic, as it is often the dynamics of people that can interfere with projects, and secondly that the principles of permaculture can be used as effectively to create vibrant, healthy and productive people and communities as they have been in landscapes.

Domesticated Animals

Domesticated animals are often incorporated into site design.

Common Practices

Agroforestry

Agroforestry is an integrated approach of using the interactive benefits from combining trees and shrubs with crops and/or livestock. It combines agricultural and forestry technologies to create more diverse, productive, profitable, healthy and sustainable land-use systems. In agroforestry systems, trees or shrubs are intentionally used within agricultural systems, or non-timber forest products are cultured in forest settings.

Forest gardening is a term permaculturalists use to describe systems designed to mimic natural forests. Forest gardens, like other permaculture designs, incorporate processes and relationships that the designers understand to be valuable in natural ecosystems. The terms forest garden and food forest are used interchangeably in the permaculture literature. Numerous permaculturists are proponents of forest gardens, such as Graham Bell, Patrick Whitefield, Dave Jacke, Eric Toensmeier and Geoff Lawton. Bell started building his forest garden in 1991 and wrote the book *The Permaculture Garden* in 1995, Whitefield wrote the book *How to Make a Forest Garden* in 2002, Jacke and Toensmeier co-authored the two volume book set *Edible Forest Gardening* in 2005, and Lawton presented the film *Establishing a Food Forest* in 2008.

Tree Gardens, such as Kandyan tree gardens, in South and Southeast Asia, are often hundreds of years old. Whether they derived initially from experiences of cultivation and forestry, as is the case in agroforestry, or whether they derived from an understanding of forest ecosystems, as is the case for permaculture systems, is not self-evident. Many studies of these systems, especially those that predate the term permaculture, consider these systems to be forms of agroforestry. Permaculturalists who include existing and ancient systems of polycropping with woody species as examples of food forests may obscure the distinction between permaculture and agroforestry.

Food forests and agroforestry are parallel approaches that sometimes lead to similar designs.

Hügelkultur

Hügelkultur is the practice of burying large volumes of wood to increase soil water retention. The porous structure of wood acts as a sponge when decomposing underground. During the rainy season, masses of buried wood can absorb enough water to sustain crops through the dry season. This technique has been used by permaculturalists Sepp Holzer, Toby Hemenway, Paul Wheaton, and Masanobu Fukuoka.

Natural Building

A natural building involves a range of building systems and materials that place major emphasis on sustainability. Ways of achieving sustainability through natural building focus on durability and the use of minimally processed, plentiful or renewable resources, as well as those that, while recycled or salvaged, produce healthy living environments and maintain indoor air quality.

The basis of natural building is the need to lessen the environmental impact of buildings and other supporting systems, without sacrificing comfort, health, or aesthetics. To be more sustainable, natural building uses primarily abundantly available, renewable, reused, or recycled materials. In addition to relying on natural building materials, the emphasis on the architectural design is heightened. The orientation of a building, the utilization of local climate and site conditions, the emphasis on natural ventilation through design, fundamentally lessen operational costs and positively impact the environment. Building compactly and minimizing the ecological footprint is common, as are on-site handling of energy acquisition, on-site water capture, alternate sewage treatment, and water reuse.

Rainwater Harvesting

Rainwater harvesting is the accumulating and storing of rainwater for reuse before it reaches the aquifer. It has been used to provide drinking water, water for livestock, water for irrigation, as well as other typical uses. Rainwater collected from the roofs of houses and local institutions can make an important contribution to the availability of drinking water. It can supplement the subsoil water level and increase urban greenery. Water collected from the ground, sometimes from areas which are especially prepared for this purpose, is called stormwater harvesting.

Greywater is wastewater generated from domestic activities such as laundry, dishwashing, and bathing, which can be recycled on-site for uses such as landscape irrigation and constructed wetlands. Greywater is largely sterile, but not potable (drinkable). Greywater differs from water from the toilets which is designated sewage or blackwater, to indicate it contains human waste. Blackwater is septic or otherwise toxic and cannot easily be reused. There are, however, continuing efforts to make use of blackwater or human waste. The most notable is for composting through a process known as humanure; a combination of the words human and manure. Additionally, the methane in humanure can be collected and used similar to natural gas as a fuel, such as for heating or cooking, and is commonly referred to as biogas. Biogas can be harvested from the human waste and the remainder still used as humanure. Some of the simplest forms of humanure use include a composting toilet or an outhouse or dry bog surrounded by trees that are heavy feeders which can be coppiced for wood fuel. This process eliminates the use of a standard toilet with plumbing.

Sheet Mulching

In agriculture and gardening, mulch is a protective cover placed over the soil. Any material or combination can be used as mulch, such as stones, leaves, cardboard, wood chips, gravel, etc., though in permaculture mulches of organic material are the most common because they perform more functions. These include: absorbing rainfall, reducing evaporation, providing nutrients, increasing organic matter in the soil, feeding and creating habitat for soil organisms, suppressing weed growth and seed germination, moderating diurnal temperature swings, protecting against frost, and reducing erosion. Sheet mulching is an agricultural no-dig gardening technique that attempts to mimic natural processes occurring within forests. Sheet mulching mimics the leaf cover that is found on forest floors. When deployed properly and in combination with other Permacultural principles, it can generate healthy, productive and low maintenance ecosystems.

Sheet mulch serves as a "nutrient bank," storing the nutrients contained in organic matter and slowly making these nutrients available to plants as the organic matter slowly and naturally breaks down. It also improves the soil by attracting and feeding earthworms, slaters and many other soil micro-organisms, as well as adding humus. Earthworms "till" the soil, and their worm castings are among the best fertilizers and soil conditioners. Sheet mulching can be used to reduce or eliminate undesirable plants by starving them of light, and can be more advantageous than using herbicide or other methods of control.

Intensive Rotational Grazing

Grazing has long been blamed for much of the destruction we see in the environment. However, it has been shown that when grazing is modeled after nature, the opposite effect can be seen. Also known as cell grazing, managed intensive rotational grazing (MIRG) is a system of grazing in which ruminant and non-ruminant herds and/or flocks are regularly and systematically moved to fresh pasture, range, or forest with the intent to maximize the quality and quantity of forage growth. This disturbance is then followed by a period of rest which allows new growth. MIRG can be used with cattle, sheep, goats, pigs, chickens, rabbits, geese, turkeys, ducks, and other animals depending on the natural ecological community that is being mimicked. Sepp Holzer and Joel Salatin have shown how the disturbance caused by the animals can be the spark needed to start ecological succession or prepare ground for planting. Allan Savory's holistic management technique has been likened to "a permaculture approach to rangeland management". One variation on MIRG that is gaining rapid popularity is called eco-grazing. Often used to either control invasives or re-establish native species, in eco-grazing the primary purpose of the animals is to benefit the environment and the animals can be, but are not necessarily, used for meat, milk or fiber.

Keyline Design

Keyline design is a technique for maximizing beneficial use of water resources of a piece of land developed in Australia by farmer and engineer P. A. Yeomans. The *Keyline* refers to a specific topographic feature linked to water flow which is used in designing the drainage system of the site.

Fruit Tree Management

Some proponents of permaculture advocate no, or limited, pruning. One advocate of this approach is Sepp Holzer who used the method in connection with Hügelkultur berms. He has successfully grown several varieties of fruiting trees at altitudes (approximately 9,000 feet (2,700 m)) far above their normal altitude, temperature, and snow load ranges. He notes that the Hügelkultur berms kept and/or generated enough heat to allow the roots to survive during alpine winter conditions. The point of having unpruned branches, he notes, was that the longer (more naturally formed) branches bend over under the snow load until they touched the ground, thus forming a natural arch against snow loads that would break a shorter, pruned, branch.

Masanobu Fukuoka, as part of early experiments on his family farm in Japan, experimented with no-pruning methods, noting that he ended up killing many fruit trees by simply letting them go, which made them become convoluted and tangled, and thus unhealthy. Then he realised this is the difference between natural-form fruit trees and the process of change of tree form that results from abandoning previously-pruned unnatural fruit trees. He concluded that the trees should be raised all their lives without pruning, so they form healthy and efficient branch patterns that follow their natural inclination. This is part of his implementation of the Tao-philosophy of Wú wéi translated in part as no-action (against nature), and he described it as no unnecessary pruning, nature farming or "do-nothing" farming, of fruit trees, distinct from non-intervention or literal no-pruning. He ultimately achieved yields comparable to or exceeding standard/intensive practices of using pruning and chemical fertilisation.

Trademark and Copyright Issues

There has been contention over who, if anyone, controls legal rights to the word *permaculture*: is it trademarked or copyrighted? and if so, who holds the legal rights to the use of the word? For a long time Bill Mollison claimed to have copyrighted the word, and his books said on the copyright page, "The contents of this book and the word PERMACULTURE are copyright." These statements were largely accepted at face-value within the permaculture community. However, copyright law does not protect names, ideas, concepts, systems, or methods of doing something; it only protects the expression or the description of an idea, not the idea itself. Eventually Mollison acknowledged that he was mistaken and that no copyright protection existed for the word *permaculture*.

In 2000, Mollison's US based Permaculture Institute sought a service mark (a form of trademark) for the word *permaculture* when used in educational services such as conducting classes, seminars, or workshops. The service mark would have allowed Mollison and his two Permaculture Institutes (one in the US and one in Australia) to set enforceable guidelines regarding how permaculture could be taught and who could teach it, particularly with relation to the PDC, despite the fact that he had instituted a system of certification of teachers to teach the PDC in 1993. The service mark failed and was abandoned in 2001. Also in 2001 Mollison applied for trademarks in Australia for the terms "Permaculture Design Course" and "Permaculture Design". These applications were both withdrawn in 2003. In 2009 he sought a trademark for "Permaculture: A Designers' Manual" and "Introduction to Permaculture", the names of two of his books. These applications were withdrawn in 2011. There has never been a trademark for the word *permaculture* in Australia.

Criticisms

General Criticisms

In 2011, Owen Hablutzel argued that "permaculture has yet to gain a large amount of specific mainstream scientific acceptance," and that "the sensitiveness to being perceived and accepted on scientific terms is motivated in part by a desire for permaculture to expand and become increasingly relevant."

In his books *Sustainable Freshwater Aquaculture* and *Farming in Ponds and Dams*, Nick Romanowski expresses the view that the presentation of aquaculture in Bill Mollison's books is unrealistic and misleading.

Agroforestry

Greg Williams argues that forests cannot be more productive than farmland because the net productivity of forests decline as they mature due to ecological succession. Proponents of permaculture respond that this is true only if one compares data between woodland forest and climax vegetation, but not when comparing farmland vegetation with woodland forest. For example, ecological succession generally results in a forest's productivity rising after its establishment only until it reaches the *woodland state* (67% tree cover), before declining until *full maturity*.

Holzer Permaculture

The Holzer Permaculture is a branch of permaculture developed independently from the mainstream permaculture in Austria by Sepp Holzer. It is particularly noteworthy because it grew out of practical application and was relatively detached from the scientific community.

Intro

Sepp Holzer started reorganising his father's property according to ecological patterns in the early 1960s after he took over the farm. As an adolescent he conducted layman experiments with plants native to the area and learned from his own observations.

Since having taken over his father's property, he has expanded it from 24 to 45 hectares. according to his methods together with his wife.

His expanded farm now spans over 45 hectares of forest gardens, including 70 ponds, and is said to be the most consistent example of permaculture worldwide. In the past he has experimented with many different animals. As a result of these experiments, there is a huge role for animals in the Holzer Permaculture.

He has created some of the world's best examples of using ponds as reflectors to increase solar gain for passive solar heating of structures, and of using the microclimate created by rock outcrops to effectively change the hardiness zone for nearby plants. He has also done original work in the use of Hügelkultur and natural branch development instead of pruning to allow fruit trees to survive high altitudes and harsh winters.

Comparison to Regular Permaculture

It is difficult to make out differences between the methods and practices of Sepp Holzer in contrast to the more scientific and theoretical permacultural mainstream. Nevertheless, here are some major points to consider:

- His designs are mostly aimed at raising temperatures and creating micro-climates with rocks, ponds and living wind barriers, in an area with 4 °C on the average and -20 °C in the winter. The ponds he makes do not contain any pond liner. Instead, he makes the ponds watertight by sifting the fine from the coarse soil in the earth pile dug up with the excavator. The excavator is then used to pile only the coarse soil up into walls which are then tampered down using the excavator's bucket. The bottom of the pond, he makes watertight by vibrating the excavator's bucket when the pond has been filled with 30-40cm of water.

- Another aspect was the necessity of creating terraces on his farm's hillsides leading him to the use of heavy machinery (like excavators). Many of the terraces he construct are also given a humus storage ditch (placed in between the terraces).

- Famous is his Hügelkultur technique, which is basically the use of raised beds in which he uses bulky material such as tree trunks. On his farm, he made a pick-your-own area where visitors can pick their own produce from the raised beds and then pay for it at a counter upon leaving the area.

- He uses animal labor alongside human labor, working his farm with only two people. He optimizes the natural patterns of animal behavior to reduce human or machine-driven labor. As an example: he uses swine to "plow" new beds for sowing. This is a very effective way of digging, as the only thing he has to do is to throw some corn and fruit on the spot he wants dug up. A couple of days later, he can bring the pigs back to their enclosure and plant new plants in the bed. Holzer is able to successfully grow his plants without using any fertilizer. The animals he uses are all heirloom races, which are hardy and require (almost) no maintenance. Examples of the races of swine he uses are Mangalitza, Swabian Hall swine, Duroc, Turopolje, ... He also keeps quail, capercaillie, hazel grouse, wisent, scottish higland cattle, hungarian steppe cattle, Dahomey miniature cattle, American bison, yak, water buffalo. He has the animals live outside, in paddocks/shelters and has them share space with orchards and forests. Many of the fruit trees in the orchards (especially those on very sloped terrain) are used exclusively to feed the animals. They collect it as it falls to the ground (those trees are hence not being picked/ nor is produce of it sold)

- He also does not prune fruit trees much nor does he cut the lower branches on fruit trees (as this can hurt the tree -due to the lignification not being able to complete before frost and the fact that the unpruned fruit trees survives snow loads that will break pruned trees.-. He also says that leaving the branches on protects the tree against browsing by animals. He does however makes a point of using deep-rooted pioneer plants such as lupins, sweet clover, lucerne and broom. These crops are said to aerate the soil and make sure no water is left standing near the tree. No use is made of wire meshes as protection against voles, since he states they are not efficient in preventing damage from voles using this technique anyway. He makes no use of contemporary fruit tree cultivars, but only uses (very strong/hardy) old, local (heirloom) cultivars. In addition to (old cultivars of) regular fruit trees, he also plants much fruit tree species that are specific to use in phytotherapy or which can only function as animal feed (i.e. crab apple, wild pear, wild cherry, blackthorn, rowan, wild service, service tree, cornelian cherry, snowy mespilus, ...). Similar to mainstream permaculture, he makes no use of chemical fertiliser or pesticides, at all. It should be noted that due to the altitude he's at, his trees bear fruit later, meaning he can sell it after most (sealevel) farmers sold their produce. Due to this (and the fact that he produces heirloom fruit varieties, and not regular varieties), he is often able to get a better price for it. In some cases, customers (like distilleries) are even willing to pick the fruit themselves, eliminating the labour expense for him.

- He grows many old cereals on his farm, such as einkorn, emmer, black emmer, spelt, fichtelgebirgshafer, wild rye, black oats, naked oats, barley, Siberian grain (secale cereale, russian cultivar), tauernroggen, ...

- He also makes much use of green manure crops (like stinging nettle, phacelia, yellow, white and narrow-leaved lupin, garden pea, grass pea, fodder & Kidney vetch, yellow, subterranean, Crimson, Persian, Egyptian, red & white sweet clover, Birdsfoot trefoil, lucerne, black medick, Sainfoin, Serradella, fiddleneck, sunflowers, Jerusalem artichoke, Gold-of-pleasure, ...)and grows these crops extensively on his farm. He leaves them standing in autumn, rather than digging them in. He instead relies on natural decay of the plants. He often relies on the natural spreading of the seeds of the crops for their re-sowing.

- Another aspect is the abandonment of other horticultural principles such as intercropping plants with very high and very low pH requirements (for example, Rhododendron with roses). Instead, Holzer mixes thirty or more different types of seeds in a bucket and tosses the mix richly onto a larger area.

The Krameterhof

Situated in Ramingstein on the slopes of Mount Schwarzenberg his farm (Krameterhof) lies at varying elevations ranging from 1100 to 1500 metres above sea level.

The exceptionally harsh climatic conditions in the area are generally considered inappropriate for farming. Nevertheless, he has managed to cultivate a variety of crops and even exotic plants like Kiwis and Sweet Chestnut.

The Krameterhof is less an operational enterprise, in terms of crop-yield (although it does provide numerous sorts of produce for the community), and more a fully functional showcase or research station for permaculture.

Endangered livestock species and rare alpine- and cultural plant species are integrated into the farm.

Farmer-managed Natural Regeneration

Farmer-managed natural regeneration (FMNR) is a low-cost, sustainable land-restoration technique used to combat poverty and hunger amongst poor subsistence farmers in developing countries by increasing food and timber production, and resilience to climate extremes. It involves the systematic regeneration and management of trees and shrubs from tree stumps, roots and seeds.

FMNR is especially applicable, but not restricted to, the dryland tropics. As well as returning degraded croplands and grazing lands to productivity, it can be used to restore degraded forests, thereby reversing biodiversity loss and reducing vulnerability to climate change. FMNR can also play an important role in maintaining not-yet-degraded landscapes in a productive state, especially when combined with other sustainable land management practices such as conservation agriculture on cropland and holistic management on rangelands.

FMNR adapts centuries-old methods of woodland management, called coppicing and pollarding, to produce continuous tree-growth for fuel, building materials, food and fodder without the need for frequent and costly replanting. On farmland, selected trees are trimmed and pruned to maximise growth while promoting optimal growing conditions for annual crops (such as access to water and sunlight). When FMNR trees are integrated into crops and grazing pastures there is an increase in crop yields, soil fertility and organic matter, soil moisture and leaf fodder. There is also a decrease in wind and heat damage, and soil erosion.

In the Sahel region of Africa, FMNR has become a potent tool in increasing food security, resilience and climate change adaptation in poor, subsistence farming communities where much of sub-Saharan Africa's poverty exists. FMNR is also being promoted in East Timor, Indonesia and Myanmar.

FMNR complements the evergreen agriculture, conservation agriculture and agroforestry movements. It is considered a good entry point for resource-poor and risk-averse farmers to adopt a low-cost and low-risk technique. This in turn has acted as a stepping stone to greater agricultural intensification as farmers become more receptive to new ideas.

Background

Throughout the developing world, immense tracts of farmland, grazing lands and forests have become degraded to the point they are no longer productive. Deforestation continues at an alarming rate. In Africa's drier regions, 74 percent of rangelands and 61 percent of rain-fed croplands are damaged by moderate to very severe desertification. In some African countries deforestation rates exceed planting rates by 300:1.

Degraded land has an extremely detrimental effect on the lives of subsistence farmers who depend on it for their food and livelihoods. Subsistence farmers often make up to 70-80 percent of the population in these regions and they regularly suffer from hunger, malnutrition and even famine as a consequence.

In the Sahel region of Africa, a band of savannah which runs across the continent immediately south of the Sahara Desert, large tracts of once-productive farmland are turning to desert. In tropical regions across the world, where rich soils and good rainfall would normally assure bountiful harvests and fat livestock, some environments have become so degraded they are no longer productive. Severe famines across the African Sahel in the 1970s and 80s led to a global response, and stopping desertification became a top priority. Conventional methods of raising exotic and indigenous tree species in nurseries were used – planting out, watering, protecting and weeding. However, despite investing millions of dollars and thousands of hours labour, there was little overall impact. Conventional approaches to reforestation in such harsh environments faced insurmountable problems and were costly and labour-intensive. Once planted out, drought, sand storms, pests, competition from weeds and destruction by people and animals negated efforts. Low levels of community ownership were another inhibiting factor.

Existing indigenous vegetation was generally dismissed as 'useless bush', and it was often cleared to make way for exotic species. Exotics were planted in fields containing living and sprouting stumps of indigenous vegetation, the presence of which was barely acknowledged, let alone seen as important.

This was an enormous oversight. In fact, these living tree stumps are so numerous they constitute a vast 'underground forest' just waiting for some care to grow and provide multiple benefits at little or no cost –and each stump can produce between 10 and 30 stems each. During the process of traditional land preparation, farmers saw the stems as weeds and slashed and burnt them before sowing their food crops. The net result was a barren landscape for much of the year with few mature trees remaining. To the casual observer, the land was turning to desert. Most concluded that there were no trees present and that the only way to reverse the problem was through tree planting.

Meanwhile, established indigenous trees continued to disappear at an alarming rate. In Niger, from the 1930s until 1993, forestry laws took tree ownership and responsibility for the care of trees out of the hands of the people; and even though ineffective and uneconomic, reforestation through conventional tree planting seemed to be the only way to address desertification at the time.

Birth and Spread

In the early 1980s, in the Maradi region of the Republic of Niger, the missionary organisation, Serving in Mission (SIM), was unsuccessfully attempting to reforest the surrounding districts using conventional means. In 1983, SIM began experimenting and promoting FMNR amongst about 10 farmers. During the severe famine of 1984, a food-for-work program was introduced that saw some 70,000 people exposed to FMNR and its practice on around 12,500 hectares of farmland. From 1985 to 1999, FMNR continued to be promoted locally and nationally as exchange visits and training days were organised for various NGOs, government foresters, Peace Corps Volunteers and farmer and civil society groups. Additionally, SIM project staff and farmers visited numerous locations across Niger to provide training.

By 2004 it was ascertained that FMNR was being practiced on over five million hectares or 50 percent of Niger's farmland – an average reforestation rate of 250,000 hectares per year over a 20-year period. This transformation prompted Senior Fellow of the World Resources Institute, Chris Reij to comment that "this is probably the largest positive environmental transformation in the Sahel and perhaps all of Africa".

Also in 2004, World Vision Australia and World Vision Ethiopia initiated a forestry-based carbon sequestration project as a potential means to stimulate community development while engaging in environmental restoration. An innovative partnership with the World Bank, the Humbo Community-based Natural Regeneration Project involved the regeneration of 2,728 hectares of degraded native forests. This brought social, economic and ecological benefits to the participating communities. Within two years, communities were collecting wild fruits, firewood and fodder, and reported that wildlife had begun to return and erosion and flooding had been reduced. In addition, the communities are now receiving payments for the sale of carbon credits through the Clean Development Mechanism (CDM) of the Kyoto protocol.

Following the success of the Humbo project, FMNR spread to the Tigray region of northern Ethiopia where 20,000 hectares have been set aside for regeneration, including 10-hectare FMNR model sites for research and demonstration in each of 34 sub-districts. In addition, the Government of Ethiopia has committed to reforest 15 million hectares of degraded land using FMNR as part of a climate change and renewable energy plan to become carbon neutral by 2025.

In Talensi, northern Ghana, FMNR is being practiced on 2,000-3,000 hectares and new projects, initiated by World Vision, are introducing FMNR into three new districts. In the Kaffrine and Diourbel regions of Senegal, FMNR has spread across 50,000 hectares in four years. World Vision is also promoting FMNR in Indonesia, Myanmar and East Timor. There are also examples of both independently promoted and spontaneous FMNR movements occurring. In Burkina Faso, for example, an increasing part of the country is being transformed into agroforestry parkland. And in Mali, an ageing agroforestry parkland of about 6 million hectares is showing signs of regeneration.

Key Principles

FMNR depends on the existence of living tree stumps or roots in crop fields, grazing pastures, woodlands or forests. Each season bushy growth will sprout from the stumps/roots often appearing like small shrubs. Continuous grazing by livestock, regular burning and/or regular harvesting for fuel wood results in these 'shrubs' never attaining tree stature. On farmland, standard practice has been for farmers to slash this regrowth in preparation for planting crops, but with a little attention this growth can be turned into a valuable resource without jeopardising, but in fact, enhancing crop yields.

For each stump, a decision is made as to how many stems will be chosen to grow. The tallest and straightest stems are selected and the remaining stems culled. Best results are obtained when the farmer returns regularly to prune any unwanted new stems and side branches as they appear. Farmers can then grow other crops between and around the trees. When farmers want wood they can cut the stem(s) they want and leave the rest to continue growing. The remaining stems will increase in size and value each year, and will continue to protect the environment. Each time a stem is harvested, a younger stem is selected to replace it.

Various naturally occurring tree species can be used which may also provide berries, fruits and nuts or have medicinal qualities. In Niger, commonly used species include: Strychnos spinosa, Balanites aegyptiaca, Boscia senegalensis, Ziziphus spp., Annona senegalensis, Poupartia birrea and Faidherbia albida. However, the most important determinants are whatever species are locally available, their ability to re-sprout after cutting, and the value local people place on those species.

Faidherbia albida, also known as the 'fertiliser tree', is popular for intercropping across the Sahel as it fixes nitrogen into the soil, provides fodder for livestock, and shade for crops and livestock. By shedding its leaves in the wet season, Faidherbia provides beneficial light shade to crops when high temperatures would otherwise damage crops or retard growth. Leaf fall contributes useful nutrients and organic matter to the soil.

The practice of FMNR is not confined to croplands. It is being practised on grazing land and in degraded communal forests as well. When there are no living stumps, seeds of naturally occurring species are used. In reality, there is no fixed way of practising FMNR and farmers are free to choose which species they will leave, the density of trees they prefer, and the timing and method of pruning.

FMNR in Practice

FMNR depends on the existence of living tree stumps, tree roots and seeds to be re-vegetated. These can be in crop fields, grazing lands or degraded forests. New stems, which sprout from these stumps and tree roots, can be selected and pruned for improved growth. Sprouting tree stumps and roots may look like shrubs and are often ignored or even slashed by farmers or foresters. However, with culling of excess stems and by selecting and pruning of the best stems, the regrowth has enormous potential to rapidly grow into trees.

Seemingly treeless fields may contain seeds and living tree stumps and roots which have the ability to sprout new stems and regenerate trees. Even this 'bare' millet field in West Africa contains hundreds of living stumps per hectare which are buried beneath the surface like an underground forest.

Step 1. Do not automatically slash all tree growth, but survey your farm noting how many and what species of trees are present.

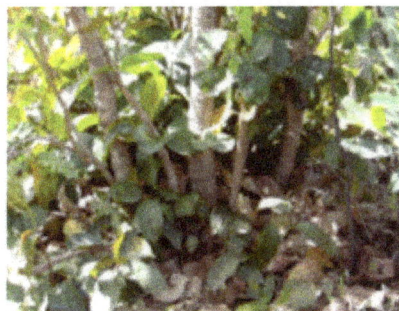

Step 2. Select the stumps which will be used for regeneration.

Step 3. Select the best five or so stems and cull unwanted ones. This way, when you want wood you can cut the stem(s) that are needed and leave the rest to continue growing. These remaining stems will increase in size and value each year, and will continue to protect the environment and provide other useful materials and services such as fodder, humus, habitat for useful pest predators and protection from the wind and sun. Each time one stem is harvested, a younger stem is selected to replace it.

Tag selected stems with a coloured rag or paint. Work with the whole community to draw up and agree on laws which will protect the trees being pruned and respect each person's rights. Where possible, include government forestry staff and local authorities in planning and decision making.

Benefits of FMNR

FMNR can restore degraded farmlands, pastures and forests by increasing the quantity and value of woody vegetation, by increasing biodiversity and by improving soil structure and fertility through leaf litter and nutrient cycling. The reforestation also retards wind and water erosion; it creates wind-breaks which decrease soil moisture evaporation, and protects crops and livestock against searing winds and temperatures. Often, dried up springs reappear and the water table rises towards historic levels; insect eating predators including insects, spiders and birds return, helping to keep crop pests in check; the trees can be a source of edible berries and nuts; and over time the biodiversity of plant and animal life is increased. FMNR can be used to combat deforestation and desertification and can also be an important tool in maintaining the integrity and productivity of land that is not yet degraded.

Trials, long-running programs and anecdotal data indicate that FMNR can at least double crop yields on low fertility soils. In the Sahel, high numbers of livestock and an eight-month dry season can mean that pastures are completely depleted before the rains commence. However, with the presence of trees, grazing animals can make it through the dry season by feeding on tree leaves and seed pods of some species, at a time when no other fodder is available. In north east Ghana, more grass became available with the introduction of FMNR because communities worked together to prevent bush fires from destroying their trees.

Well designed and executed FMNR projects can act as catalysts to empower communities as they negotiate land ownership or user rights for the trees in their care. This assists with self-organisation, and with the development of new agriculture-based micro-enterprises (e.g. selling firewood, timber and handcrafts made from timber or woven grasses).

Conventional approaches to reversing desertification, such as funding tree planting, rarely spread beyond the project boundary once external funding is withdrawn. By comparison, FMNR is cheap, rapid, locally led and implemented. It uses local skills and resources – the poorest farmers can learn by observation and teach their neighbours. Given an enabling environment, or at least the absence of a 'disabling' environment, FMNR can be done at scale and spread well beyond the original target area without ongoing government or NGO intervention.

World Vision evaluations of FMNR conducted in Senegal and Ghana in 2011 and 2012 found that households practising FMNR were less vulnerable to extreme weather shocks such as drought and damaging rain and wind storms.

The following table summarises FMNR's benefits which fit the sustainable development model of economic, social and environmental benefits:

Economic benefits	Social benefits	Environmental benefits
Increased crop yields (often double or triple)	Increased food security and nutrition (including native fruits, nuts and seeds)	Reduced erosion
Increased fodder from edible leaves and seed pods, and in some cases increased pasture growth	Less distance for women and children to travel to collect firewood	Reduced soil-moisture evaporation due to wind breaks shading and mulching
Higher livestock productivity and survival	Community capacity building to deal with local, regional and national governments and regulators	Increased soil fertility
Reduced impact from floods and drought – trees provide alternative income and livelihood sources making impacts less severe and recovery faster	Improved governance through clarification of tree ownership laws and regulations	Improved soil structure through greater quantities of organic matter
Increased income generation through diversification (e.g. timber and fuel wood) and intensification of activities	Education and training in farming and marketing	Increased water infiltration and groundwater recharge
Economic flow-on effects such as employment and greater purchasing capacity	Reduced need for migration by young people and men to cities	Increased biodiversity, environmental restoration and tree cover
Increased economic activity creates opportunities, e.g. development of new business models such as cooperatives	Higher incomes result in better opportunities for medical treatment, children's education, nutrition and clothing, etc.	Enhanced resilience to climate change
	Empowerment for community members to live independently with hope for the future	

Source: compiled by R Francis, Project Manager FMNR, World Vision Australia from Brown et al, Garrity et al, Haglund et al, McGahuey & Winterbottom, Reij et al, Rinaudo, World Recources Institute.

Key Success Factors and Constraints

While there are numerous accounts of the uptake and spread of FMNR independent of aid and development agencies, the following factors have been found to be beneficial for its introduction and spread:

- Awareness creation of FMNR's potential.

- Capacity building through workshops and exchange visits.

- Awareness of the devastating effects of deforestation. The adoption of FMNR is more likely when communities acknowledge their situation and the need to take action. This perception of need can be supported by education.

- An FMNR champion/facilitator from within the community who encourages, challenges and trains peers. This is critical during the first three to five years, and continues to be important for up to 10 years. Regular site visits also ensure early detection and remedial action on resistance and threats to FMNR through deliberate damage to trees and theft.

- The buy-in of all stakeholders including their agreement on any by-laws created for FMNR and the consequences for infringements. Stakeholders include FMNR practitioners, local, regional and national government departments of agriculture and forestry, men, women, youth, marginalised groups (including nomadic herders), cultivators and commercial interests.

- Stakeholder buy-in is also important to create a critical mass of FMNR adopters in order to change social attitudes from a position of apathy or active participation in deforestation to one of proactive sustainable tree management through FMNR.

- Government support through the creation of favourable policies, positive reinforcement of actions facilitating the spread of FMNR, and disincentives for actions working against the spread of FMNR. FMNR practitioners need to be confident that they will benefit from their labours (either private or community ownership of trees, or legally binding user rights).

- Reinforcement of existing organisational structures (farmers clubs, development groups, traditional leadership structures) or establishment of new structures which will provide a framework for communities to practise FMNR on a local, district or region-wide basis.

- A communications strategy which includes education in schools, radio programs and engagement with religious and traditional leaders to become advocates.

- Establishment of a legal, transparent and accessible market for FMNR wood and non-timber forest products, enabling practitioners to benefit financially from their activities.

Brown et al. suggest that the two main reasons why FMNR has spread so widely in Niger are attitudinal change by the community of what constitutes good land management practices, and farmers' ownership of trees. Farmers need the assurance that they will benefit from their labour. Giving farmers either outright ownership of the trees they protect, or tree-user rights, has made it possible for large-scale farmer-led reforestation to take place.

Current and Future Directions

Over nearly 30 years, FMNR has changed the farming landscape in some of the poorest countries in the world, including parts of Niger, Burkina Faso, Mali and Senegal, providing subsistence farmers with the methods necessary to become more food secure and resilient against severe weather events.

The 2011–12 food crisis in East Africa gave a stark reminder of the importance of addressing root causes of hunger. In the 2011 State of the World Report, Bunch concludes that four major factors – lack of sustainable fertile land, loss of traditional fallowing, cost of fertiliser and climate change – are coming together all at once in a sort of "perfect storm" that will almost surely result in an African famine of unprecedented proportions, probably within the next four to five years. It will most heavily affect the lowland, semi-arid to sub-humid areas of Africa (including the Sahel, parts of eastern Africa, plus a band from Malawi across to Angola and Namibia); and unless the world does something dramatic, 10 to 30 million people could die from famine between 2015 and 2020. Restoration of degraded land through FMNR is one way of addressing these major contributors to hunger.

In recent years FMNR has come to the attention of global development agencies and grass roots movements alike. The World Bank, World Resources Institute, World Agroforestry Center, USAID and the Permaculture movement are amongst those either actively promoting or advocating for the uptake of FMNR and FMNR has received recognition from a number of quarters including:

- In 2010, FMNR won the Interaction 4 Best Practice and Innovation Initiative award in recognition of high technical standards and effectiveness in addressing the food security and livelihood needs of small producers in the areas of natural resource management and agro forestry.

- In 2011, FMNR won the World Vision International Global Resilience Award for the most innovative initiative in the area of resilient development practice and natural environment and climate issues.

- In 2012 FMNR won the Arbor Day Award for Education Innovation.

In April 2012, World Vision Australia – in partnership with the World Agroforestry Center and World Vision East Africa – held an international conference in Nairobi called Beating Famine to analyse and plan how to improve food security for the world's poor through the use of FMNR and Evergreen Agriculture. The conference was attended by more than 200 participants, including world leaders in sustainable agriculture, five East African ministers of agriculture and the environment, ambassadors and other government representatives from Africa, Europe and Australia, and leaders from non-government and international organisations.

Two major outcomes of the conference were:

1. The establishment of a global FMNR network of key stakeholders to promote, encourage and initiate the scale-up of FMNR globally.

2. Country, regional and global level plans as a basis for inter-organisation collaboration for FMNR scale-up.

The conference acted as a catalyst for media coverage of FMNR in some of the world's leading outlets and a noticeable increase in momentum for an FMNR global movement. This heightened awareness of FMNR has created an opportunity for it to spread exponentially worldwide.

World Vision and the World Agroforestry Centrer are currently exploring opportunities for conducting conferences and workshops in new regions where FMNR is not yet established in order to stimulate further awareness and adoption.

Keyline Design

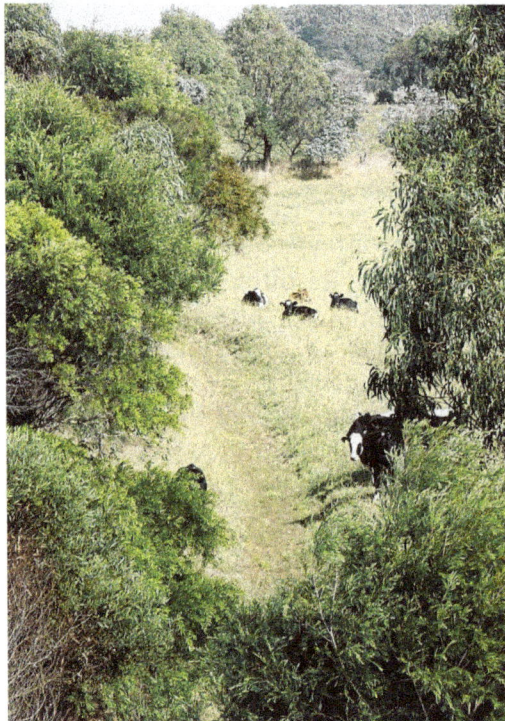

A keyline irrigation channel

Keyline design is a technique for maximizing beneficial use of water resources of a piece of land. The *Keyline* refers to a specific topographic feature linked to water flow. Beyond that however, Keyline can be seen as a collection of design principles, techniques and systems for development of rural and urban landscapes.

Keyline design was developed in Australia by farmer and engineer P. A. Yeomans, and described and explained in his books *The Keyline Plan*, *The Challenge of Landscape*, *Water For Every Farm*, and *The City Forest*.

Application

P. A. Yeomans published the first book on Keyline in 1954. Yeomans described a system of amplified contour ripping to control rainfall runoff and enable fast flood irrigation of undulating land without the need for terracing.

Keyline designs include irrigation dams equipped with through-the-wall lockpipe systems to gravity feed irrigation, stock water, and yard water. Graded earth channels may be interlinked to broaden the catchment areas of high dams, conserve the height of water, and transfer rainfall runoff into the most efficient high dam sites. Roads follow both ridge lines and water channels to provide easier movement across the land.

Keyline Scale of Permanence

The backbone of Yeomans' keyline design system, the outcome of fifteen years of adaptive experimentation, is Yeomans' Keyline Scale of Permanence (KSOP), which identifies typical farms' elements ordered according to their degree of permanence:

- Climate
- Landshape
- Water Supply
- Roads/Access
- Trees
- Structures
- Subdivision Fences
- Soil

Keyline considers these elements when planning the placement of water storage, roads, trees, buildings and fences. On undulating land, a keyline approach involves identifying several features namely ridges and valleys and the natural water courses seeking optimum water storage sites and potential interconnecting channels.

The water lines identified from the land-form subsequently provide optimal locations for the various less permanent elements (roads, fences, trees, and buildings) to optimize the natural potential of the landscape.

Rancho San Ricardo, México

Keypoint

In a smooth grassy valley, a location called the keypoint can be found where the lower and flatter portion of a primary valley floor suddenly steepens. The **keyline** of this primary valley is revealed by pegging a contour line through the keypoint, within the valley shape. All the points on the line are at the same elevation as the keypoint. Contour plowing parallel to the Keyline, both above and below will automatically become "off-contour" but the developing pattern will tend to drift rainwater runoff away from the valley centre and incidentally, prevent erosion.

Keyline pattern cultivation on ridge shapes is done parallel to any suitable contour but only working on the upper side of the contour guide line. This automatically develops a pattern of off-contour cultivation in which all the rip marks left in the soil will slope down towards the centre of the ridge shape. This pattern of cultivation allows more time for water to soak in. Keyline pattern cultivation also enables controlled flood irrigation of undulating land, which further assists in the fast development of deep biologically fertile soil, which results in improving soil nutrition and health.

In many countries, including Australia, it is important to get optimum absorption of rainfall and keyline cultivation does this as well as delaying the potentially damaging concentration of runoff. Yeomans' technique differs from traditional contour plowing in several important respects. Random contour plowing also becomes off contour but usually with the opposite effect on runoff water causing it to quickly shed off ridge shapes and be concentrated in valleys. The limitations of the traditional system of soil conservation, with its "safe disposal" approach to farm water was an important motivational factor in the development of the keyline system.

Applications

David Holmgren, one of the founders of Permaculture, used Yeoman's keyline principle extensively in the formulation of Permaculture concepts and the design of sustainable human settlements and organic farms.

Darren J. Doherty has extensive experience across the world in keyline project design, development, management & training.

A topographical example can be seen on (37°09′33″S 144°15′08″E37.159154°S 144.252248°E).

Keyline also includes concepts for rapid soil fertility enhancement and these concepts are explored in *Priority One* by P. A. Yeomans' son Allan. Yeomans and his sons were also instrumental in the design and production of special plows and cultivating equipment for use in conjunction with the keyline methodology.

Managed Intensive Rotational Grazing

In agriculture, Managed intensive rotational grazing (MIRG), also known as cell grazing, mob grazing and holistic managed planned grazing, describes a variety of closely related systems of forage use in which ruminant and non-ruminant herds and/or flocks are regularly and systematically moved to fresh rested areas with the intent to maximize the quality and quantity of forage growth.

One primary goal of MIRG is to have a vegetative cover over all grazed areas at all times, and to prevent the complete removal of all vegetation from the grazed areas ("bare dirt")

MIRG can be used with cattle, sheep, goats, pigs, chickens, turkeys, ducks and other animals. The herds graze one portion of pasture, or a paddock, while allowing the others to recover. The length of time a paddock is grazed will depend on the size of the herd and the size of the paddock. Resting grazed lands allows the vegetation to renew energy reserves, rebuild shoot systems, and deepen root systems, with the result being long-term maximum biomass production. MIRG is especially effective because grazers do better on the more tender younger plant stems. MIRG also leave parasites behind to die off minimizing or eliminating the need for de-wormers. Pasture systems alone can allow grazers to meet their energy requirements, and with the increased productivity of MIRG systems, the grazers obtain the majority of their nutritional needs without the supplemental feed sources that are required in continuous grazing systems.

One key element of this style of animal husbandry is that either each grazed area must contain all elements needed for the animals (water source, for instance) or the feed or water source must be moved each time the animals are moved. Having fixed feeding or watering stations defeats the rotational aspect, leading to degradation of the ground around the water supply or feed supply if additional feed is provided to the animals. Special care must be taken to ensure that high use areas do not become areas where mud, parasites or diseases are spread or communicated

Herd Health and Welfare

Animal Health Risks

Bloat is a common problem in grazing systems for ruminants, although not for pigs or poultry, that if left untreated can lead to animal death. This problem occurs when foam producing compounds in plants are digested by cows, causing foam to form in the rumen of the animal and ultimately prohibiting animals from expelling gas. The risk of bloat can be mitigated by seeding non-bloating legumes with the grasses. Animals are especially susceptible to bloat if they are moved to new pasture sources when they are particularly hungry. It is therefore important to ensure that the herd is eating enough at the end of a rotation when forage will be more scarce, limiting the potential for animals to gorge themselves when turned onto new paddocks.

Several problems are related to shade in pasture areas. Although shade provides relief from heat and reduces the risk of heat stress, animals tend to congregate in these areas which leads to nutrient loading, uneven grazing, and potential soil erosion. Taller shade trees move the shade area around as the day progresses minimizing this problem.

Animal Health Benefits and Animal Welfare

Herd health benefits arise from animals having access to both space and fresh air. Freedom of movement within a paddock results in increased physical fitness, which limits the potential for injuries and abrasion, and reduces the potential of exposure to high levels of harmful disease-causing microorganisms and insects.

In a concentrated animal feeding operation (CAFO), it is considered normal for a large number of animals to continuously occupy a very small area. The aisles that animals use to move around are consequently constantly coated with a moist layer of manure and urine from the many animals, leading to ailments such as foot rot due to the constant wet exposure. The manure and urine is usually just scraped off into gutters below slatted surfaces, and the surfaces and gutters are rarely washed completely clean, so molds, bacteria, and insects can grow and thrive in the potentially infectious waste. Feeding areas in a CAFO are also rarely stripped bare and washed with a disinfectant, as this would reduce the time available for animals to eat continuously, so molds and bacteria are also able to become established where the animals eat. The close confinement and lack of general environmental cleanliness leads to easier spread of infection and increased sickness, requiring the regular feeding of antibiotics to keep the confined animals healthy but which also leads to antibiotic resistance for bacteria constantly present in the CAFO.

By comparison, with managed grazing, the animals are able to spread out and exist in a natural environment more suited for their natural growth and development. As the animals move to a new paddock, wastes are left behind and allowed to decay without the animals nearby. The animals experience less disease without the need for regular antibiotic dosing, and fewer foot ailments.

Weed Control

In general, a well managed rotational grazing system has rather low pasture weed establishment because the majority of niches are already filled with established forage species, making it hard for competing weeds to emerge and become established. The use of multiple species in the co-grazing helps to minimize weeds. Established forage plants in rotational grazing pasture systems are healthy and unstressed due to the "rest" period, enhancing the competitive advantage of the forage. Additionally, in comparison to cash grain crop production, many plants which would be considered weeds are not problematic in perennial pasture. Many of these plants are actually nutritious to grazers and control of these plants is therefore not necessary in management intensive rotational systems. However, certain species such as thistles and various other weeds, are indigestible and possibly poisonous to grazers. These plant species will not be grazed by the herd and can be recognized for their prevalence in pasture systems.

A key step in managing weeds in any pasture system is identification. Once the undesired species in a pasture system are identified an integrated approach of management can be implemented to control weed populations. It is important to recognize that no single approach to weed management will result in weed free pastures; therefore, various cultural, mechanical, and chemical control methods can be combined in an integrated weed management plan. Cultural controls include avoiding spreading manure contaminated with weed seeds, cleaning equipment after working in weed infested areas, and managing weed problems in fencerows and other areas near pastures. Mechanical controls such as repeated mowing, clipping, and hand weeding can also be used to effectively manage weed infestations by weakening the plant. These methods should be implemented when weed flower buds are closed or just starting to open to prevent seed production. Although these methods for managing weeds greatly reduces reliance on herbicides, weeds problems may still persist in managed grazing systems and the use of herbicides may become necessary. Use of herbicides may restrict use of a pasture for some length of time, depending on the type and amount of the chemical used. Frequently weeds in pasture systems are patchy and therefore spot treatment of herbicides may be used as a least cost methods of chemical control.

Nutrient Availability and Soil Fertility

If pasture systems are seeded with more than 40% legumes, commercial nitrogen fertilization is unnecessary for adequate plant growth. Legumes are able to fix atmospheric nitrogen, thus providing nitrogen for themselves and surrounding plants. Although grazers remove nutrient sources from the pasture system when they feed on forage sources, the majority of the nutrients consumed by the herd are returned to the pasture system through manure. At a relatively high stocking rate, or high ratio of animals per hectare, manure will be evenly distributed across the pasture system. The nutrient content in these manure sources should be adequate to meet plant requirements, making commercial fertilization unnecessary. Management intensive rotational grazing systems are often associated with increased soil fertility which arises because manure is a rich source of organic matter that increases the health of soil. In addition, these pasture system are less susceptible to erosion because the land base has continuous ground cover throughout the year.

High levels of fertilizers entering waterways are a pertinent environmental concern associated with agricultural systems. However, management intensive rotational grazing systems effectively reduce the amount of nutrients that move off-farm which have the potential to cause environmental degradation. These systems are fertilized with on-farm sources, and are less prone to leaching as compared to commercial fertilizers. Additionally, the system is less prone to excess nutrient fertilization, so the majority of nutrients put into the system by manure sources are utilized for plant growth. Permanent pasture systems also have deeper, well established forage root systems which are more efficient at taking up nutrients from within the soil profile.

Socio-cultural-economic Considerations

Although milk yields are often lower in MIRG systems, net farm income per cow is often greater as compared to confinement operations. This is due to the additional costs associated with herd health and purchased feeds are greatly reduced in management intensive rotational grazing systems. Additionally, a transition to management intensive rotational grazing is associated with low start-up and maintenance costs. Another consideration is that while production per cow is less, the amount of cows per acre on the pasture can increase. The net effect is more productivity per acre at less cost.

The main costs associated with transitioning to management intensive rotational grazing are purchasing fencing, fencers, and water supply materials. If a pasture was continuously grazed in the past, likely capital has already been invested in fencing and a fencer system. Cost savings to graziers can also be recognized when one considers that many of the costs associated with livestock operations are transmitted to the grazers. For example, the grazers actively harvest their own sources of food for the portion of the year where grazing is possible. This translates into lower costs for feed production and harvesting, which are fuel intensive endeavors. MIRG systems rely on the grazers to produce fertilizer sources via their excretion. There is also no need for collection, storage, transportation, and application of manure, which are also all fuel intensive. Additionally, external fertilizer use contributes to other costs such as labor, purchasing costs.

It can also be demonstrated that management intensive rotational grazing system also result in time savings because the majority of work which might otherwise require human labor is transmitted to the herd.

Environmental Considerations

Many pastures undergoing MIRG are less susceptible to soil erosion and are associated with higher soil fertility than continuously grazed pastures, depending on the skill of the manager and the management system he is using. As a result, the paddocks require fewer commercial inputs, which have been associated with negative environmental impacts. In addition, because these systems tend to be more resilient and stable they are more capable of responding to changing environmental conditions and perturbations while not compromising productivity.

Human Nutrition

Animals raised on pasture have shown major differences in the nutritional quality of the products they produce for human consumption.

Criticism

Managed intensive rotational grazing paints a wide brush over many different managed grazing systems. Managers have found that rotational grazing systems can work for diverse management purposes, but scientific experiments have demonstrated that rotational grazing systems do not always necessarily work for specific ecological purposes. This controversy stems from two main categorical differences in rotational grazing, prescribed management and adaptive management. The performance of rangeland grazing strategies are similarly constrained by several ecological variables establishing that differences among them are dependent on the effectiveness of those management models. Depending on the management model, plant production has been shown to be equal or greater in continuous compared to rotational grazing in 87% of the experiments.

Natural Farming

Natural farming is an ecological farming approach established by Masanobu Fukuoka (1913–2008), a Japanese farmer and philosopher, introduced in his 1975 book *The One-Straw Revolution*. Fukuoka described his way of farming as 自然農法 (*shizen nōhō*) in Japanese. It is also referred to as "the Fukuoka Method", "the natural way of farming" or "do-nothing farming". The title refers not to lack of effort, but to the avoidance of manufactured inputs and equipment. Natural farming is related to fertility farming, organic farming, sustainable agriculture, agroforestry, ecoagriculture and permaculture but should be distinguished from biodynamic agriculture.

The system exploits the complexity of living organisms that shape each particular ecosystem. Fukuoka saw farming both as a means of producing food and as an aesthetic or spiritual approach to life, the ultimate goal of which was, "the cultivation and perfection of human beings". He suggested that farmers could benefit from closely observing local conditions. Natural farming is a closed system, one that demands no human-supplied inputs and mimics nature.

Fukuoka's ideas challenged conventions that are core to modern agro-industries, instead promoting an approach that takes advantage of the local environment. Natural farming differs from conventional organic farming, which Fukuoka considered to be another modern technique that disturbs nature.

Fukuoka claimed that his approach prevents water pollution, biodiversity loss and soil erosion, while providing ample amounts of food.

Principles

Fukuoka distilled natural farming into five principles:

1. No tillage

2. No fertilizer

3. No pesticides or herbicides

4. No weeding

5. No pruning

Though many of his plant varieties and practices relate specifically to Japan and even to local conditions in subtropical western Shikoku, his philosophy and the governing principles of his farming systems have been applied from Africa to the temperate northern hemisphere. In India, natural farming is often referred to as "Rishi Kheti". In India natural farming or rishi kheti includes ancient vedic principles of farming including use of animal waste and herbs for controlling pests and promoting growth. The rishi 's or Indian sages use cow products like buttermilk, milk, curd and its waste urine for preparing growth promoters. The Rishi or Vedic farming is regarded as non-violent farming without any usage of chemical fertilizer and pesticides. They obtain high quality natural or organic produce having medicinal values.Today still small number of farmers in Madhya Pradesh, Punjab, Maharashtra and Andhra Pradesh, Tamil Nadu use this farming in India.

Principally, natural farming minimises human labour and adopts, as closely as practical, nature's production of foods such as rice, barley, daikon or citrus in biodiverse agricultural ecosystems. Without plowing, seeds germinate well on the surface if site conditions meet the needs of the seeds placed there. Fukuoka used the presence of spiders in his fields as a key performance indicator of sustainability.}

Fukuoka specifies that the ground remain covered by weeds, white clover, alfalfa, herbaceous legumes, and sometimes deliberately sown herbaceous plants. Ground cover is present along with grain, vegetable crops and orchards. Chickens run free in orchards and ducks and carp populate rice fields.

Periodically ground layer plants including weeds may be cut and left on the surface, returning their nutrients to the soil, while suppressing weed growth. This also facilitates the sowing of seeds in the same area because the dense ground layer hides the seeds from animals such as birds.

For summer rice and winter barley grain crops, ground cover enhances nitrogen fixation. Straw from the previous crop mulches the topsoil. Each grain crop is sown before the previous one is harvested by broadcasting the seed among the standing crop. Later, this method was reduced to a single direct seeding of clover, barley and rice over the standing heads of rice. The result is a denser crop of smaller, but highly productive and stronger plants.

Fukuoka's practice and philosophy emphasised small scale operation and challenged the need for mechanised farming techniques for high productivity, efficiency and economies of scale. While his family's farm was larger than the Japanese average, he used one field of grain crops as a small-scale example of his system.

Climax Ecosystems

In ecology, climax ecosystems are mature ecosystems that have reached a high degree of stability, productivity and diversity. Natural farmers attempt to mimic those virtues, creating a comparable climax ecosystem, and employ advanced techniques such as intercropping, companion planting and integrated pest management.

No-till

Natural farming recognizes soils as a fundamental natural asset. Ancient soils possess physical and chemical attributes that render them capable of generating and supporting life abundance. It can be argued that tilling actually degrades the delicate balance of a climax soil:

1. Tilling may destroy crucial physical characteristics of a soil such as *water suction*, its ability to send moisture upwards, even during dry spells. The effect is due to pressure differences between soil areas. Furthermore, tilling most certainly destroys soil horizons and hence disrupts the established flow of nutrients. A study suggests that reduced tillage preserves the crop residues on the top of the soil, allowing organic matter to be formed more easily and hence increasing the total organic carbon and nitrogen when compared to conventional tillage. The increases in organic carbon and nitrogen increase aerobic, facultative anaerobic and anaerobic bacteria populations.

2. Tilling over-pumps oxygen to local soil residents, such as bacteria and fungi. As a result, the chemistry of the soil changes. Biological decomposition accelerates and the microbiota mass increases at the expense of other organic matter, adversely affecting most plants, including trees and vegetables. For plants to thrive a certain quantity of organic matter (around 5%) must be present in the soil.

3. Tilling uproots all the plants in the area, turning their roots into food for bacteria and fungi. This damages their ability to aerate the soil. Living roots drill millions of tiny holes in the soil and thus provide oxygen. They also create room for beneficial insects and annelids (the phylum of worms). Some types of roots contribute directly to soil fertility by funding a mutualistic relationship with certain kinds of bacteria (most famously the rhizobium) that can fix nitrogen.

Fukuoka advocated avoiding any change in the natural landscape. This idea differs significantly from some recent permaculture practice that focuses on permaculture design, which may involve the change in landscape. For example, Sepp Holzer, an Austrian permaculture farmer, advocates the creation of terraces on slopes to control soil erosion. Fukuoka avoided the creation of terraces in his farm, even though terraces were common in China and Japan in his time. Instead, he prevented soil erosion by simply growing trees and shrubs on slopes.

Fertility farming

In 1951, Newman Turner advocated the practice of "fertility farming", a system featuring the use of a cover crop, no tillage, no chemical fertilizers, no pesticides, no weeding and no composting.

Although Turner was a commercial farmer and did not practice random seeding of seed balls, his "fertility farming" principles share similarities with Fukuoka's system of natural farming. Turner also advocated a "natural method" of animal husbandry.

Nature Farming

Japanese farmer and philosopher Mokichi Okada, conceived of a "no fertilizer" farming system in the 1930s that predated Fukuoka. Okada used the same Chinese characters, which are generally translated in English as "nature farming". Agriculture researcher Hu-lian Xu claims that "nature farming" is the correct literal translation of the Japanese term.

Regenerative Agriculture

Regenerative agriculture is a sub-sector practice of organic farming designed to build soil health or to regenerate unhealthy soils. The practices associated with regenerative agriculture are those identified with other approaches to organic farming, including maintaining a high percentage of organic matter in soils, minimum tillage, biodiversity, composting, mulching, crop rotation, cover crops, and green manures.

Biodiversity

V. Organic Agriculture

Hoverfly at work

In the past, regenerative farming was seen as a long-term integrated approach that proponents used to build soil health, promote nutrient retention, and encourage pest and disease resistance. Many of the practices associated with regenerative farming are management practices associated with organic agriculture. In practice, these practices can be applied in any type of horticulture and properly managed livestock with Holistic Planned Grazing (Savory & Butterfield Holistic Management - A new decision making framework) where one of the main goals is to build soil organic matter, an organic practice understood by practitioners of organic farming to have far reaching benefits for plant health and farm sustainability. When combined with the spirit of organic agriculture such practices are said to produce healthy soil, healthy food, clean water and clean air using inexpensive inputs local to the farm. Practices that minimize biota disturbance and erosion losses while incorporating carbon rich amendments and retaining the biomass of roots and shoots are encouraged in regenerative farming.

Best Practices

Foremost among best practices in regenerative farming are zero-tolerance for synthetic pesticides, fertilizers, and other inputs that disrupt soil life. On the other hand, conservation tillage, while not yet widely used in organic systems, is viewed as a regenerative organic practice integral to soil-carbon sequestration.

Contributions from the Rodale Institute

Rodale Institute, Test Garden

Recent scientific research has shown that practices inherent in the regenerative philosophy contribute to carbon sequestration by natural processes of photosynthetic removal and retention of atmospheric CO_2 in soil organic matter. At the forefront of this research is the Rodale Institute, which is one of the major proponents of regenerative agriculture. Notably, the concept of *regenerative organic agriculture* was coined by Robert Rodale prior to his untimely death in 1990. The Rodale approach defines regenerative farming as a long-term, holistic design that attempts to grow as much food using as few resources as possible in a way that revitalizes the soil rather than depleting it, while offering a solution to carbon sequestration. The mantra is the slogan: "Healthy Soil = Healthy Food = Healthy People." In Rodale's view, when coupled with the management goal of carbon sequestration, regenerative "farming becomes, once again, a knowledge intensive enterprise, rather than a chemical and capital-intensive one."

Permaculture and Regeneration Science

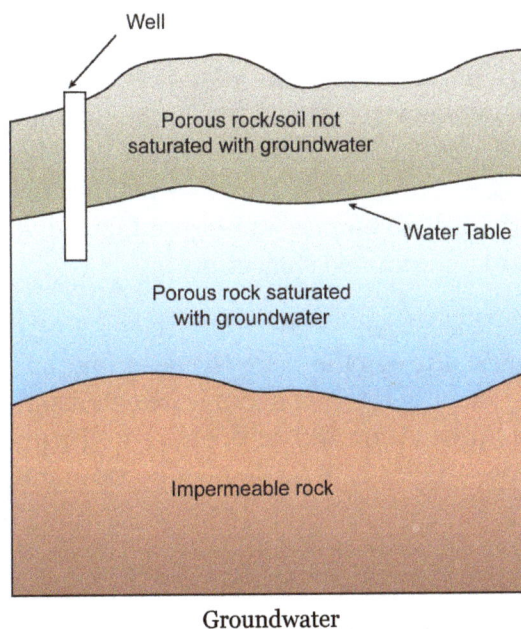

Groundwater

In permaculture, a regenerative farm is one where biological production and ecological structure are growing increasingly more complex over time, but yields continue to increase while external inputs decrease. The way that regeneration is determined in this construct is by whether the components in the system or actions taken in the system increase both biological diversity *and* biomass. The overall health of the ecological system in which the farm and its humans are guests is determined by the health of the water and soil. This is achieved by strengthening (making more resilient, redundant) three key components of regenerative ecosystem management:

1. Increased biomass

2. Increased biological activity

3. Intentional remineralization

Foundations

The foundation of regenerative agriculture has occurred over thousands of years, a span of time when most farming was organically practiced worldwide. Even then, its knowledge base was not global because its practitioners excluded traditional non-western local knowledge systems, about which we are only beginning to learn more. It is not possible to list the many contributors or contributions to this process over many centuries. Yet, some voices and systems should be illuminated. Among these, in alphabetical order, are the following:

William Albrecht (1888–1974), an agronomist at the University of Missouri for many years. Through his writings, lectures, and radio programs, Albrecht promoted the intimate relationship between healthy soil and animal nutrition, a relationship that includes humans. Feed the soil to feed the plants to feed the consumer was his mantra.

Lady Evelyn Barbara "Eve" Balfour (1899–1990) was a founding figure in the 20th century organic movement and an organic farming pioneer. She was one of the first women to study agriculture at an English university, graduating from the University of Reading, and began farming in 1920. By 1939, she had launched a privately funded experimental farm, called Haughley, to test the principles of organic farming. The initial findings of the work there and her research were published in Living Soil (1948), which has become an organic farming classic. Haughley was the first long-term comparative research project measuring results from organic and chemically based farming. What the results of her work at Haughley revealed should have led to a global push to institute sustainable agriculture throughout the British empire and beyond, but that was not to be, as the global power of post-WWII capitalism undermined such efforts.

George Washington Carver (1864–1943), inventor and scientist of Tuskegee University, was the grand-daddy of sustainable agriculture in the USA. His contributions are too many to detail here, but among them is his development of the U.S. agriculture extension system and his many inventions in polymer science. His influence on U.S. agriculture in the first half of the 20th century is not fully appreciated nor, more tragically, even known by contemporary proponents of today's biointensive farming methods. Carver's scientific research was the foundation of US agriculture policy under two generations of Secretaries of Agriculture. His death in 1948 marked the end of organic agricultural practices in commercial agriculture in the United States until recently, as petrochemical companies and Agribusiness corporations competed successfully to control the actors in the various industries.

Alan Chadwick (1909–1980) was a leading innovator of organic farming techniques and an influential educator in the field of biodynamic/French intensive gardening through much of the 20th century. A student of Rudolf Steiner, Chadwick is often cited as inspirational to the development of the "California Cuisine" movement.

Field Hamois Belgium Luc Viatour

George Chan (born 1923), former Ministry of Agro-Industry & Fisheries of Mauritius, and Mae-Wan Ho, Director of the Institute of Science in Society, are architects of Dream Farm, a model of a sustainable zero-emission, zero-waste production farm that maximizes the use of renewable energies, turns residues into food and energy resources, and eliminates the need for fossil fuels. A core principle of Dream Farm is that sustainable systems are organisms, comprising the farmer, livestock and crops. In a Dream Farm, very little input is wasted or exported to the environment; most are recycled and kept inside the system by consciously integrating food and energy production. Mr. Chan has been a pioneer in the field of sustainable recycling of waste using agro-ecological systems. He developed the Integrated Farming and Waste Management System, which involves a sustainable cycle in which matter and energy circulate through different stages, dramatically increasing yields.

Masanobu Fukuoka (1913–2008) of Japan lived a long and dutiful life in partnership with his environment. He was a farmer, activist, and author of the practices and theory of natural farming, on which he based his four core uncompromising principles: no cultivation, no (chemical) fertilizers, no weeding, and no pesticides. Among his teachings is that for a farmer to be successful, he must form a partnership with the natural environment, derive an intimate understanding of it together with the plants a farmer chooses to grow. Features of his philosophy are present in most contemporary farming practices. His "seed balls" cultivation innovation is widely used in many horticultural environments and in commercial retail products, including lawn seed.

Takao Furuno (born 1950) of Japan is the architect of the Aigamo Method, a modernization of an 800-year-old Chinese technique of using ducks to promote sustainable rice cultivation. Funuro's system is polycultural, combining rice cultivation with duck husbandry, aquaculture, and vegetable production. Together, these enterprises provide income from rice, vegetable and flower production, eggs, meat, and live ducks from the aigamo, and fish from the paddies. The aigamo ducks are a breed derived from wild and domestic ducks, whose ducklings provide the labor for cultivation, pest control, and manure for fertilization of the rice paddies. This beneficial marriage between duck and rice eliminates dependency on chemical fertilizers, pesticides, molluscicides, fossil fuels, and heavy duty equipment as costs, while sustaining a safe environment for farmers to work and simultaneously increasing net production and farm income.

John D. Hamaker (1914–1994) was a mechanical engineer, ecologist and practical visionary who from 1968 to 1994 tried to awaken the world to the crucial need to remineralize the soils and regenerate the Earth's life-support system. His motivations included a profound desire to help create a healthy, just and real civilization rooted in ecological wisdom, and his realization that malnutrition and disease followed by famine and glaciation could be ended by a total human commitment.

Julius Hensel (1844–1903), a German miller and author of Bread from Stones. In the 1890s, Hensel was an early advocate of restoring trace minerals to soil with dust from primeval stones and reported successful results with his steinmehl (stonemeal). It is said that his ideas were not accepted due to both technical limitations and financing. But, according to proponents of his method, Hensel's opposition from manufacturers of chemical fertilizers, set the stage for what would happen eventually in American agriculture following WWII when there was little opposition and the rise of the U.S. petrochemical industry. Today, Hensel's pioneer work in opposing the use of chemicals in agriculture found rebirth in the Organic Movement a half a century later. Yet, Hensel may be more modern than the most modern agricultural reformer. On the basis of theoretical chemical considerations, supported by practical tests, he claimed that his rock dust can replace not only chemical fertilizers, but all animal ones as well.

^{Sepp Holzer} (born 1942) of Austria, and his wife Veronika, have created a diverse and natural way of growing food in an unconventional way by using a terraced system of mounds on the Austrian mountainsides, referred to as hugelkultur. The mounds are built on a foundation of organic materials, a traditional way of growing in the region of The Krameterhof in Lungau, Austria, just not at 1000+ meters above sea level. Holzer's edible microclimates are considered one of the few perfectly working permaculture systems in the world. After almost 40 years of continuous production, the Holzer farms contain a complex of pond culture, terraces, water power station, thousands of fruit trees cultivated among companionable plant families, thirty different types of potatoes, many different grains, fruits, vegetables, herbs and wildflowers are growing just about everywhere — in the forest, on extremely steep hills, on rocky outcrops, on stone pathways, and around ponds, all without the use of any pesticides, herbicides or synthetic fertilizers. Their video, *Farming With Nature: A Case Study of Successful Temperate Permaculture*, is widely distributed on the Internet and a must-view for anyone looking to develop a sustainable collaborative system of food production.

Sir Albert Howard (1873–1947), an English botanist, was an agricultural advisor to the British government in charge of a colonial research farm at Indore in India. Howard has been called the father of modern composting for his refinement of a traditional Indian composting system into what is now known as the Indore method. It was at Indore that Howard documented and tested Indian organic farming techniques. Sir Howard shared this knowledge through the Soil Association in England and the Rodale Institute in the United States. In his later years, he was the editor of the influential journal, *Biodynamics Journal*.

Elaine Ingham, a microbiologist and founder of Soil Foodweb, Inc. in 1996, is recognized as a premier authority in soil microbiology and the soil food web. Through her pioneering research and lectures, Dr. Ingham has been instrumental in popularizing the importance of soil health and the growing public understanding of the soil food web in sustaining this health. Since January 2011, she has been the Chief Scientist at The Rodale Institute where she continues to study the microbial life of the soil and to give lectures on her findings.

John Ikerd, a successor to William Albrecht and professor emeritus at the University of Missouri, continues to be a staunch advocate for the "small" family farm and farmers. Today, he is an active crusader for sustainability in the US food system. His views and writings are available on his Website and at YouTube. Ikerd is author of The essentials of economic sustainability, Small Farms are Real Farms: Sustaining People through Agriculture and Sustainable Capitalism (2005).

John Jeavons, the inspiration and architect of a sustainable 8-step food production method officially known as Grow Biointensive, which combines elements of French intensive and biodynamics techniques. The Grow Biointensive approach is promoted by Ecology Action, a non-profit that operates a research mini-farm in Willits, CA and a retail store called Bountiful Gardens. Both projects promote the Grow Biointensive method teaching people in more than a hundred countries. Ecology Action's research and publications have several goals: (1) enabling small-scale farms and farmers worldwide to significantly increase food production and income by (2) utilizing predominantly local, renewable resources to decrease expenses and energy inputs (labor, land, water) and (3) building fertile topsoil at a rate 60 times faster than in nature.

Patricia Lanza was born in 1935 in Crossville, Tennessee, to teenagers George and Mamie Neal. An only child for eight years, she spent her early formative years with grandparents while her parents worked in Detroit, MI. She perfected and authored several books about lasagna gardening or sheet mulch gardening, a type of gardening perfected by Ruth Stout in the 1950s, but whose books had gone out of print. Lanza introduced a new generation of gardeners to the ultimate no-till method of growing in which little or no labor is wasted digging and amending soils. Instead, the lasagna method feeds the soil biota from above and encourages the soil food web to do the work of aerating and mixing the nutrients into the soil below.

Jacob Mittleider, architect of the Mittleider Method, a popular contemporary method of soil-less growing. Mittleider's system is being continued and refined by Jim Kennard at the Food for Everyone Foundation.

Bill Mollison and David Holmgren, architects of permaculture or "permanent agriculture", a holistic approach that combines ecological design with natural principles of horticultural production. Both men are ardent advocates for creating communities that work in harmony with nature rather than in opposition to it. Permaculture is a worldwide phenomenon whose principles and practices have been published in many of the world's languages and is rapidly being integrated into public planning projects across the globe. Video presentations about Permaculture and its practitioners are broadly distributed on the Internet and can be found at YouTube.

Maynard Murray (1910–1983) was a medical doctor and a pioneer in merging the disciplines of biology, health and agriculture from the 1930s when he began experimenting with "sea-solids"– mineral salts that remain after total sea water evaporation. Around 1940, he began to perform extensive experiments to determine when the proportions of trace minerals and other elements present in sea water were optimum for growth and health of both land and sea life. His extensive experiments demonstrated repeatedly and conclusively that plants fertilized with sea solids and animals fed sea-solid-fertilized feeds grow stronger and more resistant to disease. Murray recounts his experiments and presents his conclusions in his classic work, Sea Energy Agriculture (1976; republished in 2001). Largely ignored during his lifetime, his lifelong quest contributed greatly to our understanding of the role of trace minerals in the healthy growth of all organisms on the planet, including humans.

J.I. Rodale (1898–1971) was an early proponent of organic and regenerative farming and founder of the Rodale Institute in the United States. He is credited with launching organic gardening practices more broadly in the United States through his writings, research, and publishing enterprise. The Rodale enterprises continue to make contributions around the world through their advocacy, research, demonstrations, and publishing.

Robert Rodale (1930–1990), former CEO of The Rodale Institute, was a major advocate of regenerative agriculture, fostered the Regenerative Agriculture Association, published numerous books on the subject, funded research, established demonstration fields, sponsored practitioners in the field, and spread the knowledge system of regenerative agriculture around the globe. He coined the concept of 'regenerative organic agriculture' to distinguish it from 'sustainable' agriculture.

Bhaskar H. Save (born 1922) is creator of the highly successful Kalpavruksha ("wish-fulfilling tree") Farm in Umbergaon, India established in 1953. After practicing traditional agriculture for many years with poor results, Sri Save committed his resources to organic farming and developed a system of natural farming that Masanobu Fukuoka, the noted founder of natural farming, praised as the best example of natural farming he had witnessed anywhere. Sri Save used intensive interplanting in which short life-span vegetables (alpa-jeevi), medium life-span species (madhya-jeevi – such as banana, papaya, and custard apple), and long life-span species (deergha-jeevi–such as chikoo, coconut, mango) are combined and phased in over time until the long life-span species mature.

Ethan Roland Soloviev and Gregory Landua, cofounders of Terra Genesis International (a regenerative agriculture and supply company), published a paper in 2016 titled Levels of Regenerative Agriculture. In this paper, they describe a four-fold framework consisting of:

- Functional Regenerative Agriculture: "humans can do good through their agricultural production"

- Integrative Regenerative Agriculture: "grow the health and vitality of the whole ecosystem"

- Systemic Regenerative Agriculture: requiring personal development; "farms are woven into an ecosystem of enterprises operating in their bioregion"

- Evolutionary Regenerative Agriculture: requiring pattern understanding; "harmonize with the potential of a place," and "develop a diversity of global and local regenerative producer webs"

Rather than creating a hierarchy, Soloviev and Landua posit that each level of regenerative agriculture has its place, depending upon context and aim.

Rudolf Steiner (1861–1925), an Austrian and one of the great minds of the 20th century, is the architect of the Biodynamics method of agriculture, the first truly modern ecological food production system. Biodynamic agriculture was one of the first systems also to treat farms as unified and individual organisms or ecosystems. The system has its basis in a world-view referred to as Anthroposophy, which Steiner developed from his marriage of spiritual, cultural, and intellectual life experiences. Biodynamics emphasizes a holistic balance between the soil, plants, and animals in a closed, self-nourishing system, with an emphasis on animal waste products and composts and exclusion of artificial chemicals. Some approaches are unique, such as (1) the use of fermented herbal and mineral preparations as compost additives and field sprays and (2) the use of an astronomical sowing and planting calendar. The biodynamics method continues to have a large following worldwide, and deservedly so—Rudolf Steiner was a brilliant thinker in many ways.

Ruth Stout (1884–1980) lived a long, active and productive life. By the 1950s, she had perfected a "no-till" method of gardening that she promoted as "no work" in her writings about gardening, including two books, How to Have a Green Thumb Without an Aching Back and Gardening Without Work for the Aging, the Busy, and the Indolent. The latter volume was republished by Mother Earth News in 2011. Her work has led to other innovations in no-till practices, such as slash and mulch in the tropics.

Charles Walters (1926–2009) was an economist, journalist, farmers advocate in the first phase of his career with the National Farmer's Organization; and founder, publisher and editor of Acres U.S.A., North America's oldest publisher on production-scale organic and sustainable farming in the second phase of his extraordinary life. Walters penned hundreds of articles on the technologies of organic and sustainable agriculture and is author or co-author of several books, including *Eco-Farm, Weeds: Control Without Poisons, Unforgiven*, a book about visionary farm economist Carl Wilken, and many more. In 1970, shortly after he started Acres, Walters coined the term "eco-agriculture" because he wanted to unify the concepts of "ecological" and "economical" in the belief that unless agriculture was ecological, it could not be economical.

Keyline Irrigation, Taranaki Farm

Don Weaver, a protégé of John Hamaker, is an ecologist and gardener, who assisted Hamaker in advocating for policies and practices of soil remineralization, biosphere regeneration, and climatic stabilization. He continues to promote these causes today.

Booker T. Whatley (1915–2005), a horticulturalist and beneficiary of the George Washington Carver tradition, may be best remembered for popularizing U-Pick farms and their direct marketing approach through fee-based customer subscriptions. But, he was also among the first practitioners of sustainable agriculture to focus more directly on the economic concerns of small farmers, encouraging them to identify high value crops and enterprises that were more profitable on smaller units of land, such as shiitake mushrooms, the husbandry of small ruminants, specialty cheeses, and much more.

P.A. Yeomans (1904–1984), an Australian geologist and the architect of the Keyline design, an innovative solution to farm water management, is little known outside of Australia for his many contributions from sustainable agriculture to soil fertility to farm management in the early 20th century. His use of land topography to harvest rain water into ponds is quietly used today in the construction of swales and berms in production units ranging from backyard gardens to the monumental landscaping practices of Sepp Holzer today.

References

- Smith, Joseph Russell; Smith, John (1987). Tree Crops: A permanent agriculture. Island Press. ISBN 978-1-59726873-8.

- Holmgren, David (2002). Permaculture: Principles & Pathways Beyond Sustainability. Holmgren Design Services. p. 1. ISBN 0-646-41844-0.

- Ash, Andrew, The Ecograze Project – developing guidelines to better manage grazing country (PDF), et al., CSIRO, ISBN 0-9579842-0-0, retrieved 7 April 2013

- Nick Romanowski (2007). Sustainable Freshwater Aquaculture: The Complete Guide from Backyard to Investor. UNSW Press. p. 130. ISBN 978-0-86840-835-4.

- Yeomans, P.A. (1973). Water for Every Farm: A practical irrigation plan for every Australian property. Sydney NSW: K.G. Murray. ISBN 0-646-12954-6. ISBN 0-909325-29-4.

- Yeomans, P.A.; Yeomans, K.B. (1993). Water for Every Farm — Yeomans Keyline Plan. Keyline Designs. ISBN 0646129546. 2002 ISBN 0646418750

- Yeomans, P.A.; Yeomans, K.B. (2008). Water for Every Farm — Yeomans Keyline Plan. Keyline Designs. ISBN 1438225784. External link in |publisher= (help)

- Colin Adrien MacKinley Duncan (1996). The Centrality of Agriculture: Between Humankind and the Rest of Nature. McGill-Queen's Press - MQUP. ISBN 978-0-7735-6571-5.

- Stephen Morse; Michael Stockin (1995). People and Environment: Development for the Future. Taylor & Francis Group. ISBN 978-1-85728-283-2.

- Elpel, Thomas J. (November 1, 2002). Participating in Nature: Thomas J. Elpel's Field Guide to Primitive Living Skills. ISBN 1892784122.

- Priya Reddy; Prescott College Environmental studies (2010). Sustainable Agricultural Education: An Experiential Approach to Shifting Consciousness and Practices. Prescott College. ISBN 978-1-124-38302-6.

- Helena Norberg-Hodge; Peter Goering; John Page (1 January 2001). From the Ground Up: Rethinking Industrial Agriculture. Zed Books. ISBN 978-1-85649-994-1.

- Masanobu Fukuoka (1987). The Natural Way of Farming: The Theory and Practice of Green Philosophy. Japan Publications. ISBN 978-0-87040-613-3.

- Sylvia, D.M.; Fuhrmann, J.J.; Hartel, P.G.; Zuberer, D.A. (1999). Principles and Applications of Soil Microbiology. New Jersey: Prentice Hall. pp. 39–41. ISBN 0130941174.

- Xu, Hui-Lian (2001). NATURE FARMING In Japan (Monograph). T. C. 37/661(2), Fort Post Office, Trivandrum - 695023, Kerala, India.: Research Signpost. ISBN 81-308-0111-6. Retrieved 6 March 2011.

6

Environmental and Conservation Movements

Environmental and conservation movements have increased in number in the past couple of years. The increase in global warming and environmental degradation has caused several NGOs and governments to take steps regarding environmental degradation. Some of the movements explained in this section are conservation movement, Chipko movement and Anarcho-naturism. This section helps the reader in understanding the importance and the history of conserving our environment.

Environmental Movement

Apollo 8's *Earthrise*, 24 December 1968

The environmental movement (sometimes referred to as the ecology movement), also including conservation and green politics, is a diverse scientific, social, and political movement for addressing environmental issues. Environmentalists advocate the sustainable management of resources and stewardship of the environment through changes in public policy and individual behavior. In its recognition of humanity as a participant in (not enemy of) ecosystems, the movement is centered on ecology, health, and human rights.

The environmental movement is an international movement, represented by a range of organizations, from the large to grassroots and varies from country to country. Due to its large membership, varying and strong beliefs, and occasionally speculative nature, the environmental movement is not always united in its goals. The movement also encompasses some other movements with a more specific focus, such as the climate movement. At its broadest, the movement includes private citizens, professionals, religious devotees, politicians, scientists, nonprofit organizations and individual advocates.

History

Early awareness

Levels of air pollution rose during the Industrial Revolution, sparking the first modern environmental laws to be passed in the mid-19th century

Early interest in the environment was a feature of the Romantic movement in the early 19th century. The poet William Wordsworth had travelled extensively in the Lake District and wrote that it is a "sort of national property in which every man has a right and interest who has an eye to perceive and a heart to enjoy".

The origins of the environmental movement lay in the response to increasing levels of smoke pollution in the atmosphere during the Industrial Revolution. The emergence of great factories and the concomitant immense growth in coal consumption gave rise to an unprecedented level of air pollution in industrial centers; after 1900 the large volume of industrial chemical discharges added to the growing load of untreated human waste. Under increasing political pressure from the urban middle-class, the first large-scale, modern environmental laws came in the form of Britain's Alkali Acts, passed in 1863, to regulate the deleterious air pollution (gaseous hydrochloric acid) given off by the Leblanc process, used to produce soda ash.

Conservation Movement

Students from the forestry school at Oxford, on a visit to the forests of Saxony in the year 1892

The modern conservation movement was first manifested in the forests of India, with the practical application of scientific conservation principles. The conservation ethic that began to evolve included three core principles: that the human activity damaged the environment, that there was a civic duty to maintain the environment for future generations, and that scientific, empirically based methods should be applied to ensure this duty was carried out. Sir James Ranald Martin was prominent in promoting this ideology, publishing many medico-topographical reports that demonstrated the scale of damage wrought through large-scale deforestation and desiccation, and lobbying extensively for the institutionalization of forest conservation activities in British India through the establishment of Forest Departments. The Madras Board of Revenue started local conservation efforts in 1842, headed by Alexander Gibson, a professional botanist who systematically adopted a forest conservation program based on scientific principles. This was the first case of state management of forests in the world. Eventually, the government under Governor-General Lord Dalhousie introduced the first permanent and large-scale forest conservation program in the world in 1855, a model that soon spread to other colonies, as well the United States. In 1860, the Department banned the use shifting cultivation. Dr. Hugh Cleghorn's 1861 manual, *The forests and gardens of South India*, became the definitive work on the subject and was widely used by forest assistants in the subcontinent.

Sir Dietrich Brandis joined the British service in 1856 as superintendent of the teak forests of Pegu division in eastern Burma. During that time Burma's teak forests were controlled by militant Karen tribals. He introduced the "taungya" system, in which Karen villagers provided labour for clearing, planting and weeding teak plantations. He formulated new forest legislation and helped establish research and training institutions. The Imperial Forest School at Dehradun was founded by him.

Formation of Environmental Protection Societies

The late 19th century saw the formation of the first wildlife conservation societies. The zoologist Alfred Newton published a series of investigations into the *Desirability of establishing a 'Close-time' for the preservation of indigenous animals* between 1872 and 1903. His advocacy for legislation to protect animals from hunting during the mating season led to the formation of the Plumage League (later the Royal Society for the Protection of Birds) in 1889. The society acted as a protest group campaigning against the use of great crested grebe and kittiwake skins and feathers in fur clothing. The Society attracted growing support from the suburban middle-classes, and influenced the passage of the Sea Birds Preservation Act in 1869 as the first nature protection law in the world.

John Ruskin an influential thinker who articulated the Romantic ideal of environmental protection and conservation

For most of the century from 1850 to 1950, however, the primary environmental cause was the mitigation of air pollution. The Coal Smoke Abatement Society was formed in 1898 making it one of the oldest environmental NGOs. It was founded by artist Sir William Blake Richmond, frustrated with the pall cast by coal smoke. Although there were earlier pieces of legislation, the Public Health Act 1875 required all furnaces and fireplaces to consume their own smoke.

Systematic and general efforts on behalf of the environment only began in the late 19th century; it grew out of the amenity movement in Britain in the 1870s, which was a reaction to industrialization, the growth of cities, and worsening air and water pollution. Starting with the formation of the Commons Preservation Society in 1865, the movement championed rural preservation against the encroachments of industrialisation. Robert Hunter, solicitor for the society, worked with Hardwicke Rawnsley, Octavia Hill, and John Ruskin to lead a successful campaign to prevent the construction of railways to carry slate from the quarries, which would have ruined the unspoilt valleys of Newlands and Ennerdale. This success led to the formation of the Lake District Defence Society (later to become The Friends of the Lake District).

In 1893 Hill, Hunter and Rawnsley agreed to set up a national body to coordinate environmental conservation efforts across the country; the "National Trust for Places of Historic Interest or Natural Beauty" was formally inaugurated in 1894. The organisation obtained secure footing through the 1907 National Trust Bill, which gave the trust the status of a statutory corporation. and the bill was passed in August 1907.

An early "Back-to-Nature" movement, which anticipated the romantic ideal of modern environmentalism, was advocated by intellectuals such as John Ruskin, William Morris, and Edward Carpenter, who were all against consumerism, pollution and other activities that were harmful to the natural world. The movement was a reaction to the urban conditions of the industrial towns, where sanitation was awful, pollution levels intolerable and housing terribly cramped. Idealists championed the rural life as a mythical Utopia and advocated a return to it. John Ruskin argued that people should return to a *small piece of English ground, beautiful, peaceful, and fruitful. We will have no steam engines upon it . . . we will have plenty of flowers and vegetables . . . we will have some music and poetry; the children will learn to dance to it and sing it.*

Practical ventures in the establishment of small cooperative farms were even attempted and old rural traditions, without the "taint of manufacture or the canker of artificiality", were enthusiastically revived, including the Morris dance and the maypole.

The movement in the United States began in the late 19th century, out of concerns for protecting the natural resources of the West, with individuals such as John Muir and Henry David Thoreau making key philosophical contributions. Thoreau was interested in peoples' relationship with nature and studied this by living close to nature in a simple life. He published his experiences in the book *Walden*, which argues that people should become intimately close with nature. Muir came to believe in nature's inherent right, especially after spending time hiking in Yosemite Valley and studying both the ecology and geology. He successfully lobbied congress to form Yosemite National Park and went on to set up the Sierra Club in 1892. The conservationist principles as well as the belief in an inherent right of nature were to become the bedrock of modern environmentalism. However, the early movement in the U.S. developed with a contradiction; preservationists like John Muir wanted land and nature set aside for its own sake, and conservationists, such as Gifford Pinchot (appointed as the first Chief of the US Forest Service from 1905-1910), wanted to manage natural resources for human use.

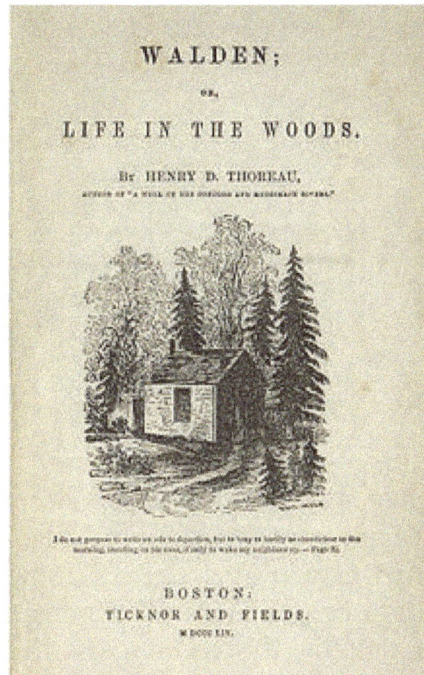

Original title page of *Walden* by Henry David Thoreau

20th Century

In the 20th century, environmental ideas continued to grow in popularity and recognition. Efforts were starting to be made to save some wildlife, particularly the American bison. The death of the last passenger pigeon as well as the endangerment of the American bison helped to focus the minds of conservationists and popularize their concerns. In 1916 the National Park Service was founded by US President Woodrow Wilson. Pioneers of the movement called for more efficient and professional management of natural resources. They fought for reform because they believed the destruction of forests, fertile soil, minerals, wildlife and water resources would lead to the downfall of society. The group that has been the most active in recent years is the climate movement.

The conservation of natural resources is the fundamental problem. Unless we solve that problem, it will avail us little to solve all others.

Theodore Roosevelt (4 October 1907)

The U.S movement did not really take off until after World War II as people began to recognize the costs of environmental negligence, disease, and widespread air and water pollution through the occurrence of several environmental disasters that occurred post-World War II. Aldo Leopold wrote "A Sand County Almanac" in the 1940s. He believed in a land ethic that recognized that maintaining the "beauty, integrity, and health of natural systems" as a moral and ethical imperative.

Another important book in the promotion of the environmental movement was Rachel Carson's "Silent Spring" about declining bird populations due to DDT, an insecticide, pollution and man's attempts to control nature through use of synthetic substances. Both of these books helped bring the issues into the public eye Rachel Carson's "Silent Spring" sold over two million copies.

Earth Day 2007 at City College, San Diego

The first Earth Day was celebrated on 22 April 1970. Its founder, former Wisconsin Senator, Gaylord Nelson was inspired to create this day of environmental education and awareness after seeing the oil spill off the coast of Santa Barbara in 1969. Greenpeace was created in 1971 as an organization that believed that political advocacy and legislation were ineffective or inefficient solutions and supported non-violent action. 1980 saw the creation of Earth First!, a group with an ecocentric view of the world – believing in equality between the rights of humans to flourish, the rights of all other species to flourish and the rights of life-sustaining systems to flourish.

In the 1950s, 1960s, and 1970s, several events illustrated the magnitude of environmental damage caused by humans. In 1954, a hydrogen bomb test at Bikini Atoll exposed the 23 man crew of the Japanese fishing vessel *Lucky Dragon 5* to radioactive fallout. In 1967 the oil tanker *Torrey Canyon* ran aground off the coast of Cornwall, and in 1969 oil spilled from an offshore well in California's Santa Barbara Channel. In 1971, the conclusion of a lawsuit in Japan drew international attention to the effects of decades of mercury poisoning on the people of Minamata.

At the same time, emerging scientific research drew new attention to existing and hypothetical threats to the environment and humanity. Among them were Paul R. Ehrlich, whose book *The Population Bomb* (1968) revived Malthusian concerns about the impact of exponential population growth. Biologist Barry Commoner generated a debate about growth, affluence and "flawed technology." Additionally, an association of scientists and political leaders known as the Club of Rome published their report *The Limits to Growth* in 1972, and drew attention to the growing pressure on natural resources from human activities.

Meanwhile, technological accomplishments such as nuclear proliferation and photos of the Earth from outer space provided both new insights and new reasons for concern over Earth's seemingly small and unique place in the universe.

In 1972, the United Nations Conference on the Human Environment was held in Stockholm, and for the first time united the representatives of multiple governments in discussion relating to the state of the global environment. This conference led directly to the creation of government environmental agencies and the UN Environment Program.

By the mid-1970s anti-nuclear activism had moved beyond local protests and politics to gain a wider appeal and influence. Although it lacked a single co-ordinating organization the anti-nuclear movement's efforts gained a great deal of attention, especially in the United Kingdom and United States. In the aftermath of the Three Mile Island accident in 1979, many mass demonstrations took place. The largest one was held in New York City in September 1979 and involved 200,000 people.

Since the 1970s, public awareness, environmental sciences, ecology, and technology have advanced to include modern focus points like ozone depletion, global climate change, acid rain, and the potentially harmful genetically modified organisms (GMOs).

United States

Beginning in the conservation movement at the beginning of the 20th century, the contemporary environmental movement's roots can be traced back to Murray Bookchin's *Our Synthetic Environment*, Paul R. Ehrlich's The Population Bomb, and Rachel Carson's *Silent Spring*. American environmentalists have campaigned against nuclear weapons and nuclear power in 1960s and 1970s, acid rain in the 1980s, ozone depletion and deforestation in the 1990s, and most recently climate change and global warming.

The United States passed many pieces of environmental legislation in the 1970s, such as the Clean Water Act, the Clean Air Act, the Endangered Species Act, and the National Environmental Policy Act. These remain as the foundations for current environmental standards.

Latin America

After the International Environmental Conference in Stockholm in 1972 Latin American officials returned with a high hope of growth and protection of the fairly untouched natural resources. Governments spent millions of dollars, and created departments and pollution standards. However, the outcomes have not always been what officials had initially hoped. Activists blame this on growing urban populations and industrial growth. Many Latin American countries have had a large inflow of immigrants that are living in substandard housing. Enforcement of the pollution standards is lax and penalties are minimal; in Venezuela, the largest penalty for violating an environmental law is 50,000 bolivar fine ($3,400) and 3 days in jail. In the 1970s or 1980s many Latin American countries were transitioning from military dictatorships to democratic governments.

Brazil

In 1992, Brazil came under scrutiny with the United Nations Conference on Environment and Development in Rio de Janeiro. Brazil has a history of little environmental awareness. It has the highest biodiversity in the world and also the highest amount of habitat destruction. One-third of the world's forests lie in Brazil, and they have the largest river, The Amazon, and the largest rainforest, the Amazon Rainforest. The people have raised funds to create state parks and increase the consciousness of people who have destroyed forests and polluted waterways. They have several organizations that have fronted the environmental movement. The Blue Wave Foundation was created in 1989 and has partnered with advertising companies to promote national education campaigns to keep Brazil's beaches clean. Funatura was created in 1986 and is a wildlife sanctuary program. Pro-Natura International is a private environmental organization created in 1986.

Europe

In 1952 the Great London Smog episode killed thousands of people and led the UK to create the first Clean Air Act in 1956. In 1957 the first major nuclear accident occurred in Windscale in northern England. The supertanker *Torrey Canyon* ran aground off the coast of Cornwall in 1967 causing the first major oil leak that killed marine life along the coast. In 1972, in Stockholm, the United Nations Conference on the Human Environment created the UN Environment Programme. The EU's environmental policy was formally founded by a European Council declaration and the first five-year environment programme was adopted. The main idea of the declaration was that prevention is better than the cure and the polluter should pay. 1979 saw the partial meltdown of Three Mile Island in the USA.

In the 1980s the green parties that were created a decade before began to have some political success.. In 1986, there was a nuclear accident in Chernobyl, Ukraine. The end of the 1980s and start of the 1990s saw the fall of communism across central and Eastern Europe, the fall of the [Berlin Wall], and the Union of East and West Germany. In 1992 there was a UN summit held in Rio de Janeiro where Agenda 21 was adopted. The Kyoto Protocol was created in 1997 which set specific targets and deadlines to reduce global greenhouse gas emissions. In the early 2000s activists believed that environmental policy concerns were overshadowed by energy security, globalism, and terrorism.

Asia

Middle East

The environmental movement is reaching the less developed world with different degrees of success. The Arab world, including the Middle East and North Africa, has different adaptations of the environmental movement. Countries on the Persian Gulf have high incomes and rely heavily on the large amount of energy resources in the area. Each country in the Arab world has varying combinations of low or high amounts of natural resources and low or high amounts of labor.

The League of Arab States has one specialized sub-committee, of 12 standing specialized subcommittee in the Foreign Affairs Ministerial Committees, which deals with Environmental Issues. Countries in the League of Arab States have demonstrated an interest in environmental issue, on paper some environmental activists have doubts about the level of commitment to environmental issues;; being a part of the world community may have obliged these countries to portray concern for the environment. Initial level of environmental awareness may be the creation of a ministry of the environment. The year of establishment of a ministry is also indicative of level of engagement. Saudi Arabia was the first to establish environmental law in 1992 followed by Egypt in 1994. Somalia is the only country without environmental law. In 2010 the Environmental Performance Index listed Algeria as the top Arab country at 42 of 163; Morocco was at 52 and Syria at 56. The Environmental Performance Index measures the ability of a country to actively manage and protect their environment and the health of their citizens. A weighted index is created by giving 50% weight for environmental health objective (health) and 50% for ecosystem vitality (ecosystem); values range from 0-100. No Arab countries were in the top quartile, and 7 countries were in the lowest quartile.

South Korea and Taiwan

South Korea and Taiwan experienced similar growth in industrialization from 1965-1990 with few environmental controls. South Korea's Han River and Nakdong River were so polluted by unchecked dumping of industrial waste that they were close to being classified as biologically dead. Taiwan's formula for balanced growth was to prevent industrial concentration and encourage manufacturers to set up in the countryside. This led to 20% of the farmland being polluted by industrial waste and 30% of the rice grown on the island was contaminated with heavy metals. Both countries had spontaneous environmental movements drawing participants from different classes. Their demands were linked with issues of employment, occupational health, and agricultural crisis. They were also quite militant; the people learned that protesting can bring results. The polluting factories were forced to make immediate improvements of the conditions or pay compensation to victims. Some were even forced to shut down or move locations. The people were able to force the government to come out with new restrictive rules on toxins, industrial waste, and air pollution. All of these new regulations caused the migration of those polluting industries from Taiwan and South Korea to China and other countries in Southeast Asia with more relaxed environmental laws.

China

China's environmental movement is characterized by spontaneous alliances that often only occur at the local level. The Chinese have realized the ability of riots and protests to have success and had led to an increase in disputes in China by 30% since 2005 to more than 50,000 events. Protests cover topics such as environmental issues, land-loss, income, and political issues. They have also grown in size from about 10 people or fewer in the mid-1990s to 52 people per incident in 2004. China has more relaxed environmental laws than other countries in Asia, so many polluting factories have relocated to China causing pollution in China. Water pollution, water scarcity, soil pollution, soil degradation, and desertification are issues currently in discussion in China. The groundwater table of the North China Plain is dropping by 1.5 m (5 ft) per year. This groundwater table occurs in the region of China that produces 40% of the country's grain.

India

Environmental and public health is an ongoing struggle within India. The first seed of an environmental movement in India was the foundation in 1964 of *Dasholi Gram Swarajya Sangh*, a labour coperative started by Chandi Prasad Bhatt. It was inaugurated by Sucheta Kriplani and founded on a land donated by Shyma Devi. This initiative was eventually followed up with the Chipko movement starting in 1974.

The most severe single event underpinning the movement was the Bhopal gas leakage on 3 December 1984. 40 tons of methyl isocyanate was released, immediately killing 2,259 people and ultimately affecting 700,000 citizens.

India has a national campaign against Coca Cola and Pepsi Cola plants due to their practices of drawing ground water and contaminating fields with sludge. The movement is characterized by local struggles against intensive aquaculture farms. The most influential part of the environmental movement in India is the anti-dam movement. Dam creation has been thought of as a way for

India to catch up with the West by connecting to the power grid with giant dams, coal or oil-powered plants, or nuclear plants. Jhola Aandolan a mass movement is conducting as fighting against polyethylene carry bags uses and promoting cloth/jute/paper carry bags to protect environment & nature. Activists in the Indian environmental movement consider global warming, sea levels rising, and glaciers retreating decreasing the amount of water flowing into streams to be the biggest challenges for them to face in the early twenty-first century.

Bangladesh

Mithun Roy Chowdhury, President, Save Nature & Wildlife (SNW), Bangladesh, insisted that the people of Bangladesh raise their voice against Tipaimukh Dam, being constructed by the Government of India. He said Tipaimukh Dam project will be another "death trap for Bangladesh like the Farakka Barrage," that would lead to an environmental disaster for 50 million people in the Meghna River basin. He said that this project will start desertification in Bangladesh.

Scope of the Movement

Before flue-gas desulfurization was installed, the air-polluting emissions from this power plant in New Mexico contained excessive amounts of sulfur dioxide

Environmental science is the study of the interactions among the physical, chemical and biological components of the environment.

- Ecology, or ecological science, is the scientific study of the distribution and abundance of living organisms and how these properties are affected by interactions between the organisms and their environment.

Primary Focus Points

The environmental movement is broad in scope and can include any topic related to the environment, conservation, and biology, as well as preservation of landscapes, flora, and fauna for a variety of purposes and uses. When an act of violence is committed against someone or some institution in the name of environmental defense it is referred to as eco terrorism.

- The conservation movement seeks to protect natural areas for sustainable consumption, as well as traditional (hunting, fishing, trapping) and spiritual use.

- Environmental conservation is the process in which one is involved in conserving the natural aspects of the environment. Whether through reforestation, recycling, or pollution control, environmental conservation sustains the natural quality of life.

- Environmental health movement dates at least to Progressive Era, and focuses on urban standards like clean water, efficient sewage handling, and stable population growth. Environmental health could also deal with nutrition, preventive medicine, aging, and other concerns specific to human well-being. Environmental health is also seen as an indicator for the state of the environment, or an early warning system for what may happen to humans

- Environmental justice is a movement that began in the U.S. in the 1980s and seeks an end to environmental racism and prevent low-income and minority communities from an unbalanced exposure to highways, garbage dumps, and factories. The Environmental Justice movement seeks to link "social" and "ecological" environmental concerns, while at the same time preventing de facto racism, and classism. This makes it particularly adequate for the construction of labor-environmental alliances.

- Ecology movement could involve the Gaia Theory, as well as Value of Earth and other interactions between humans, science, and responsibility.

- Bright green environmentalism is a currently popular sub-movement, which emphasizes the idea that through technology, good design and more thoughtful use of energy and resources, people can live responsible, sustainable lives while enjoying prosperity.

- Light green, and dark green environmentalism are yet other sub-movements, respectively distinguished by seeing environmentalism as a lifestyle choice (light greens), and promoting reduction in human numbers and/or a relinquishment of technology (dark greens)

- Deep Ecology is an ideological spinoff of the ecology movement that views the diversity and integrity of the planetary ecosystem, in and for itself, as its primary value.

- The anti-nuclear movement opposes the use of various nuclear technologies. The initial anti-nuclear objective was nuclear disarmament and later the focus began to shift to other issues, mainly opposition to the use of nuclear power. There have been many large anti-nuclear demonstrations and protests. Major anti-nuclear groups include Campaign for Nuclear Disarmament, Friends of the Earth, Greenpeace, International Physicians for the Prevention of Nuclear War, and the Nuclear Information and Resource Service.

Environmental Law and Theory

Property Rights

Many environmental lawsuits question the legal rights of property owners, and whether the general public has a right to intervene with detrimental practices occurring on someone else's land.

Environmental law organizations exist all across the world, such as the Environmental Law and Policy Center in the midwestern United States.

Citizens' Rights

One of the earliest lawsuits to establish that citizens may sue for environmental and aesthetic harms was Scenic Hudson Preservation Conference v. Federal Power Commission, decided in 1965 by the Second Circuit Court of Appeals. The case helped halt the construction of a power plant on Storm King Mountain in New York State.

Nature's Rights

Christopher D. Stone's 1972 essay, "Should trees have standing?" addressed the question of whether natural objects themselves should have legal rights. In the essay, Stone suggests that his argument is valid because many current rightsholders (women, children) were once seen as objects.

Environmental Reactivism

Numerous criticisms and ethical ambiguities have led to growing concerns about technology, including the use of potentially harmful pesticides, water additives like fluoride, and the extremely dangerous ethanol-processing plants.

NIMBY syndrome refers to public outcry caused by knee-jerk reaction to an unwillingness to be exposed to even necessary developments. Some serious biologists and ecologists created the scientific ecology movement which would not confuse empirical data with visions of a desirable future world.

Environmentalism Today

Composite images of Earth generated by NASA in 2001 (left) and 2002 (right)

Today, the sciences of ecology and environmental science, in addition to any aesthetic goals, provide the basis of unity to some of the serious environmentalists. As more information is gathered in scientific fields, more scientific issues like biodiversity, as opposed to mere aesthetics, are a concern to environmentalists. Conservation biology is a rapidly developing field.

In recent years, the environmental movement has increasingly focused on global warming as one of the top issues. As concerns about climate change moved more into the mainstream, from the connections drawn between global warming and Hurricane Katrina to Al Gore's film An Inconvenient Truth, more and more environmental groups refocused their efforts. In the United States, 2007 witnessed the largest grassroots environmental demonstration in years, Step It Up 2007, with rallies in over 1,400 communities and all 50 states for real global warming solutions.

Many religious organizations and individual churches now have programs and activities dedicated to environmental issues. The religious movement is often supported by interpretation of scriptures. Most major religious groups are represented including Jewish, Islamic, Anglican, Orthodox, Evangelical, Christian and Catholic.

Radical Environmentalism

Radical environmentalism emerged from an ecocentrism-based frustration with the co-option of mainstream environmentalism. The radical environmental movement aspires to what scolar Christopher Manes calls "a new kind of environmental activism: iconoclastic, uncompromising, discontented with traditional conservation policy, at times illegal ..." Radical environmentalism presupposes a need to reconsider Western ideas of religion and philosophy (including capitalism, patriarchy and globalization) sometimes through "resacralising" and reconnecting with nature. Greenpeace represents an organisation with a radical approach, but has contributed in serious ways towards understanding of critical issues, and has a science-oriented core with radicalism as a means to media exposure. Groups like Earth First! take a much more radical posture. Some radical environmentalist groups, like Earth First! and the Earth Liberation Front, illegally sabotage or destroy infrastructural capital.

Criticisms

Conservative critics of the movement characterize it as radical and misguided. Especially critics of the United States Endangered Species Act, which has come under scrutiny lately, and the Clean Air Act, which they said conflict with private property rights, corporate profits and the nation's overall economic growth. Critics also challenge the scientific evidence for global warming. They argue that the environmental movement has diverted attention from more pressing issues.

Conservation Movement

Much attention has been given to preserving the natural characteristics of Hopetoun Falls, Australia, while allowing ample access for visitors.

The conservation movement, also known as nature conservation, is a political, environmental and a social movement that seeks to protect natural resources including animal and plant species as well as their habitat for the future.

The early conservation movement included fisheries and wildlife management, water, soil conservation and sustainable forestry. The contemporary conservation movement has broadened from the early movement's emphasis on use of sustainable yield of natural resources and preservation of wilderness areas to include preservation of biodiversity. Some say the conservation movement is part of the broader and more far-reaching environmental movement, while others argue that they differ both in ideology and practice. Chiefly in the United States, conservation is seen as differing from environmentalism in that it aims to preserve natural resources expressly for their continued sustainable use by humans. In other parts of the world conservation is used more broadly to include the setting aside of natural areas and the active protection of wildlife for their inherent value, as much as for any value they may have for humans.

History

Early History

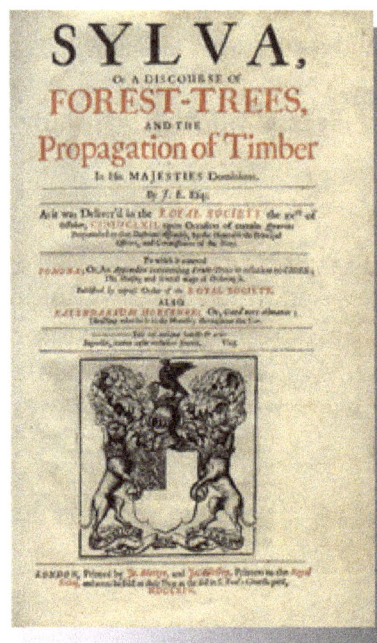

Sylva, or A Discourse of Forest-Trees and the Propagation of Timber in His Majesty's Dominions,
title page of the first edition (1664).

The conservation movement can be traced back to John Evelyn's work *Sylva*, presented as a paper to the Royal Society in 1662. Published as a book two years later, it was one of the most highly influential texts on forestry ever published. Timber resources in England were becoming dangerously depleted at the time, and Evelyn advocated the importance of conserving the forests by managing the rate of depletion and ensuring that the cut down trees get replenished.

The field developed during the 18th century, especially in Prussia and France where scientific forestry methods were developed. These methods were first applied rigorously in British India from the early-19th century. The government was interested in the use of forest produce and began managing the forests with measures to reduce the risk of wildfire in order to protect the "household" of nature, as it was then termed. This early ecological idea was in order to preserve the growth of delicate teak trees, which was an important resource for the Royal Navy. Concerns over teak depletion were raised as early as 1799 and 1805 when the Navy was undergoing a massive expansion during the Napoleonic Wars; this pressure led to the first formal conservation Act, which prohibited the felling of small teak trees. The first forestry officer was appointed in 1806 to regulate and preserve the trees necessary for shipbuilding. This promising start received a setback in the 1820s and 30s, when laissez-faire economics and complaints from private landowners brought these early conservation attempts to an end.

Origins of the Modern Conservation Movement

Conservation was revived in the mid-19th century, with the first practical application of scientific conservation principles to the forests of India. The conservation ethic that began to evolve included three core principles: that human activity damaged the environment, that there was a civic duty to maintain the environment for future generations, and that scientific, empirically based methods should be applied to ensure this duty was carried out. Sir James Ranald Martin was prominent in promoting this ideology, publishing many medico-topographical reports that demonstrated the scale of damage wrought through large-scale deforestation and desiccation, and lobbying extensively for the institutionalization of forest conservation activities in British India through the establishment of Forest Departments. Edward Percy Stebbing warned of desertification of India. The Madras Board of Revenue started local conservation efforts in 1842, headed by Alexander Gibson, a professional botanist who systematically adopted a forest conservation program based on scientific principles. This was the first case of state management of forests in the world.

These local attempts gradually received more attention by the British government as the unregulated felling of trees continued unabated. In 1850, the British Association in Edinburgh formed a committee to study forest destruction at the behest of Dr. Hugh Cleghorn a pioneer in the nascent conservation movement.

He had become interested in forest conservation in Mysore in 1847 and gave several lectures at the Association on the failure of agriculture in India. These lectures influenced the government under Governor-General Lord Dalhousie to introduce the first permanent and large-scale forest conservation program in the world in 1855, a model that soon spread to other colonies, as well the United States. In the same year, Cleghorn organised the Madras Forest Department and in 1860 the Department banned the use shifting cultivation. Cleghorn's 1861 manual, *The forests and gardens of South India*, became the definitive work on the subject and was widely used by forest assistants in the subcontinent. In 1861, the Forest Department extended its remit into the Punjab.

Sir Dietrich Brandis, a German forester, joined the British service in 1856 as superintendent of the teak forests of Pegu division in eastern Burma. During that time Burma's teak forests were controlled by militant Karen tribals. He introduced the "taungya" system, in which Karen villagers provided labour for clearing, planting and weeding teak plantations. After seven years in Burma, Brandis was appointed Inspector General of Forests in India, a position he served in for 20 years. He formulated new forest legislation and helped establish research and training institutions. The Imperial Forest School at Dehradun was founded by him.

Germans were prominent in the forestry administration of British India. As well as Brandis, Berthold Ribbentrop and Sir William P.D. Schlich brought new methods to Indian conservation, the latter becoming the Inspector-General in 1883 after Brandis stepped down. Schlich helped to establish the journal *Indian Forester* in 1874, and became the founding director of the first forestry school in England at Cooper's Hill in 1885. He authored the five-volume *Manual of Forestry* (1889–96) on silviculture, forest management, forest protection, and forest utilisation, which became the standard and enduring textbook for forestry students.

Conservation in the United States

The American movement received its inspiration from 19th century works that exalted the inherent value of nature, quite apart from human usage. Author Henry David Thoreau (1817-1862) made key philosophical contributions that exalted nature. Thoreau was interested in peoples' relationship with nature and studied this by living close to nature in a simple life. He published his experiences in the book *Walden*, which argued that people should become intimately close with nature. The ideas of Sir Brandis, Sir William P.D. Schlich and Carl A. Schenck were also very influential - Gifford Pinchot, the first chief of the USDA Forest Service, relied heavily upon Brandis' advice for introducing professional forest management in the U.S. and on how to structure the Forest Service.

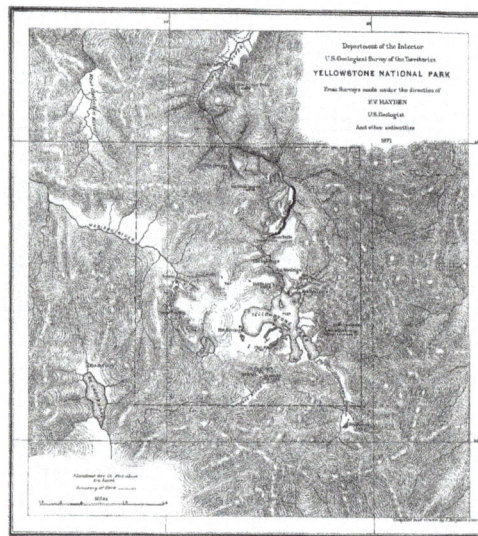

F. V. Hayden's map of Yellowstone National Park, 1871.

Both Conservationists and Preservationists appeared in political debates during the Progressive Era (the 1890s—early 1920s). There were three main positions. The laissez-faire position held that owners of private property—including lumber and mining companies, should be allowed to do anything they wished for their property.

The conservationists, led by future President Theodore Roosevelt and his close ally George Bird Grinnell, were motivated by the wanton waste that was taking place at the hand of market forces, including logging and hunting. This practice resulted in placing a large number of North American game species on the edge of extinction. Roosevelt recognized that the laissez-faire approach of the U.S. Government was too wasteful and inefficient. In any case, they noted, most of the natural resources in the western states were already owned by the federal government. The best course of action, they argued, was a long-term plan devised by national experts to maximize the long-term economic benefits of natural resources. To accomplish the mission, Roosevelt and Grinnell formed the Boone and Crockett Club in 1887. The Club was made up of the best minds and influential men of the day. The Boone and Crockett Club's contingency of conservationists, scientists, politicians, and intellectuals became Roosevelt's closest advisers during his march to preserve wildlife and habitat across North America. Preservationists, led by John Muir (1838–1914), argued that the conservation policies were not strong enough to protect the interest of the natural world because they continued to focus on the natural world as a source of economic production.

The debate between conservation and preservation reached its peak in the public debates over the construction of California's Hetch Hetchy dam in Yosemite National Park which supplies the water supply of San Francisco. Muir, leading the Sierra Club, declared that the valley must be preserved for the sake of its beauty: "No holier temple has ever been consecrated by the heart of man."

President Roosevelt put conservationist issue high on the national agenda. He worked with all the major figures of the movement, especially his chief advisor on the matter, Gifford Pinchot and was deeply committed to conserving natural resources. He encouraged the Newlands Reclamation Act of 1902 to promote federal construction of dams to irrigate small farms and placed 230 million acres (360,000 mi² or 930,000 km²) under federal protection. Roosevelt set aside more federal land for national parks and nature preserves than all of his predecessors combined.

A PRACTICAL FORESTER
(A subject that had attention all through Mr.
Roosevelt's Presidency.)
From the *Pioneer Press* (St. Paul)

Roosevelt established the United States Forest Service, signed into law the creation of five national parks, and signed the year 1906 Antiquities Act, under which he proclaimed 18 new national monuments. He also established the first 51 bird reserves, four game preserves, and 150 national forests, including Shoshone National Forest, the nation's first. The area of the United States that he placed under public protection totals approximately 230,000,000 acres (930,000 km²).

Gifford Pinchot had been appointed by McKinley as chief of Division of Forestry in the Department of Agriculture. In 1905, his department gained control of the national forest reserves. Pinchot promoted private use (for a fee) under federal supervision. In 1907, Roosevelt designated 16 million acres (65,000 km²) of new national forests just minutes before a deadline.

In May 1908, Roosevelt sponsored the Conference of Governors held in the White House, with a focus on natural resources and their most efficient use. Roosevelt delivered the opening address: "Conservation as a National Duty.".

In 1903 Roosevelt toured the Yosemite Valley with John Muir, who had a very different view of conservation, and tried to minimize commercial use of water resources and forests. Working through the Sierra Club he founded, Muir succeeded in 1905 in having Congress transfer the Mariposa Grove and Yosemite Valley to the federal government. While Muir wanted nature preserved for its own sake, Roosevelt subscribed to Pinchot's formulation, "to make the forest produce the largest amount of whatever crop or service will be most useful, and keep on producing it for generation after generation of men and trees."

Theodore Roosevelt's view on conservationism remained dominant for decades; - Franklin D. Roosevelt authorised the building of many large-scale dams and water projects, as well as the expansion of the National Forest System to buy out sub-marginal farms. In 1937, the Pittman–Robertson Federal Aid in Wildlife Restoration Act was signed into law, providing funding for state agencies to carry out their conservation efforts.

Since 1970

Environmental reemerged on the national agenda in 1970, with Republican Richard Nixon playing a major role, especially with his creation of the Environmental Protection Agency. The debates over the public lands and environmental politics played a supporting role in the decline of liberalism and the rise of modern environmentalism. Although Americans consistently rank environmental issues as "important", polling data indicates that in the voting booth voters rank the environmental issues low relative to other political concerns.

The growth of the Republican party's political power in the inland West (apart from the Pacific coast) was facilitated by the rise of popular opposition to public lands reform. Successful Democrats in the inland West and Alaska typically take more conservative positions on environmental issues than Democrats from the Coastal states. Conservatives drew on new organizational networks of think tanks, industry groups, and citizen-oriented organizations, and they began to deploy new strategies that affirmed the rights of individuals to their property, protection of extraction rights, to hunt and recreate, and to pursue happiness unencumbered by the federal government at the expense of resource conservation.

Areas of Concern

Deforestation and overpopulation are issues affecting all regions of the world. The consequent destruction of wildlife habitat has prompted the creation of conservation groups in other countries, some founded by local hunters who have witnessed declining wildlife populations first hand. Also, it was highly important for the conservation movement to solve problems of living conditions in the cities and the overpopulation of such places.

Boreal Forest and the Arctic

The idea of incentive conservation is a modern one but its practice has clearly defended some of the sub Arctic wildernesses and the wildlife in those regions for thousands of years, especially by indigenous peoples such as the Evenk, Yakut, Sami, Inuit and Cree. The fur trade and hunting by these peoples have preserved these regions for thousands of years. Ironically, the pressure now upon them comes from non-renewable resources such as oil, sometimes to make synthetic clothing which is advocated as a humane substitute for fur. Similarly, in the case of the beaver, hunting and fur trade were thought to bring about the animal's demise, when in fact they were an integral part of its conservation. For many years children's books stated and still do, that the decline in the beaver population was due to the fur trade. In reality however, the decline in beaver numbers was because of habitat destruction and deforestation, as well as its continued persecution as a pest (it causes flooding). In Cree lands however, where the population valued the animal for meat and fur, it continued to thrive. The Inuit defend their relationship with the seal in response to outside critics.

Latin America (Bolivia)

The Izoceño-Guaraní of Santa Cruz Department, Bolivia is a tribe of hunters who were influential in establishing the Capitania del Alto y Bajo Isoso (CABI). CABI promotes economic growth and survival of the Izoceno people while discouraging the rapid destruction of habitat within Bolivia's Gran Chaco. They are responsible for the creation of the 34,000 square kilometre Kaa-Iya del Gran Chaco National Park and Integrated Management Area (KINP). The KINP protects the most biodiverse portion of the Gran Chaco, an ecoregion shared with Argentina, Paraguay and Brazil. In 1996, the Wildlife Conservation Society joined forces with CABI to institute wildlife and hunting monitoring programs in 23 Izoceño communities. The partnership combines traditional beliefs and local knowledge with the political and administrative tools needed to effectively manage habitats. The programs rely solely on voluntary participation by local hunters who perform self-monitoring techniques and keep records of their hunts. The information obtained by the hunters participating in the program has provided CABI with important data required to make educated decisions about the use of the land. Hunters have been willing participants in this program because of pride in their traditional activities, encouragement by their communities and expectations of benefits to the area.

Africa (Botswana)

In order to discourage illegal South African hunting parties and ensure future local use and sustainability, indigenous hunters in Botswana began lobbying for and implementing conservation practices in the 1960s. The Fauna Preservation Society of Ngamiland (FPS) was formed in 1962 by the husband and wife team: Robert Kay and June Kay, environmentalists working in conjunction with the Batawana tribes to preserve wildlife habitat.

The FPS promotes habitat conservation and provides local education for preservation of wildlife. Conservation initiatives were met with strong opposition from the Botswana government because of the monies tied to big-game hunting. In 1963, BaTawanga Chiefs and tribal hunter/adventurers in conjunction with the FPS founded Moremi National Park and Wildlife Refuge, the first area to be set aside by tribal people rather than governmental forces. Moremi National Park is home to a variety of wildlife, including lions, giraffes, elephants, buffalo, zebra, cheetahs and antelope, and covers an area of 3,000 square kilometers. Most of the groups involved with establishing this protected land were involved with hunting and were motivated by their personal observations of declining wildlife and habitat.

Chipko Movement

The Chipko movement or chipko andolan was primarily a forest conservation movement in India that began in 1973 and went on to become a rallying point for many future environmental movements all over the world; it created a precedent for non-violent protest started in India. It occurred at a time when there was hardly any environmental movement in the developing world, and its success meant that the world immediately took notice of this non-violent movement, which was to inspire in time many such eco-groups by helping to slow down the rapid deforestation, expose vested interests, increase ecological awareness, and demonstrate the viability of people power. Above all, it stirred up the existing civil society in India, which began to address the issues of tribal and marginalized people. Today, beyond the eco-socialism hue, it is being seen increasingly as an ecofeminism movement. Although many of its leaders were men, women were not only its backbone, but also its mainstay, because they were the ones most affected by the rampant deforestation, which led to a lack of firewood and fodder as well as water for drinking and irrigation. Over the years they also became primary stakeholders in a majority of the afforestation work that happened under the Chipko movement.

In 1987, the Chipko Movement was awarded the Right Livelihood Award. The chipko aandolan is a movement that practised the Gandhian methods of Satyagraha where both male and female activists played vital roles, including Gaura Devi, Sudesha Devi, Bachni Devi and Chandi Prasad Bhatt.

History

The Chipko movement started in the early 18th century in the state of Rajasthan. Amrita Devi with 84 villagers risked their lives to protect the forest trees from being felled on the order of the *maharaja* (king).

In Khejarli village, 363 Bishnois died in 1730 AD while protecting green Khejri that are considered sacred by the community. Gradually, a rising awareness of the ecological crisis, which came from an immediate loss of livelihood caused by it, resulted in the growth of political activism in the region.

Background

In India the forest cover started deteriorating at an alarming rate, resulting in hardships for those involved in labour-intensive fodder and firewood collection. This also led to a deterioration in the soil conditions, and erosion in the area. As water sources dried up in the hills, water shortages became widespread. Subsequently, communities gave up raising livestock, which added to the problems of malnutrition of many regions in India. This crisis was heightened by the fact that forest conservation policies, like the Indian Forest Act, 1927, traditionally restricted the access of local communities to the forests, resulting in scarce farmlands in an over-populated and extremely poor area, despite all of its natural wealth. Thus the sharp decline in the local agrarian economy lead to a migration of people into the plains in search of jobs, leaving behind several depopulated villages in the 1960s.

The year 1964 saw the establishment of *Dasholi Gram Swarajya Sangh* (DGSS) ("Dasholi Society for Village Self-Rule"), set up by Gandhian social worker Chandi Prasad Bhatt in Gopeshwar, and inspired by Jayaprakash Narayan and the Sarvodaya movement, with an aim to set up small industries using the resources of the forest. Their first project was a small workshop making farm tools for local use. Its name was later changed to DGSS from the original *Dasholi Gram Swarajya Mandal* (DGSM) in the 1980s. Here they had to face restrictive forest policies, a hangover of colonial era still prevalent, as well as the "contractor system", in which these pieces of forest land were commodified and auctioned to big contractors, usually from the plains, who brought along their own skilled and semi-skilled laborers, leaving only the menial jobs like hauling rocks for the hill people, and paying them next to nothing. On the other hand, the hill regions saw an influx of more people from the outside, which only added to the already strained ecological balance.

Hastened by increasing hardships, the Garhwal Himalayas soon became the centre for a rising ecological awareness of how reckless deforestation had denuded much of the forest cover, resulting in the devastating Alaknanda River floods of July 1970, when a major landslide blocked the river and affected an area starting from Hanumanchatti, near Badrinath to 350 km downstream till Haridwar, further numerous villages, bridges and roads were washed away. Thereafter, incidences of landslides and land subsidence became common in an area which was experiencing a rapid increase in civil engineering projects.

Beginnings and Organization

Soon villagers, especially women, began to organize themselves under several smaller groups, taking up local causes with the authorities, and standing up against commercial logging operations that threatened their livelihoods. In October 1971, the Sanga workers held a demonstration in Gopeshwar to protest against the policies of the Forest Department. More rallies and marches were held in late 1972, but to little effect, until a decision to take direct action was taken. The first such occasion occurred when the Forest Department turned down the Sangh's annual request for ten Ash Trees for its farm tools workshop, and instead awarded a contract for 300 trees to Simon Company, a sporting goods manufacturer in distant Allahabad, to make tennis rackets. In March 1973, the lumbermen arrived at Gopeshwar, and after a couple of weeks, they were confronted at village Mandal on April 24, 1973, where about hundred villagers and DGSS workers were beating drums and shouting slogans, thus forcing the contractors and their lumbermen to retreat. This was the first confrontation of the movement, The contract was eventually cancelled and awarded to the Sangh instead. By now, the issue had grown beyond the mere procurement of an annual quota of three ash trees, and encompassed a growing concern over commercial logging and the government's forest policy, which the villagers saw as unfavourable towards them. The Sangh also decided to resort to tree-hugging, or Chipko, as a means of non-violent protest.

But the struggle was far from over, as the same company was awarded more ash trees, in the Phata forest, 80 km away from Gopeshwar. Here again, due to local opposition, starting on June 20,

1973, the contractors retreated after a stand-off that lasted a few days. Thereafter, the villagers of Phata and Tarsali formed a vigil group and watched over the trees until December, when they had another successful stand-off, when the activists reached the site in time. The lumbermen retreated leaving behind the five ash trees felled.

The final flash point began a few months later, when the government announced an auction scheduled in January 1974, for 2,500 trees near Reni village, overlooking the Alaknanda River. Bhatt set out for the villages in the Reni area, and incited the villagers, who decided to protest against the actions of the government by hugging the trees. Over the next few weeks, rallies and meetings continued in the Reni area.

On March 25, 1974, the day the lumbermen were to cut the trees, the men of the Reni village and DGSS workers were in Chamoli, diverted by state government and contractors to a fictional compensation payment site, while back home labourers arrived by the truckload to start logging operations. A local girl, on seeing them, rushed to inform Gaura Devi, the head of the village *Mahila Mangal Dal*, at Reni village (Laata was her ancestral home and Reni adopted home). Gaura Devi led 27 of the village women to the site and confronted the loggers. When all talking failed, and the loggers started to shout and abuse the women, threatening them with guns, the women resorted to hugging the trees to stop them from being felled. This went on into late hours. The women kept an all-night vigil guarding their trees from the cutters until a few of them relented and left the village. The next day, when the men and leaders returned, the news of the movement spread to the neighbouring Laata and others villages including Henwalghati, and more people joined in. Eventually, only after a four-day stand-off, the contractors left.

Aftermath

The news soon reached the state capital, where then state Chief Minister, Hemwati Nandan Bahuguna, set up a committee to look into the matter, which eventually ruled in favour of the villagers. This became a turning point in the history of eco-development struggles in the region and around the world.

The struggle soon spread across many parts of the region, and such spontaneous stand-offs between the local community and timber merchants occurred at several locations, with hill women demonstrating their new-found power as non-violent activists. As the movement gathered shape under its leaders, the name Chipko Movement was attached to their activities. According to Chipko historians, the term originally used by Bhatt was the word "angalwaltha" in the Garhwali language for "embrace", which later was adapted to the Hindi word, *Chipko*, which means to stick. Over the next five years, the movement spread to many districts in the region, and within a decade throughout the Uttarakhand Himalayas. Larger issues of ecological and economic exploitation of the region were raised. The villagers demanded that no forest-exploiting contracts should be given to outsiders and local communities should have effective control over natural resources like land, water, and forests. They wanted the government to provide low-cost materials to small industries and ensure development of the region without disturbing the ecological balance. The movement took up economic issues of landless forest workers and asked for guarantees of minimum wage. Globally Chipko demonstrated how environment causes, up until then considered an activity of the rich, were a matter of life and death for the poor, who were all too often the first ones to be devastated by an environmental tragedy. Several scholarly studies were made in the aftermath of the movement. In 1977, in another area, women tied sacred threads, Raksha Bandhan, around trees earmarked for felling in a Hindu tradition which signifies a bond between brother and sisters.

Women's participation in the Chipko agitation was a very novel aspect of the movement. The forest contractors of the region usually doubled up as suppliers of alcohol to men. Women held sustained agitations against the habit of alcoholism and broadened the agenda of the movement to cover other social issues. The movement achieved a victory when the government issued a ban on felling of trees in the Himalayan regions for fifteen years in 1980 by then Prime Minister Indira Gandhi, until the green cover was fully restored. One of the prominent Chipko leaders, Gandhian Sunderlal Bahuguna, took a 5,000-kilometre trans-Himalaya foot march in 1981–83, spreading the Chipko message to a far greater area. Gradually, women set up cooperatives to guard local forests, and also organized fodder production at rates conducive to local environment. Next, they joined in land rotation schemes for fodder collection, helped replant degraded land, and established and ran nurseries stocked with species they selected.

Participants

Surviving participants of the first all-woman Chipko action at Reni village in 1974 on left jen wadas, reassembled thirty years later.

One of Chipko's most salient features was the mass participation of female villagers. As the backbone of Uttarakhand's Agrarian economy, women were most directly affected by environmental degradation and deforestation, and thus related to the issues most easily. How much this participation impacted or derived from the ideology of Chipko has been fiercely debated in academic circles.

Despite this, both female and male activists did play pivotal roles in the movement including Gaura Devi, Sudesha Devi, Bachni Devi, Chandi Prasad Bhatt, Sundarlal Bahuguna, Govind Singh Rawat, Dhoom Singh Negi, Shamsher Singh Bisht and Ghanasyam Raturi, the Chipko poet, whose songs echo throughout the Himalayas. Out of which, Chandi Prasad Bhatt was awarded the Ramon Magsaysay Award in 1982, and Sundarlal Bahuguna was awarded the Padma Vibhushan in 2009.

Legacy

In Tehri district, Chipko activists would go on to protest limestone mining in the Doon Valley (Dehra Dun) in the 1980s, as the movement spread through the Dehradun district, which had earlier seen deforestation of its forest cover leading to heavy loss of flora and fauna. Finally quarrying was banned after years of agitation by Chipko activists, followed by a vast public drive for afforestation, which turned around the valley, just in time. Also in the 1980s, activists like Bahuguna protested against construction of the Tehri dam on the Bhagirathi River, which went on for the next two decades, before founding the *Beej Bachao Andolan*, the Save the Seeds movement, that continues to the present day.

Over time, as a United Nations Environment Programme report mentioned, Chipko activists started "working a socio-economic revolution by winning control of their forest resources from the hands of a distant bureaucracy which is only concerned with the selling of forestland for making urban-oriented products.". The Chipko movement became a benchmark for socio-ecological movements in other forest areas of Himachal Pradesh, Rajasthan and Bihar; in September 1983, Chipko inspired a similar, Appiko movement in Karnataka state of India, where tree felling in the Western Ghats and Vindhyas was stopped. In Kumaon region, Chipko took on a more radical tone, combining with the general movement for a separate Uttarakhand state, which was eventually achieved in 2000.

In recent years, the movement not only inspired numerous people to work on practical programmes of water management, energy conservation, afforestation, and recycling, but also encouraged scholars to start studying issues of environmental degradation and methods of conservation in the Himalayas and throughout India.

On March 26, 2004, Reni, Laata, and other villages of the Niti Valley celebrated the 30th anniversary of the Chipko Movement, where all the surviving original participants united. The celebrations started at Laata, the ancestral home of Gaura Devi, where Pushpa Devi, wife of late Chipko Leader Govind Singh Rawat, Dhoom Singh Negi, Chipko leader of Henwalghati, Tehri Garhwal, and others were celebrated. From here a procession went to Reni, the neighbouring village, where the actual Chipko action took place on March 26, 1974. This marked the beginning of worldwide methods to improve the present situation.

Anarcho-naturism

Anarcho-naturism (also anarchist naturism and naturist anarchism) appeared in the late 19th century as the union of anarchist and naturist philosophies. "In many of the alternative communities established in Britain in the early 1900s nudism, anarchism, vegetarianism and free love were accepted as part of a politically radical way of life. In the 1920s the inhabitants of the anarchist community at Whiteway, near Stroud in Gloucestershire, shocked the conservative residents of the area with their shameless nudity." Mainly it had importance within individualist anarchist circles in Spain, France, Portugal, and Cuba.

Anarcho-naturism advocates vegetarianism, free love, nudism, hiking and an ecological world view within anarchist groups and outside them. Anarcho-naturism also promotes an ecological worldview, small ecovillages, and most prominently nudism as a way to avoid the artificiality of the industrial mass society of modernity. Naturist individualist anarchists see the individual in their biological, physical and psychological aspects and tried to eliminate social determinations.

History

Early Influences

An important early influence on anarchist naturism was the thought of Henry David Thoreau, Leo Tolstoy and Elisee Reclus.

Thoreau was an American author, poet, naturalist, tax resister, development critic, surveyor, historian, philosopher, and leading transcendentalist. He is best known for his book *Walden*, a reflection upon simple living in natural surroundings, and his essay, *Civil Disobedience*, an argument for individual resistance to civil government in moral opposition to an unjust state. His thought is an early influence on green anarchism, but with an emphasis on the individual experience of the natural world, influencing later naturist currents. Simple living as a rejection of a materialist lifestyle and self-sufficiency were Thoreau's goals, and the whole project was inspired by transcendentalist philosophy. "Many have seen in Thoreau one of the precursors of ecologism and anarcho-primitivism represented today in John Zerzan. For George Woodcock this attitude can be also motivated by certain idea of resistance to progress and of rejection of the growing materialism which is the nature of American society in the mid-19th century." John Zerzan himself included the text "Excursions" (1863) by Thoreau in his edited compilation of anti-civilization writings called *Against civilization: Readings and reflections* from 1999.

France

For the influential French anarchist Élisée Reclus, naturism "was at the same time a physical means of revitalization, a rapport with the body completely different from the hypocrisy and taboos which prevailed at the time, a more convivial way to see life in society, and an incentive to respect the planet. Thus naturism develops in France, in particular under the influence of Élizée Reclus, at the end of the 19th century and beginning of the 20th century among anarchistic communities resulting from utopian socialism."

In France, later important propagandists of anarcho-naturism include Henri Zisly and Émile Gravelle who collaborated in *La Nouvelle Humanité, Le Naturien, Le Sauvage, L'Ordre Naturel,* and *La Vie Naturelle*. Their ideas were important in individualist anarchist circles in France as well as Spain, where Federico Urales (pseudonym of Joan Montseny) promoted the ideas of Gravelle and Zisly in *La Revista Blanca* (1898–1905). Zisly's political activity, "primarily aimed at supporting a return to 'natural life' through writing and practical involvement, stimulated lively confrontations within and outside the anarchist environment. Zisly vividly criticized progress and civilization, which he regarded as 'absurd, ignoble, and filthy.' He openly opposed industrialization, arguing that machines were inherently authoritarian, defended nudism, advocated a non-dogmatic and non-religious adherence to the 'laws of nature,' recommended a lifestyle based on limited needs and self-sufficiency, and disagreed with vegetarianism, which he considered 'anti-scientific.'"

Richard D. Sonn comments on the influence of naturist views in the wider French anarchist movement:

In her memoir of her anarchist years that was serialized in *Le Matin* in 1913, Rirette Maîtrejean made much of the strange food regimens of some of the *compagnons*. [...] She described the "tragic bandits" of the Bonnot gang as refusing to eat meat or drink wine, preferring plain water. Her humorous comments reflected the practices of the "naturist" wing of individualist anarchists who favored a simpler, more "natural" lifestyle centered on a vegetarian diet. In the 1920s, this wing was expressed by the journal *Le Néo-Naturien, Revue des Idées Philosophiques et Naturiennes*. Contributors condemned the fashion of smoking cigarettes, especially by young women; a long article of 1927 actually connected cigarette smoking with cancer! Others distinguished between vegetarians, who foreswore the eating of meat, from the stricter "vegetalians," who ate nothing but vegetables. An anarchist named G. Butaud, who made this distinction, opened a restaurant called the Foyer Végétalien in the nineteenth arrondissement in 1923. Other issues of the journal included vegetarian recipes. In 1925, when the young anarchist and future detective novelist Léo Malet arrived in Paris from Montpellier, he initially lodged with anarchists who operated another vegetarian restaurant that served only vegetables, with neither fish nor eggs. Nutritional concerns coincided with other means of encouraging health bodies, such as nudism and gymnastics. For a while in the 1920s, after they were released from jail for antiwar and birth-control activities, Jeanne and Eugène Humbert retreated to the relative safety of the "integral living" movement that promoted nude sunbathing and physical fitness, which were seen as integral aspects of health in the Greek sense of *gymnos*, meaning nude. This back-to-nature, primitivist current was not a monopoly of the left; the same interests were echoed by right-wing Germans in the interwar era. In France, however, these proclivities were mostly associated with anarchists, insofar as they suggested an ideal of self-control and the rejection of social taboos and prejudices.

—*Richard D. Sonn,*

Henri Zisly

Henri Zisly (born in Paris, November 2, 1872; died in 1945) was a French individualist anarchist and naturist. He participated alongside Henri Beylie and Émile Gravelle in many journals such as *La Nouvelle Humanité* and *La Vie Naturelle*, which promoted anarchist-naturism. In 1902, he was one of the main initiators, alongside Georges Butaud and Sophie Zaïkowska, of the cooperative *Colonie de Vaux* established in Essômes-sur-Marne, in l'Aisne.

Zisly's political activity, "primarily aimed at supporting a return to 'natural life' through writing and practical involvement, stimulated lively confrontations within and outside the anarchist environment. Zisly vividly criticized progress and civilization, which he regarded as 'absurd, ignoble, and filthy.' He openly opposed industrialization, arguing that machines were inherently authoritarian, defended nudism, advocated a non-dogmatic and non-religious adherence to the 'laws of nature,' recommended a lifestyle based on limited needs and self-sufficiency, and disagreed with vegetarianism, which he considered 'anti-scientific.'"

Spain

This relationship between anarchism and naturism was quite important at the end of the 1920s in Spain:

> The linking role played by the Sol y Vida group was very important. The goal of this group was to take trips and enjoy the open air. The Naturist athenaeum, *Ecléctico*, in Barcelona, was the base from which the activities of the group were launched. First *Etica* and then *Iniciales*, which began in 1929, were the publications of the group, which lasted until the Spanish Civil War. We must be aware that the naturist ideas expressed in them matched the desires that the libertarian youth had of breaking up with the conventions of the bourgeoisie of the time. That is what a young worker explained in a letter to *Iniciales*. He writes it under the odd pseudonym of *silvestre del campo* (wild man in the country). "I find great pleasure in being naked in the woods, bathed in light and air, two natural elements we cannot do without. By shunning the humble garment of an exploited person, (garments which, in my opinion, are the result of all the laws devised to make our lives bitter), we feel there no others left but just the natural laws. Clothes mean slavery for some and tyranny for others. Only the naked man who rebels against all norms, stands for anarchism, devoid of the prejudices of outfit imposed by our money-oriented society."

Isaac Puente, an influential Spanish anarchist during the 1920s and 1930s and an important propagandist of anarcho-naturism, was a militant of both the CNT anarcho-syndicalist trade union and Iberian Anarchist Federation. He published the book *El Comunismo Libertario y otras proclamas insurreccionales y naturistas* (en:*Libertarian Communism and other insurrectionary and naturist proclaims*) in 1933, which sold around 100,000 copies, and wrote the final document for the Extraordinary Confederal Congress of Zaragoza of 1936 which established the main political line for the CNT for that year. Puente was a doctor who approached his medical practice from a naturist point of view. He saw naturism as an integral solution for the working classes, alongside Neo-Malthusianism, and believed it concerned the living being while anarchism addressed the social being. He believed capitalist societies endangered the well-being of humans from both a socioeconomic and sanitary viewpoint, and promoted anarcho-communism alongside naturism as a solution.

This ecological tendency in Spanish anarchism was strong enough as to call the attention of the CNT–FAI in Spain. Daniel Guérin in *Anarchism: From Theory to Practice* reports:

Spanish anarcho-syndicalism had long been concerned to safeguard the autonomy of what it called "affinity groups." There were many adepts of naturism and vegetarianism among its members, especially among the poor peasants of the south. Both these ways of living were considered suitable for the transformation of the human being in preparation for a libertarian society. At the Saragossa congress the members did not forget to consider the fate of groups of naturists and nudists, "unsuited to industrialization." As these groups would be unable to supply all their own needs, the congress anticipated that their delegates to the meetings of the confederation of communes would be able to negotiate special economic agreements with the other agricultural and industrial communes. On the eve of a vast, bloody, social transformation, the CNT did not think it foolish to try to meet the infinitely varied aspirations of individual human beings.

— *Daniel Guérin,*

Cuba

The historian Kirwin R. Schaffer in his study of Cuban anarchism reports anarcho-naturism as "[a] third strand within the island's anarchist movement" alongside anarcho-communism and anarcho-syndicalism. Naturism offered a global alternative health and lifestyle movement. Naturists focused on redefining one's life to live simply, to eat cheap but nutritious vegetarian diets, and to raise one's own food if possible. The countryside was posited as a romantic alternative to urban living, and some naturists even promoted what they saw as the healthful benefits of nudism. Globally, the naturist movement counted anarchists, liberals, and socialists as its followers. However, in Cuba a particular "anarchist" dimension evolved led by people like Adrián del Valle, who spearheaded the Cuban effort to shift naturism's focus away from only individual health to naturism having a "social emancipatory" function."

Schaffer reports the influence that anarcho-naturism had outside naturist circles. So "[f]or instance, nothing inherently prevented an anarcho-syndicalist in the Havana restaurant workers' union from supporting the alternative health care programs of the anarcho-naturists and seeing those alternative practices as 'revolutionary.'". "Anarcho-naturists promoted a rural ideal, simple living, and being in harmony with Nature as ways to save the laborers from the increasingly industrialized character of Cuba. Besides promoting an early twentieth-century "back-to-the-land" movement, they used these romantic images of Nature to illustrate how far removed a capitalist industrialized Cuba had departed from an anarchist view of natural harmony." The main propagandizer in Cuba of anarcho-naturism was the Catalonia born "Adrián del Valle (aka Palmiro de Lidia)...Over the following decades, Del Valle became a constant presence in not only the anarchist press that proliferated in Cuba but also mainstream literary publications...From 1912 to 1913 he edited the freethinking journal *El Audaz*. Then he began his largest publishing job by helping to found and edit the monthly alternative health magazine that followed the anarcho-naturist line *Pro-Vida*.

Other Countries

Naturism also met anarchism in the United Kingdom. "In many of the alternative communities established in Britain in the early 1900s nudism, anarchism, vegetarianism and free love were accepted as part of a politically radical way of life. In the 1920s the inhabitants of the anarchist community at Whiteway, near Stroud in Gloucestershire, shocked the conservative residents of the area with their shameless nudity."

In Italy, during the IX Congress of the Italian Anarchist Federation in Carrara in 1965, a group decided to split off from this organization and created the *Gruppi di Iniziativa Anarchica*. In the seventies it mostly comprised "veteran individualist anarchists with an pacifism orientation, naturism, etc,...".

Criticisms

American anarcho-syndicalist Sam Dolgoff shows some of the criticism that some people on the other anarchist currents at the time had for anarcho-naturist tendencies. "Speaking of life at the Stelton Colony of New York in the 1930s, noted with disdain that it, "like other colonies, was infested by vegetarians, naturists, nudists, and other cultists, who sidetracked true anarchist goals." One resident "always went barefoot, ate raw food, mostly nuts and raisins, and refused to use a tractor, being opposed to machinery, and he didn't want to abuse horses, so he dug the earth himself." Such self-proclaimed anarchists were in reality "ox-cart anarchists," Dolgoff said, "who opposed organization and wanted to return to a simpler life." In an interview with Paul Avrich before his death, Dolgoff also grumbled, "I am sick and tired of these half-assed artists and poets who object to organization and want only to play with their belly buttons".

Environmental Movement in the United States

1970s U.S. postage stamp block

In the United States today, the organized environmental movement is represented by a wide range of organizations sometimes called non-governmental organizations or NGOs. These organizations exist on local, national, and international scales. Environmental NGOs vary widely in political views and in the amount they seek to influence the environmental policy of the United States and other governments. The environmental movement today consists of both large national groups and also many smaller local groups with local concerns. Some resemble the old U.S. conservation movement - whose modern expression is The Nature Conservancy, Audubon Society and National Geographic Society - American organizations with a worldwide influence.

Scope of the Movement

- The early Conservation movement, which began in the late 19th century, included fisheries and wildlife management, water, soil conservation and sustainable forestry. Today it includes sustainable yield of natural resources, preservation of wilderness areas and biodiversity.

- The modern Environmental movement, which began in the 1960s with concern about air and water pollution, became broader in scope to including all landscapes and human activities.

- Environmental health movement dating at least to Progressive Era (1890s - 1920s) urban reforms including clean water supply, more efficient removal of raw sewage and reduction in crowded and unsanitary living conditions. Today Environmental health is more related to nutrition, preventive medicine, aging well and other concerns specific to the human body's well-being.

- Sustainability movement which started in the 1980s focused on Gaia theory, value of Earth and other interrelations between human sciences and human responsibilities. Its spinoff Deep Ecology was more spiritual but often claimed to be science.

- Environmental justice is a movement that began in the U.S. in the 1980s and seeks an end to environmental racism. Often, low-income and minority communities are located close to highways, garbage dumps, and factories, where they are exposed to greater pollution and environmental health risk than the rest of the population. The Environmental Justice movement seeks to link "social" and "ecological" environmental concerns, while at the same time keeping environmentalists conscious of the dynamics in their own movement, i.e. racism, sexism, homophobia, classicism, and other malaises of dominant culture.

As public awareness and the environmental sciences have improved in recent years, environmental issues have broadened to include key concepts such as "sustainability" and also new emerging concerns such as ozone depletion, global warming, acid rain, land use and biogenetic pollution.

Environmental movements often interact or are linked with other social movements, e.g. for peace, human rights, and animal rights; and against nuclear weapons and/or nuclear power, endemic diseases, poverty, hunger, etc.

Some US colleges are now going green by signing the "President's Climate Commitment," a document that a college President can sign to enable said colleges to practice environmentalism by switching to solar power, etc.

Membership of selected US environmental organizations (000s)					
	1971	**1981**	**1992**	**1997**	**2004**
Sierra Club (1892)	124	246	615	569	736
National Audubon Society (1905)	115	400	600	550	550
National Parks Conservation Association (1919)	49	27	230	375	375
Izaak Walton League (1922)	54	48	51	42	45
Wilderness Society (1935)	62	52	365	237	225
National Wildlife Federation (1936)	540	818	997	650	650
Defenders of Wildlife (1947)	13	50	77	215	463
The Nature Conservancy (1951)	22	80	545	865	972
WWF-US (1961)	n.a.	n.a.	970	1,200	1,200
Environmental Defense Fund (1967)	20	46	175	300	350
Friends of the Earth (US) (1969)	7	25	30	20	35
Natural Resources Defense Council (1970)	5	40	170	260	450
Greenpeace USA (1972)	n.a.	n.a.	2,225	400	250

History

Early European settlers to the United States brought from Europe the concept of the commons. In the colonial era, access to natural resources was allocated by individual towns, and disputes over fisheries or land use were resolved at the local level. Changing technologies, however, strained traditional ways of resolving disputes of resource use, and local governments had limited control over powerful special interests. For example, the damming of rivers for mills cut off upriver towns from fisheries; logging and clearing of forest in watersheds harmed local fisheries downstream. In New England, many farmers became uneasy as they noticed clearing of forest changed stream flows and a decrease in bird population which helped control insects and other pests. These concerns become widely known with the publication of Man and Nature (1864) by George Perkins Marsh. The environmental impact method of analysis is generally the main mode for determining what issues the environmental movement is involved in. This model is used to determine how to proceed in situations that are detrimental to the environment by choosing the way that is least damaging and has the fewest lasting implications.

Conservation Movement

Conservation first became a national issue during the progressive era's conservation movement (1890s - 1920s). The early national conservation movement shifted emphasis to scientific management which favored larger enterprises and control began to shift from local governments to the states and the federal government.(Judd) Some writers credit sportsmen, hunters and fishermen with the increasing influence of the conservation movement. In the 1870s sportsman magazines such as American Sportsmen, Forest and Stream, and Field and Stream are seen as leading to the growth of the conservation movement.(Reiger) This conservation movement also urged the establishment of state and national parks and forests, wildlife refuges, and national monuments intended to preserve noteworthy natural features. Conservation groups focus primarily on an issue that's origins are routed in general expansion. As Industrialization became more prominent as well as the increasing trend towards Urbanization the conservative environmental movement began. Contrary to popular belief conservation groups are not against expansion in general, instead they are concerned with efficiency with resources and land development.

Progressive Era

Theodore Roosevelt and his close ally George Bird Grinnell, were motivated by the wanton waste that was taking place at the hand of market hunting. This practice resulted in placing a large number of North American game species on the edge of extinction. Roosevelt recognized that the laissez-faire approach of the U.S. Government was too wasteful and inefficient. In any case, they noted, most of the natural resources in the western states were already owned by the federal government. The best course of action, they argued, was a long-term plan devised by national experts to maximize the long-term economic benefits of natural resources. To accomplish the mission, Roosevelt and Grinnell formed the Boone and Crockett Club in 1887. The Club was made up of the best minds and influential men of the day. The Boone and Crockett Club's contingency of conservationists, scientists, politicians, and intellectuals became Roosevelt's closest advisers during his march to preserve wildlife and habitat across North America. As president, Theodore Roosevelt became a prominent conservationist, putting the issue high on the national agenda. He worked with all the major figures of the movement, especially his chief advisor on the matter, Gifford Pinchot. Roosevelt was deeply committed to conserving natural resources, and is considered to be the nation's first conservation President. He encouraged the Newlands Reclamation Act of 1902 to promote federal construction of dams to irrigate small farms and placed 230 million acres (360,000 mi^2 or 930,000 km^2) under federal protection. Roosevelt set aside more Federal land for national parks and nature preserves than all of his predecessors combined.

Roosevelt established the United States Forest Service, signed into law the creation of five National Parks, and signed the 1906 Antiquities Act, under which he proclaimed 18 new U.S. National Monuments. He also established the first 51 Bird Reserves, four Game Preserves, and 150 National Forests, including Shoshone National Forest, the nation's first. The area of the United States that he placed under public protection totals approximately 230,000,000 acres (930,000 km²).

Gifford Pinchot had been appointed by McKinley as chief of Division of Forestry in the Department of Agriculture. In 1905, his department gained control of the national forest reserves. Pinchot promoted private use (for a fee) under federal supervision. In 1907, Roosevelt designated 16 million acres (65,000 km²) of new national forests just minutes before a deadline.

In May 1908, Roosevelt sponsored the Conference of Governors held in the White House, with a focus on natural resources and their most efficient use. Roosevelt delivered the opening address: "Conservation as a National Duty."

In 1903 Roosevelt toured the Yosemite Valley with John Muir, who had a very different view of conservation, and tried to minimize commercial use of water resources and forests. Working through the Sierra Club he founded, Muir succeeded in 1905 in having Congress transfer the Mariposa Grove and Yosemite Valley to the National Park Service. While Muir wanted nature preserved for the sake of pure beauty, Roosevelt subscribed to Pinchot's formulation, "to make the forest produce the largest amount of whatever crop or service will be most useful, and keep on producing it for generation after generation of men and trees." Muir and the Sierra Club vehemently opposed the damming of the Hetch Hetchy Valley in Yosemite in order to provide water to the city of San Francisco. Roosevelt and Pinchot supported the dam, as did President Woodrow Wilson. The Hetch Hetchy dam was finished in 1923 and is still in operation, but the Sierra Club still wants to tear it down.

Other influential conservationists of the Progressive Era included George Bird Grinnell (a prominent sportsmen who founded the Boone and Crockett Club), the Izaak Walton League and John Muir, the founder of the Sierra Club in 1892. Conservationists organized the National Parks Conservation Association, the Audubon Society, and other groups that still remain active.

New Deal

Franklin Delano Roosevelt (1933–45), like his cousin Theodore Roosevelt, was an ardent conservationist. He used numerous programs of the departments of Agriculture and Interior to end wasteful land-use, mitigate the effects of the Dust Bowl, and efficiently develop natural resources in the West. One of the most popular of all New Deal programs was the Civilian Conservation Corps (1933–1943), which sent two million poor young men to work in rural and wilderness areas, primarily on conservation projects.

Post 1945

After World War II increasing encroachment on wilderness land evoked the continued resistance of conservationists, who succeeded in blocking a number of projects in the 1950s and 1960s, including the proposed Bridge Canyon Dam that would have backed up the waters of the Colorado River into the Grand Canyon National Park.

The Inter-American Conference on the Conservation of Renewable Natural Resources met in 1948 as a collection of nearly 200 scientists from all over the Americans forming the trusteeship principle that:

"No generation can exclusively own the renewable resources by which it lives. We hold the commonwealth in trust for prosperity, and to lessen or destroy it is to commit treason against the future"

Beginning of the Modern Movement

Earth Day Flag

During the 1950s, 1960s and 1970s, several events occurred which raised the public awareness of harm to the environment caused by man. In 1954, the 23 man crew of the Japanese fishing vessel Lucky Dragon was exposed to radioactive fallout from a hydrogen bomb test at Bikini Atoll, in 1969, an ecologically catastrophic oil spill from an offshore well in California's Santa Barbara Channel, Barry Commoner's protest against nuclear testing, Rachel Carson's book Silent Spring, Paul R. Ehrlich's The Population Bomb all added anxiety about the environment. Pictures of Earth from space emphasized that the earth was small and fragile.

As the public became more aware of environmental issues, concern about air pollution, water pollution, solid waste disposal, dwindling energy resources, radiation, pesticide poisoning (particularly as described in Rachel Carson's influential Silent Spring, 1962), noise pollution, and other environmental problems engaged a broadening number of sympathizers. That public support for environmental concerns was widespread became clear in the Earth Day demonstrations of 1970.

Unlike the Progressive Era's conservation movement (1890s - 1920s), which was largely elitist consisting of largely of wealthy, politically powerful men, the modern environmental movement was a social movement with more popular support. The environmental movement borrowed tactics from both the successful civil rights movement and the protests against the Vietnam war.

Wilderness Preservation

In the modern wilderness preservation movement, important philosophical roles are played by the writings of John Muir who had been activist in the late 19th and early 20th century. Along with Muir perhaps most influential in the modern movement is Henry David Thoreau who published Walden in 1854. Also important was forester and ecologist Aldo Leopold, one of the founders of the Wilderness Society in 1935, who wrote a classic of nature observation and ethical philosophy, *A Sand County Almanac*, published in 1949. Other philosophical foundations were established by Ralph Waldo Emerson and Thomas Jefferson.

There is also a growing movement of campers and other people who enjoy outdoor recreation activities to help preserve the environment while spending time in the wilderness.

Anti-nuclear Movement

The anti-nuclear movement in the United States consists of more than 80 anti-nuclear groups which have acted to oppose nuclear power or nuclear weapons, or both, in the United States. These groups include the Abalone Alliance, Clamshell Alliance, Institute for Energy and Environmental Research, Nuclear Information and Resource Service, and Physicians for Social Responsibility. The anti-nuclear movement has delayed construction or halted commitments to build some new nuclear plants, and has pressured the Nuclear Regulatory Commission to enforce and strengthen the safety regulations for nuclear power plants.

Anti-nuclear protests reached a peak in the 1970s and 1980s and grew out of the environmental movement. Campaigns which captured national public attention involved the Calvert Cliffs Nuclear Power Plant, Seabrook Station Nuclear Power Plant, Diablo Canyon Power Plant, Shoreham Nuclear Power Plant, and Three Mile Island. On June 12, 1982, one million people demonstrated in New York City's Central Park against nuclear weapons and for an end to the cold war arms race. It was the largest anti-nuclear protest and the largest political demonstration in American history. International Day of Nuclear Disarmament protests were held on June 20, 1983 at 50 sites across the United States. There were many Nevada Desert Experience protests and peace camps at the Nevada Test Site during the 1980s and 1990s.

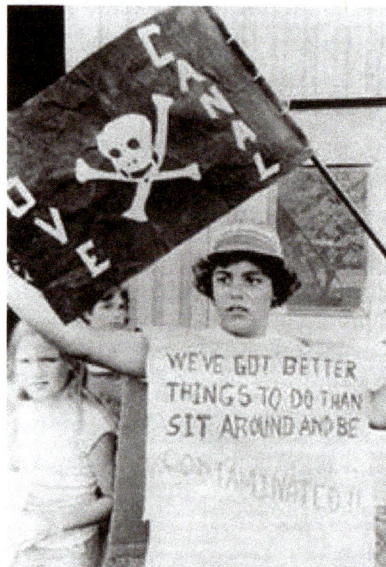

Protest about the Love Canal contamination by a resident, ca. 1978

More recent campaigning by anti-nuclear groups has related to several nuclear power plants including the Enrico Fermi Nuclear Power Plant, Indian Point Energy Center, Oyster Creek Nuclear Generating Station, Pilgrim Nuclear Generating Station, Salem Nuclear Power Plant, and Vermont

Yankee Nuclear Power Plant. There have also been campaigns relating to the Y-12 Nuclear Weapons Plant, the Idaho National Laboratory, proposed Yucca Mountain nuclear waste repository, the Hanford Site, the Nevada Test Site, Lawrence Livermore National Laboratory, and transportation of nuclear waste from the Los Alamos National Laboratory.

Some scientists and engineers have expressed reservations about nuclear power, including: Barry Commoner, S. David Freeman, John Gofman, Arnold Gundersen, Mark Z. Jacobson, Amory Lovins, Arjun Makhijani, Gregory Minor, Joseph Romm and Benjamin K. Sovacool. Scientists who have opposed nuclear weapons include Linus Pauling and Eugene Rabinowitch.

Antitoxics Groups

Antitoxics groups are a subgroup that is affiliated with the Environmental Movement in the United States, that is primarily concerned with the effects that cities and their by products have on humans. This aspect of the movement is a self-proclaimed "movement of housewives". Concern around the issues of ground water contamination and air pollution rose in the early 1980s and individuals involved in antitoxics groups claim that they are concerned for the health of their families. A prominent case can be seen in the Love Canal Homeowner's association (LCHA); in this case a housing development was built on a site that had been used for toxic dumping by the Hooker Chemical Company. As a result of this dumping the residents had symptoms of skin irritation, Lois Gibbs, a resident of the development, started a grassroots campaign for reparations. Eventual success led to the government having to purchase homes that were sold in the development.

Federal Legislation in the 1970s

Prior to the 1970s the protection of basic air and water supplies was a matter mainly left to each state. During the 1970s, primary responsibility for clean air and water shifted to the federal government. Growing concerns, both environmental and economic, from cites and towns as well as sportsman and other local groups, and senators such as Maine's Edmund S. Muskie, led to passage of extensive legislation, notably the Clean Air Act of 1970 and the Water Pollution Control Act Amendments of 1972. Other legislation included National Environmental Policy Act (NEPA), signed into law in 1970, which established a United States Environmental Protection Agency and a Council on Environmental Quality; the Marine Protection, Research, and Sanctuaries Act of 1972; the Endangered Species Act of 1973, the Safe Drinking Water Act (1974), the Resource Conservation and Recovery Act (1976), the Water Pollution Control Act Amendments of 1977, which became known as the Clean Water Act, and the Comprehensive Environmental Response, Compensation, and Liability Act, commonly known as the Superfund Act (1980). These laws regulated public drinking water systems, toxic substances, pesticides, and ocean dumping; and protected wildlife, wilderness, and wild and scenic rivers. Moreover, the new laws provide for pollution research, standard setting, contaminated site cleanup, monitoring, and enforcement.

The creation of these laws led to a major shift in the environmental movement. Groups such as the Sierra Club shifted focus from local issues to becoming a lobby in Washington and new groups, for example, the Natural Resources Defense Council and Environmental Defense, arose to influence politics as well. (Larson)

Renewed Focus on Local Action

In the 1980s President Ronald Reagan sought to curtail scope of environmental protection taking steps such as appointing James G. Watt who was called one of the most "blatantly anti-environmental political appointees". The major environmental groups responded with mass mailings which led to increased membership and donations. The large environmental organization increasingly relied on ties within Washington, D.C. to advance their environmental agenda. At the same time membership in environmental groups became more suburban and urban. Groups such as animal rights, and the gun control lobby became linked with environmentalism while sportsmen, farmers and ranchers were no longer influential in the movement.

When industry groups lobbied to weaken regulation and a backlash against environmental regulations, the so-called wise use movement gained importance and influence. The wise use movement and anti-environmental groups were able to portray environmentalist as out of touch with mainstream values. (Larson)

"Post-environmentalism"

In 2004, with the environmental movement seemingly stalled, some environmentalists started questioning whether "environmentalism" was even a useful political framework. According to a controversial essay titled "The Death of Environmentalism " (Michael Shellenberger and Ted Nordhaus, 2004) American environmentalism has been remarkably successful in protecting the air, water, and large stretches of wilderness in North America and Europe, but these environmentalists have stagnated as a vital force for cultural and political change.

Shellenberger and Nordhaus wrote, "Today environmentalism is just another special interest. Evidence for this can be found in its concepts, its proposals, and its reasoning. What stands out is how arbitrary environmental leaders are about what gets counted and what doesn't as 'environmental.' Most of the movement's leading thinkers, funders, and advocates do not question their most basic assumptions about who we are, what we stand for, and what it is that we should be doing." Their essay was followed by a speech in San Francisco called "Is Environmentalism Dead?" by former Sierra Club President, Adam Werbach, who argued for the evolution of environmentalism into a more expansive, relevant and powerful progressive politics. Werbach endorsed building an environmental movement that is more relevant to average Americans, and controversially chose to lead Wal-Mart's effort to take sustainability mainstream.

These "post-environmental movement" thinkers argue that the ecological crises the human species faces in the 21st century are qualitatively different from the problems the environmental movement was created to address in the 1960s and 1970s. They argue that climate change and habitat destruction are global and more complex, therefore demanding far deeper transformations of the economy, the culture and political life. The consequence of environmentalism's outdated and arbitrary definition, they argue, is political irrelevancy.

These "politically neutral" groups tend to avoid global conflicts and view the settlement of inter-human conflict as separate from regard for nature - in direct contradiction to the ecology movement and peace movement which have increasingly close links: while Green Parties, Greenpeace, and groups like the ACTivist Magazine regard ecology, biodiversity, and an end to non-human extinction as an absolute basis for peace, the local groups may not, and see a high degree of global competition and conflict as justifiable if it lets them preserve their own local uniqueness. However, such groups tend not to "burn out" and to sustain for long periods, even generations, protecting the same local treasures.

Local groups increasingly find that they benefit from collaboration, e.g. on consensus decision making methods, or making simultaneous policy, or relying on common legal resources, or even sometimes a common glossary. However, the differences between the various groups that make up the modern environmental movement tend to outweigh such similarities, and they rarely co-operate directly except on a few major global questions. In a notable exception, over 1,000 local groups from around the country united for a single day of action as part of the Step It Up 2007 campaign for real solutions to global warming.

Groups such as The Bioregional Revolution are calling on the need to bridge these differences, as the converging problems of the 21st century they claim compel us to unite and to take decisive action. They promote bioregionalism, permaculture, and local economies as solutions to these problems, overpopulation, global warming, global epidemics, and water scarcity, but most notably to "peak oil"—the prediction that we are likely to reach a maximum in global oil production which could spell drastic changes in many aspects of our everyday lives.

Environmental Rights

Many environmental lawsuits turn on the question of who has standing; are the legal issues limited to property owners, or does the general public have a right to intervene? Christopher D. Stone's 1972 essay, "Should trees have standing?" seriously addressed the question of whether natural objects themselves should have legal rights, including the right to participate in lawsuits. Stone suggested that there was nothing absurd in this view, and noted that many entities now regarded as having legal rights were, in the past, regarded as "things" that were regarded as legally rightless; for example, aliens, children and women. His essay is sometimes regarded as an example of the fallacy of hypostatization.

One of the earliest lawsuits to establish that citizens may sue for environmental and aesthetic harms was Scenic Hudson Preservation Conference v. Federal Power Commission, decided in 1965 by the Second Circuit Court of Appeals. The case helped halt the construction of a power plant on Storm King Mountain in New York State.

Role of Science

Conservation biology is an important and rapidly developing field.

One way to avoid the stigma of an "ism" was to evolve early anti-nuclear groups into the more scientific Green Parties, sprout new NGOs such as Greenpeace and Earth Action, and devoted groups to protecting global biodiversity and preventing global warming and climate change. But in the process, much of the emotional appeal, and many of the original aesthetic goals were lost. Nonetheless, these groups have well-defined ethical and political views, backed by science.

Criticisms of the Environmental Movement

Some people are skeptical of the environmental movement and feel that it is more deeply rooted in politics than science. Although there have been serious debates about climate change and effects of some pesticides and herbicides that mimic animal sex steroids, science has shown that some of the claims of environmentalists have credence.

Claims made by environmentalists may be perceived as veiled attacks on industry and globalization rather than legitimate environmental concerns. Detractors note that a significant number of environmental theories and predictions have been inaccurate and suggest that the regulations recommended by environmentalists will more likely harm society rather than help nature.

DDT

Specific examples include when Rachel Carson, in her book *Silent Spring*, suggested that the pesticide DDT caused cancer and drastically harmed ecosystems. DDT is highly toxic to aquatic life, including crawfish, daphnids, sea shrimp and many species of fish. However, DDT is also used to control malaria.

Prominent novelist and Harvard Medical School graduate Michael Crichton appeared before the U.S. Senate Committee on Environment and Public Works to address such concerns and recommended the employment of double-blind experimentation in environmental research. Crichton suggested that because environmental issues are so political in nature, policy makers need neutral, conclusive data to base their decisions on, rather than conjecture and rhetoric, and double-blind experiments are the most efficient way to achieve that aim.

A consistent theme acknowledged by both supporters and critics (though more commonly vocalized by critics) of the environmental movement is that we know very little about the Earth we live in. Most fields of environmental studies are relatively new, and therefore what research we have is limited and does not date far enough back for us to completely understand long-term environmental trends. This has led a number of environmentalists to support the use of the precautionary principle in policy making, which ultimately asserts that we don't know how certain actions may affect the environment, and because there is reason to believe they may cause more harm than good we should refrain from such actions.

Elitist

In the December 1994 *Wild Forest Review*, Alexander Cockburn and Jeffrey St. Clair wrote "The mainstream environmental movement was elitist, highly paid, detached from the people, indifferent to the working class, and a firm ally of big government....The environmental movement is now accurately perceived as just another well-financed and cynical special interest group, its rancid infrastructure supported by Democratic Party operatives and millions in grants from corporate foundations."

Wilderness Myth

Historian and President of the American Historical Association William Cronon has criticized the modern environmental movement for having a romantic idealizations of wilderness. Cronon writes "wilderness serves as the unexamined foundation on which so many of the quasi-religious values of modern environmentalism rest." Cronon claims that "to the extent that we live in an urban-industrial civilization but at the same time pretend to ourselves that our real home is in the wilderness, to just that extent we give ourselves permission to evade responsibility for the lives we actually lead."

Similarly Michael Pollan has argued that the wilderness ethic leads people to dismiss areas whose wildness is less than absolute. In his book Second Nature, Pollan writes that "once a landscape is no longer 'virgin' it is typically written off as fallen, lost to nature, irredeemable."

Debates within the Movement

Within the environmental movement an ideological debate has taken place between those with an ecocentric view point and an anthropocentric view point. The anthropocentric view has been seen as the conservationist approach to the environment with nature viewed, at least in part, as resource to be used by man. In contrast to the conservationist approach the ecocentric view, associated with John Muir, Henry David Thoreau and William Wordsworth referred to as the preservationist movement. This approach sees nature in a more spiritual way. Many environmental historians consider the split between John Muir and Gifford Pinchot. During the preservation / conservation debate the term preservationist become to be seen as a pejorative term.

While the ecocentric view focused on biodiversity and wilderness protection the anthropocentric view focus on urban pollution and social justice. Some environmental writers, for example William Cronon have criticized the ecocentric view as have a dualist view as man being separate from nature. Critics of the anthropocentric view point contend that the environmental movement has been taken over by so-called leftist with an agenda beyond environmental protection.

Several books after the middle of the 20th century contributed to the rise of American environmentalism (as distinct from the longer-established conservation movement), especially among college and university students and the more literate public. One was the publication of the first textbook on ecology, *Fundamentals of Ecology,* by Eugene Odum and Howard Odum, in 1953. Another was the appearance of the best-seller *Silent Spring* by Rachel Carson, in 1962. Her book brought about a whole new interpretation on pesticides by exposing their harmful effects in nature. From this book many began referring to Carson as the "mother of the environmental movement". Another influential development was a 1965 lawsuit, Scenic Hudson Preservation Conference v. Federal Power Commission, opposing the construction of a power plant on Storm King Mountain, which is said to have given birth to modern United States environmental law. The wide popularity of *The Whole Earth Catalogs*, starting in 1968, was quite influential among the younger, hands-on, activist generation of the 1960s and 1970s. Recently, in addition to opposing environmental degradation and protecting wilderness, an increased focus on coexisting with natural biodiversity has appeared, a strain that is apparent in the movement for sustainable agriculture and in the concept of Reconciliation Ecology.

Environmentalism and Politics

Environmentalists became much more influential in American politics after the creation or strengthening of numerous U.S. environmental laws, including the Clean Air Act and Clean Water Act and the formation of the US Environmental Protection Agency, or EPA in 1970. These successes were followed by the enactment of a whole series of laws regulating waste (Resource Conservation and Recovery Act), toxic substances (Toxic Substances Control Act), pesticides (FI-FRA: Federal Insecticide, Fungicide, and Rodenticide Act), clean-up of polluted sites (Superfund), protection of endangered species (Endangered Species Act), and more.

Fewer environmental laws have been passed in the last decade as corporations and other conservative interests have increased their influence over American politics. Corporate cooperation against environmental lobbyists has been organized by the Wise Use group. At the same time, many environmentalists have been turning toward other means of persuasion, such as working with business, community, and other partners to promote sustainable development.

Much environmental activism is directed towards conservation, as well as the prevention or elimination of pollution. However, conservation movements, ecology movements, peace movements, green parties, green- and eco-anarchists often subscribe to very different ideologies, while supporting the same goals as those who call themselves "environmentalists". To outsiders, these groups or factions can appear to be indistinguishable.

As human population and industrial activity continue to increase, environmentalists often find themselves in serious conflict with those who believe that human and industrial activities should not be overly regulated or restricted, such as some libertarians.

Environmentalists often clash with others, particularly "corporate interests," over issues of the management of natural resources, like in the case of the atmosphere as a "carbon dump", the focus of climate change, and global warming controversy. They usually seek to protect commonly owned or unowned resources for future generations.

Those who take issue with new untested technologies are more precisely known, especially in Europe, as political ecologists. They usually seek, in contrast, to preserve the integrity of existing ecologies and ecoregions, and in general are more pessimistic about human "management".

Radical Environmentalism

While most environmentalists are mainstream and peaceful, a small minority are more radical in their approach. Adherents of radical environmentalism and ecological anarchism are involved in direct action campaigns to protect the environment. Some campaigns have employed controversial tactics including sabotage, blockades, and arson, while most use peaceful protests such as marches, tree-sitting, and the like. There is substantial debate within the environmental movement as to the acceptability of these tactics, but almost all environmentalists condemn violent actions that can harm humans.

References

- Jan Marsh (1982). Back to the Land: The Pastoral Impulse in England, 1880-1914. Quartet Books. ISBN 9780704322769.

- Chapman, Roger (2010). Culture wars: an encyclopedia of issues, viewpoints, and voices. M.E. Sharpe, Inc. p. 162. ISBN 0-7656-1761-7.

- Greg Barton (2002). Empire Forestry and the Origins of Environmentalism. Cambridge University Press. p. 48. ISBN 9781139434607.

- Hijacking Chipko Political ecology: a critical introduction, by Paul Robbins. Published by Wiley-Blackwell, 2004. ISBN 1-4051-0266-7. Page 194.

- The women of Chipko Staying alive: women, ecology, and development, by Vandana Shiva, Published by Zed Books, 1988. ISBN 0-86232-823-3. Page 67

- The Chipko Movement Politics in the developing world: a concise introduction, by Jeffrey Haynes. Published by Wiley-Blackwell, 2002. ISBN 0-631-22556-0. Page 229.

- Chipko Movement The Future of the Environment: The Social Dimensions of Conservation and Ecological Alternatives, by David C. Pitt. Published by Routledge, 1988. ISBN 0-415-00455-1. Page 112.

- Singh, Mahesh Prasad; Singh, J. K.; Mohanka, Reena (2007-01-01). Forest Environment and Biodiversity. Daya Publishing House. p. 157. ISBN 9788170354215.

- Starting.. Of myths and movements: rewriting Chipko into Himalayan history, by Haripriya Rangan. Published by Verso, 2000. ISBN 1-85984-305-0. Page 4-5.

- Ecological crisis Water Wars: Privatization, Pollution and Profit, by Vandana Shiva. Published by Pluto Press, 2002. ISBN 0-7453-1837-1. Page 3.

- From Chipko to Uttaranchal: Haripriya Ranjan Liberation ecologies: environment, development, social movements, by Richard Peet, Michael Watts. Published by Routledge, 1996. ISBN 0-415-13362-9. Page 205-206.

- Shaffer, Kirwin R. (2005). Anarchism and countercultural politics in early twentieth-century Cuba. Gainsville: University Press of Florida. ISBN 0813027918.

- Diez, Xavier (2007). El anarquismo individualista en España (1923-1938). Barcelona: Virus. ISBN 978-84-96044-87-6. Retrieved 2016-07-13.

- Harvey E. Klehr (1988-01-01). Far Left of Center: The American Radical Left Today. Transaction Publishers. p. 150. ISBN 978-0-88738-875-0.

- J. Samuel Walker (2006-01-28). Three Mile Island: A Nuclear Crisis in Historical Perspective. University of California Press. p. 10. ISBN 978-0-520-24683-6.

Permissions

Index